The distribution of the various nuclear species in the cosmos, ranging from very common elements (hydrogen and helium), through moderately common elements like oxygen, magnesium and iron, to rare ones like gold and thorium, is the result of many different processes in the past history of the universe. It reveals much about the nature of the Big Bang, the density of baryonic matter, the details of the stellar evolution and nucleosynthesis. It also tells us about the formation and evolution of stars and galaxies. This book gives a concise account of a wide range of overlapping fields, including thermonuclear reactions, abundance measurements in astronomical sources, cosmological element production, stellar evolution and nucleosynthesis, light element production by cosmic rays and the effects of galactic processes on the distribution of the elements. This last theme, often referred to as 'galactic chemical evolution', is developed in an analytical treatment based on work in which the author has played a major part, and this is used to examine critically the assumptions underlying the many numerical models of recent years and to provide a clear understanding of what can really be deduced from the observations. Based on a lecture course given at Copenhagen University, the book is mainly aimed at advanced undergraduates and beginning graduate students with a background in either physics or astronomy, but it is also likely to be helpful to professional scientists.

Nucleosynthesis and Chemical Evolution of Galaxies

Nucleosynthesis and Chemical Evolution of Galaxies

B. E. J. PAGEL, FRS
Professor of Astrophysics, NORDITA, Copenhagen

CAMBRIDGE
UNIVERSITY PRESS

PUBLISHED BY THE PRESS SYNDICATE OF THE UNIVERSITY OF CAMBRIDGE
The Pitt Building, Trumpington Street, Cambridge CB2 1RP, United Kingdom

CAMBRIDGE UNIVERSITY PRESS
The Edinburgh Building, Cambridge CB2 2RU, United Kingdom
40 West 20th Street, New York, NY 10011-4211, USA
10 Stamford Road, Oakleigh, Melbourne 3166, Australia

First published 1997

Printed in the United Kingdom at the University Press, Cambridge

Typeset in 10/13pt Times

A catalogue record for this book is available from the British Library

Library of Congress cataloguing in publication data

Pagel, B. E. J. (Bernard Ephraim Julius)
    Nucleosynthesis and chemical evolution of galaxies / Bernard E. J. Pagel
        p.   cm.
    Includes bibliographical references.
    ISBN 0 521 55061 0
    1. Galaxies–Evolution.   2. Stars–Evolution.   3. Cosmochemistry.
    4. Nucleosynthesis.   I. Title.
    QB857.5.E96P34   1997
    523.1′12–dc21   97-5994CIP

ISBN 0 521 55061 0 hardback
ISBN 0 521 55958 8 paperback

Contents

Abbreviations

A & A	*Astronomy & Astrophysics*
ABB	after the Big Bang
AGB	asymptotic giant branch
AGN	active galactic nucleus
AMR	age–metallicity relation
amu	atomic mass unit
Ap. J.	*Astrophysical Journal*
AQ	Astrophysical Quantities, by C.W. Allen
Astr. J.	*Astronomical Journal*
B^2FH	E.M. & G.R. Burbidge, Fowler & Hoyle
BABI	basaltic achondrite
BBNS	Big Bang nucleosynthesis
BCG	blue compact galaxy
BDM	baryonic dark matter
B.E.	binding energy
BSG	blue supergiant
CC1	carbonaceous chondrite type 1
CDM	cold dark matter
CM	centre of mass
CNO	carbon, nitrogen & oxygen
CO	carbon & oxygen
CP	chemically peculiar
DDO	Toronto intermediate-band system
3DHO	three-dimensional harmonic oscillator
E-AGB	early AGB
ELS	Eggen, Lynden-Bell & Sandage
EW	equivalent width
FIP	first ionization potential
FUN	fractionation and unknown nuclear
GCE	galactic chemical evolution
GCR	galactic cosmic ray
GT	Gamow–Teller
GUT	grand unification theory
HB	horizontal branch
HR	Hertzsprung–Russell
HST	Hubble Space Telescope
IGM	intergalactic medium
IMF	initial mass function
IMS	intermediate-mass stars

IR	infra-red
IRAS	Infra-Red Astronomy Satellite
ISM	interstellar medium
ISW	infinite square well
IUE	International Ultraviolet Explorer
KBH	Kulkarni, Blitz & Heiles
K.E.	kinetic energy
LEP	Large Electron–Positron Collider
LINER	low-ionization emission line region
LMC	Large Magellanic Cloud
LTE	local thermodynamic equilibrium
MEMMU	Milne, Eddington, Minnaert, Menzel & Unsöld
MNRAS	*Monthly Notices of the Royal Astronomical Society*
MS	main sequence
MWB	microwave background radiation
NDM	non-baryonic dark matter
ORS	Oliver, Rowan-Robinson & Saunders
PDMF	present-day mass function
PN	planetary nebula
PP	Partridge & Peebles
QM	quantum mechanics
QSO	quasi-stellar object
RGB	red-giant branch
RSG	red supergiant
SBBN	standard Big Bang nucleosynthesis
SFR	star formation rate
SGB	subgiant branch
SIAL	aluminium-rich silicate
SIMA	magnesium-rich silicate
SMC	Small Magellanic Cloud
SN	supernova
SNIa	Type Ia supernova
SNII	Type II supernova
SNR	supernova remnant
snu	solar neutrino unit
SZ	Searle & Zinn
TAMS	terminal main sequence
TP-AGB	thermal pulse AGB
UBV	Johnson UBV broad-band system
UV	ultra-violet
WIMPS	weakly interacting massive particles
WR	Wolf–Rayet
ZAHB	zero-age horizontal branch
ZAMS	zero-age main sequence

Preface

This book is based on a lecture course given at Copenhagen University in the past few years to a mixed audience of advanced undergraduates, graduate students and some senior colleagues with backgrounds in either physics or astronomy. It is intended to cover a wide range of interconnected topics including thermonuclear reactions, cosmic abundances, primordial synthesis of elements in the Big Bang, stellar evolution and nucleosynthesis. There is also a (mainly analytical) treatment of factors governing the distribution of element abundances in stars, gas clouds and galaxies and related observational data are presented.

Some of the content of the course is a concise summary of fairly standard material concerning abundance determinations in stars, cold gas and ionized nebulae, cosmology, stellar evolution and nucleosynthesis that is available in much more detail elsewhere, notably in the books cited in the reading list or in review articles; here I have attempted to concentrate on giving up-to-date information, often in graphical form, and to give the simplest possible derivations of well-known results (e.g. exponential distribution of exposures in the main s-process). The section on Chemical Evolution of Galaxies deals with a rapidly growing subject in a more distinctive way, based on work in which I and some colleagues have been engaged over the years. The problem in this field is that uncertainties arising from problems in stellar and galactic evolution are compounded. Observational results are accumulating at a rapid rate and numerical models making a variety of often arbitrary assumptions are proliferating, leading to a jungle of more or less justifiable inferences that are often forgotten in the next instant paper. The analytical formalism on which I have been working on and off since my paper with the late B.E. Patchett in 1975, and to which very significant contributions have also been made by D.D. Clayton, M.G. Edmunds, F.G.A. Hartwick, R.B. Larson, D. Lynden-Bell, W.L.W. Sargent, L. Searle, B.M. Tinsley and others, is designed not only to keep the computations simple but also to introduce some order into the subject and provide the reader with an insight into what actually are the important factors in chemical evolution models, whether analytical or numerical, and which are the major uncertainties.

The book should be considered basically as a textbook suitable for beginning graduate students with a background in either physics or astronomy, but it is hoped that parts of it will also be useful to professional scientists. For this purpose, I have tried to keep the text as expository as possible with a minimum of references, but added notes at the ends of some chapters to provide a guide to the literature.

I should like to take this opportunity to thank Donald Lynden-Bell for arousing my interest in this subject and for his continued encouragement and stimulation over the years; and likewise Michael Edmunds, whose collaboration in both observational and theoretical projects has been a source of pleasure as well as (one hopes) insight. Thanks are due to them, and also to Sven Åberg, Chris Pethick and the late Roger Tayler, for helpful comments on successive versions. Finally, I warmly thank Elisabeth Grothe for her willing and expert work on the diagrams.

Bernard Pagel

1 Introduction and overview

1.1 General introduction

The existence and distribution of the chemical elements and their isotopes is a consequence of nuclear processes that have taken place in the past in the Big Bang and subsequently in stars and in the interstellar medium (ISM) where they are still ongoing. These processes are studied theoretically, experimentally and observationally. Theories of cosmology, stellar evolution and interstellar processes are involved, as are laboratory investigations of nuclear and particle physics, cosmochemical studies of elemental and isotopic abundances in the Earth and meteorites and astronomical observations of the physical nature and chemical composition of stars, galaxies and the interstellar medium.

Fig. 1.1 shows a general scheme or 'creation myth' which summarizes our general ideas of how the different nuclear species (loosely referred to hereinafter as 'elements') came to be created and distributed in the observable universe. Initially — in the first few minutes after the Big Bang — universal cosmological nucleosynthesis at a temperature of the order of 10^9 K created all the hydrogen and deuterium, some ^{3}He, the major part of ^{4}He and some ^{7}Li, leading to primordial mass fractions $X \simeq 0.76$ for hydrogen, $Y \simeq 0.24$ for helium and $Z = 0.00$ for all heavier elements (sometimes loosely referred to by astronomers as 'metals'!). The existence of the latter in our present-day world is the result of nuclear reactions in stars followed by more or less violent expulsion of the products when the stars die, as first set out in plausible detail by E.M. & G.R. Burbidge, W.A. Fowler and F. Hoyle (usually abbreviated to B^{2}FH) in a classic article in *Rev. Mod. Phys.* in 1957 and independently by A.G.W. Cameron in an Atomic Energy of Canada report in the same year.

Estimates of primordial abundances from the Big Bang, based on astrophysical and cosmochemical observations and arguments, lead to a number of interesting deductions which will be described in Chapter 4. In particular, there are reasons to believe that, besides the luminous matter that we observe in the form of stars, gas and dust, there is dark matter detectable only from its gravitational effects (including gravitational lensing). This dark matter, in turn, has two quite distinct components: one is ordinary baryonic matter, consisting of protons, neutrons and electrons, which happens not to shine at any detectable wavelength (and may have primordial chemical composition); there is evidence for such baryonic dark matter (BDM) in the form of substellar sized objects from gravitational microlensing events in the halo of our Galaxy, and other

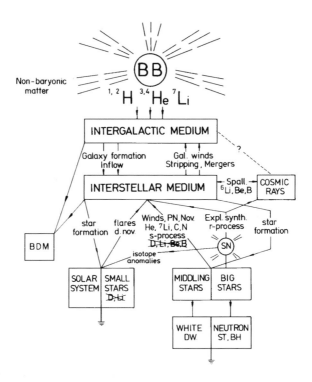

Fig. 1.1. A scenario for cosmic chemical evolution, adapted from Pagel (1981).

forms may also exist as cold molecular gas and as rarefied ionized intergalactic gas. The other kind of dark matter could consist of some combination of massive neutrinos (if any of these have mass, which is not known at present) with more exotic non-baryonic dark matter (NDM), consisting of some kind of particles envisaged in extensions of the 'Standard Model' of particle physics. The most popular candidates are so-called cold dark matter (CDM) particles that decoupled from radiation very early on in the history of the universe and now have an exceedingly low kinetic temperature which favours the formation of structures on the sort of scales that are observed in galaxy red-shift surveys. The seeds of those structures would have been quantum fluctuations imprinted very early (maybe $\sim 10^{-35}$ s) after the Big Bang during a so-called inflationary era of super-rapid expansion preceding the kind of expanding universe that we experience now, which later enabled pockets of mainly dark matter slightly denser than the average to separate out from the general expansion by their own gravity.

The first few minutes are followed by 'dark ages' lasting of the order of 10^5 years during which the universe was radiation-dominated and the baryonic gas, consisting almost entirely of hydrogen and helium, was ionized and consequently opaque to radiation. But expansion was accompanied by cooling, and when the temperature was down to a few thousand K (at a red-shift of a few thousand), matter began to dominate and first helium and then hydrogen became neutral by recombination.[1] The universe became transparent, background radiation was scattered for the last time (and is now received as black-body radiation with a temperature of 2.7 K) and eventually gas began to settle in the interiors of the pre-existing dark-matter halos, resulting in the formation of galaxies, stars and groups and clusters of galaxies. The epoch and mode of galaxy formation are not well known, but the compact ultra-luminous objects known as quasars or quasi-stellar objects (QSOs), and intergalactic gas clouds with spectral lines of heavy elements seen in absorption on the sight-lines to quasars, are known with red-shifts up to about 5, corresponding to an era when the expanding universe was only 1/6 of its present size. The emission-line spectra of quasars indicate a large heavy-element abundance (solar or more), suggesting prior stellar activity, and numerical simulations of structure formation suggest that galaxies might form at red-shifts up to 10 or so, corresponding to an age of the universe between about 500 Myr and 1.5 Gyr according to what kind of cosmological model applies to our actual universe. The first number refers to the so-called Einstein–de Sitter model, preferred by many scientists, in which there is just enough gravitating matter to eventually slow down the expansion to zero velocity, while the second refers to a so-called open universe that goes on expanding for ever. In the most popular models, galaxies form by cooling and collapse of baryonic gas contained in non-gaseous (and consequently non-dissipative) dark-matter halos, with complications caused by mergers that may take place both in early stages and much later; these mergers (or tidal interactions) may play a significant role in triggering star formation at the corresponding times.

Galaxies thus have a mixture of stars and diffuse interstellar medium. (The diffuse ISM generally consists of gas and dust, but will often be loosely referred to hereinafter as 'gas'.) Most observed galaxies belong to a sequence first established in the 1920s and 30s by Edwin Hubble. So-called early-type galaxies according to this classification are the ellipticals, which are spheroidal systems consisting of old stars and relatively little gas or dust detectable at optical, infra-red or radio wavelengths; they do, however, contain substantial amounts of very hot X-ray emitting gas. Later-type galaxies have a rotating disk-like component surrounding an elliptical-like central bulge, with the relative brightness of the disk increasing along the sequence. The disks display a spiral structure, typical of spiral galaxies such as our own Milky Way system, and the proportion of cool gas relative to stars increases along the sequence, together with the relative rate of formation of new stars from the gas. At the end of the Hubble sequence are irregular galaxies, such as the Magellanic Clouds (the nearest galaxies outside our own), and outside the sequence there is a variety of small systems known as dwarf spheroidals, dwarf irregulars and blue compact and H II (i.e. ionized hydrogen) galaxies. The last three classes actually overlap (being partly classified by the method of discovery) and they are dominated by the light of young stars and (in the case of H II and some blue compact galaxies) the gas ionized by those young stars (see

---

[1] Since this was the first time electrons were captured by protons and α-particles it might be more appropriate to talk about 'combination' rather than 'recombination'!

Chapter 3). There are also radio galaxies and so-called Seyfert galaxies with bright nuclei that have spectra resembling those of quasars; these are collectively known as active galactic nuclei (AGNs), and are associated with large ellipticals and early-type spirals. The increase in the proportion of cool gas along the Hubble sequence could be due to differences in age, the most gas-poor galaxies being the oldest, or to differences in the rates at which gas has been converted into stars in the past or added or removed by interaction with other galaxies and the intergalactic medium (IGM), probably some combination of all of these.

The figure illustrates very schematically some possible interactions between galaxies and the intergalactic medium. The diffuse IGM is a somewhat ghostly entity which may recently have been detected in the form of absorption of light from distant quasars by the He^+ Lyman-α line at 304 Å. This has been spread out by red-shift into a continuum shortward of 304 Å in the quasar's rest frame, and then also red-shifted to the middle ultra-violet range above 1200 Å accessible to the Hubble Space Telescope (HST); this so-called Gunn–Peterson effect (Gunn & Peterson 1965) was first predicted for neutral hydrogen, but has not yet been detected in that case, presumably because the hydrogen is too highly ionized. More definite evidence for intergalactic gas is found in the form of distributed X-ray emission from clusters of galaxies, where heavy elements are also present with an abundance of the order of 1/3 solar. All other evidence for intergalactic gas comes from some form of clouds, e.g. those producing the absorption-line systems in spectra of quasars, which exhibit a wide range of column densities and chemical compositions, usually with low or very low heavy-element abundances. These may represent primitive galactic halos (tenuous ellipsoidal outer extensions of disk galaxies), disks or building blocks thereof. Neutral gas is detected in disk and blue compact galaxies from the hyperfine 21 cm transition of neutral atomic hydrogen H I, but very few isolated intergalactic H I clouds have ever been detected. The interactions between the ISM and IGM include expulsion of diffuse material from galaxies in the form of galactic winds, stripping of (especially the outer) layers of the gas content of galaxies by ram pressure in an intra-cluster medium when the galaxy is a member of a cluster and/or inflow of intergalactic gas into galaxies, which latter may be indicated by observations of high-velocity H I clouds at high galactic latitudes in our own Milky Way system.

The figure also gives a schematic illustration of the complex interactions between the ISM and stars. Stars inject energy, recycled gas and nuclear reaction products ('ashes of nuclear burning') enriching the ISM from which other generations of stars form later. This leads to an increase in the heavy-element content of both the ISM and newly formed stars; the subject of 'galactic chemical evolution' (GCE) is really all about these processes. On the other hand, nuclear products may be lost from the ISM by galactic winds or diluted by inflow of relatively unprocessed material. The heavy-element content of the intra-cluster X-ray gas in rich clusters like Coma (the nearest rich cluster of galaxies, in the constellation Coma Berenices) is thought to result from winds from the constituent galaxies (see Chapter 11), or possibly from the destruction of dwarf galaxies in the cluster.

The effects of different sorts of stars on the ISM depend on their (initial) mass and on whether they are effectively single stars or interacting binaries; some of the latter are believed to be the progenitors of Type Ia supernovae (SN Ia) which are important contributors to iron-group elements in the Galaxy. Big stars, with initial mass above about $10M_\odot$ ($M_\odot$ is the mass of the Sun), have short lives (~ 10 Myr), they emit partially burned material in the form of stellar winds and those that are not too massive eventually explode as Type II (or related Types Ib and Ic) supernovae ejecting elements up to the iron group with a sprinkling of heavier elements. (Supernovae are classed as Type I or II according to whether they respectively lack, or have, lines of hydrogen in their spectra, but all except Type Ia seem to be associated with massive stars of short lifetime which undergo core collapse leading to a neutron star remnant.) The most massive stars of all are expected to collapse into black holes, with or without a prior supernova explosion; the upper limit for core collapse supernovae is uncertain, but it could be somewhere in the region of $50M_\odot$.

Middle-sized stars, between about 1 and $10M_\odot$, undergo complicated mixing processes and mass loss in advanced stages of evolution, culminating in the ejection of a planetary nebula while the core becomes a white dwarf. Such stars are important sources of fresh carbon, nitrogen and heavy elements formed by the slow neutron capture (s-) process (see Chapter 6). Finally, small stars below $1M_\odot$ have lifetimes comparable to the age of the universe and contribute little to chemical enrichment or gas recycling and increasingly merely serve to lock up material.

The result of all these processes is that the Sun was born 4.7 Gyr ago with mass fractions $X \simeq 0.70$, $Y \simeq 0.28$, $Z \simeq 0.02$. These abundances (with perhaps a slightly lower value of Z) are also characteristic of the local ISM and young stars. The material in the solar neighbourhood is about 15 per cent 'gas' (including dust which is about 1 per cent by mass of the gas) and about 85 per cent stars or compact remnants thereof; these are white dwarfs (mainly), neutron stars and black holes.

1.2   Some basic facts of nuclear physics

(1) An atomic nucleus consists of Z protons and N neutrons, where Z is the *atomic number* defining the charge of the nucleus, the number of electrons in the neutral atom and hence the chemical element, and $Z + N = A$, the *mass number* of the nuclear species. Protons and neutrons are referred to collectively as *nucleons*. Different values of A or N for a given element lead to different *isotopes*, while nuclei with the same A and different Z are referred to as *isobars*. A given nuclear species is usually symbolised by the chemical symbol with Z as an (optional) lower and A as an upper prefix, e.g. $^{56}_{26}$Fe.

Stable nuclei occupy a 'β-stability valley' in the Z, N plane (see Fig. 1.2), where one can imagine energy (or mass) being plotted along a third axis perpendicular to the paper. Various processes, some of which are shown in the figure, transform one nucleus into another. Thus, under normal conditions, a

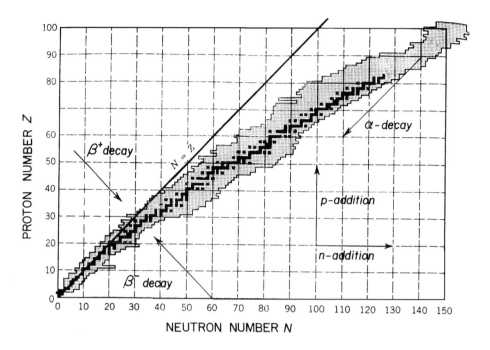

Fig. 1.2. Chart of the nuclides, in which Z is plotted against N. Stable nuclei are shown in dark shading and known radioactive nuclei in light shading. Arrows indicate directions of some simple nuclear transformations. After K.S. Krane, *Introductory Nuclear Physics*, ©1988 by John Wiley & Sons. Reproduced by permission of John Wiley & Sons, Inc.

nucleus outside the valley undergoes spontaneous decays, while in accelerators, stars and the early universe nuclei are transformed into one another by various reactions.

(2) The binding energy per nucleon varies with A along the stability valley as shown in Fig. 1.3, and this has the following consequences:

(a) Since the maximum binding energy per nucleon is possessed by ^{62}Ni, followed closely by ^{56}Fe, energy is released by either fission of heavier or fusion of lighter nuclei. The latter process is the main source of stellar energy, with the biggest contribution (7 MeV per nucleon) coming from the conversion of hydrogen into helium (H-burning).

(b) Some nuclei are more stable than others, e.g. the α-particle nuclei ^{4}He, ^{12}C, ^{16}O, ^{20}Ne, ^{24}Mg, ^{28}Si, ^{32}S, ^{36}Ar, ^{40}Ca. Nuclei with a couple of A-values (5 and 8) are violently unstable, owing to the nearby helium peak. Others are stable but only just: examples are D, ^{6,7}Li, ^{9}Be and ^{10,11}B, which are destroyed by thermonuclear reactions at relatively low temperatures.

(3) Nuclear reactions involving charged particles (p, α etc.) require them to have enough kinetic energy to get through in spite of the electrostatic repulsion of the

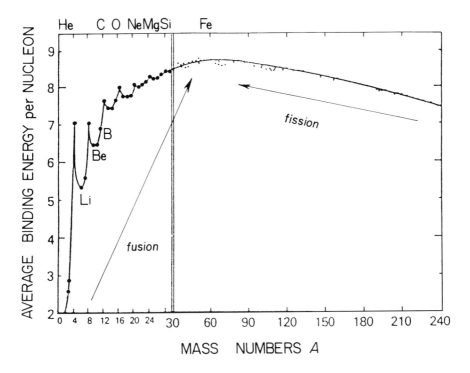

Fig. 1.3. Binding energy per nucleon as a function of mass number. Adapted from Rolfs & Rodney (1988).

target nucleus (the 'Coulomb barrier'); the greater the charges, the greater the energy required. In the laboratory, the energy is supplied by accelerators, and analogous processes are believed to occur in reactions induced in the ISM by cosmic rays (see Chapter 9). In the interiors of stars, the kinetic energy exists by virtue of high temperatures (leading to *thermonuclear* reactions) and when one fuel (e.g. hydrogen) runs out, the star contracts and becomes hotter, eventually allowing a more highly charged fuel such as helium to 'burn'.

There is no Coulomb barrier for neutrons, but free neutrons are unstable so that they have to be generated *in situ*, which again demands high temperatures.

1.3   The local abundance distribution

Fig. 1.4 shows the 'local galactic' abundances of isobars, based on a combination of elemental and isotopic determinations in the Solar System with data from nearby stars and emission nebulae. These are sometimes referred to as 'cosmic abundances', but because there are significant variations among stars and between and across galaxies this term is best avoided. The curve shows a number of features that give clues to the origin of the various elements:

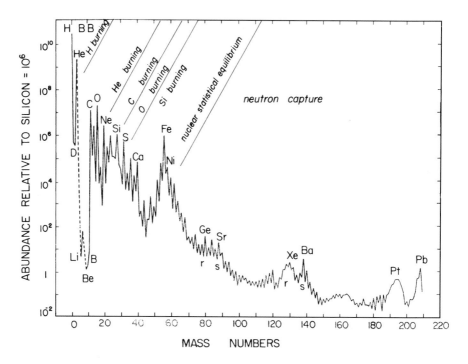

Fig. 1.4. The 'local galactic' abundance distribution of nuclear species, normalised to 10^6 ^{28}Si atoms, adapted from Cameron (1982).

(1) Hydrogen is by far the most abundant element, followed fairly closely by helium. This is mainly formed in the Big Bang, with some topping up of the primordial helium abundance ($Y_P \simeq 0.24$) by subsequent H-burning in stars ($Y \simeq 0.28$ here and now). Only a small fraction ($Z \simeq 0.02$) of material in the ISM and in newly formed stars has been subjected to high enough temperatures and densities to burn helium.

(2) The fragile nuclei Li, Be and B are very scarce, being destroyed in the harsh environment of stellar interiors although some of them may also be created there. (The lithium abundance comes from measurements in meteorites; it is still lower in the solar photosphere because of destruction by mixing with hotter layers below.) On the other hand, they are much more abundant in primary cosmic rays as a result of $\alpha - \alpha$ fusion and spallation reactions between p, α and (mainly) CNO nuclei at high energies. The latter may also account for the production of some of these species in the ISM. Some ^{7}Li (about 10 per cent) also comes from the Big Bang.

(3) Although still more fragile than Li, Be and B, deuterium is vastly more abundant (comparable to Si and S). D is virtually completely destroyed in gas recycled through stars and there are no plausible mechanisms for creating it there in such quantities, but it is nicely accounted for by the Big Bang. The observed

abundance is a lower limit to the primordial abundance. The light helium isotope ^{3}He is the main product when deuterium is destroyed and can also be freshly produced in stars. It has comparable abundance to D and some is made in the Big Bang, but in this case the interpretation of the observed abundance is more complicated.

(4) Nuclei from carbon to calcium show a fairly regular downward progression modulated by odd:even and shell effects in nuclei which affect their binding energy (see Chapter 2). These arise from successive stages in stellar evolution when the exhaustion of one fuel is followed by gravitational contraction and heating enabling the previous ashes to 'burn'. The onset of carbon burning, which leads to Mg and nearby elements, is accompanied by a drastic acceleration of stellar evolution due to neutrino emission. This occurs because at the high densities and temperatures required for such burning, much of the energy released comes out in the form of neutrinos which directly escape from the interior of the star, relative to photons for which it takes of the order of 10^6 yrs for their energy to reach the surface. It is this rapid loss of energy that speeds up the stellar evolution (cf. Chapter 5).

(5) The iron-group elements show an approximation to nuclear statistical equilibrium at a temperature of several $\times 10^9$ K or several tenths of an MeV leading to the iron peak. This was referred to by B^{2}FH as the 'e-process' referring to thermodynamic equilibrium. Because the neutrinos escape, complete thermodynamic equilibrium does not apply here, but one has an approach to statistical equilibrium in which forward and reverse nuclear reactions balance. The iron peak is thought to result from explosive nucleosynthesis which may occur in one or other of two typical situations. One of these involves the shock that emerges from the core of a massive star that has collapsed into a neutron star; heat from the shock ignites the overlying silicon and oxygen layers in a Type II (or related) supernova outburst. Another possible cause is the sudden ignition of carbon in a white dwarf that has accreted enough material from a companion to bring it over the Chandrasekhar mass limit (supernova Type Ia). In either case, the quasi-equilibrium is frozen at a distribution characteristic of the highest temperature reached in the shock before the material is cooled by expansion, but the initial neutron–proton ratio is another important parameter. Calculations and observations both indicate that the dominant product is actually ^{56}Ni, the 'doubly magic' most stable nucleus with equal numbers of protons and neutrons, which later decays into ^{56}Fe.

(6) Once iron-group elements have been produced, all nuclear binding energy has been extracted and further gravitational contraction merely leads to photodisintegration of nuclei with a catastrophic consumption of energy. This is one of several mechanisms that either drive or accelerate core collapse. By contrast, the SN Ia explosion of a white dwarf simply leads to disintegration of the entire star. Consequently charged-particle reactions do not lead to significant nucleosynthesis beyond the iron group and the majority of heavier nuclei result

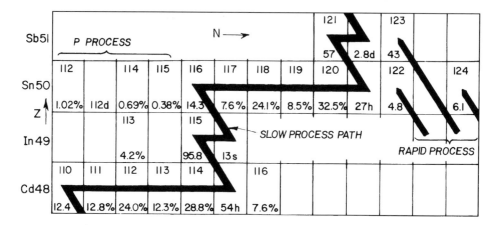

Fig. 1.5. Paths of the r-, s- and p-processes in the neighbourhood of the tin isotopes. Numbers in the boxes give mass numbers and percentage abundance of the isotope for stable species, and β-decay lifetimes for unstable ones. ^{116}Sn is an s-only isotope, shielded from the r-process by ^{116}Cd. After D.D. Clayton, W.A. Fowler, T.E. Hull & B.A. Zimmerman, *Ann. Phys.*, **12**, 331, ©1961 by Academic Press, Inc. Courtesy Don Clayton.

from neutron captures. One or more such captures on a seed nucleus (in practice mostly ^{56}Fe) lead to the production of a β-unstable nucleus (e.g. ^{59}Fe) and the eventual outcome then depends on the relative time-scales for neutron addition and β-decay. In the slow or s-process, neutrons are added so slowly that most unstable nuclei have time to undergo β-decay and nuclei are built up along the stability valley in Fig. 1.2 up to ^{209}Bi, whereafter α-decay terminates the process. In the rapid or r-process, the opposite is the case and many neutrons are added under conditions of very high temperature and neutron density, building unstable nuclei up to the point where (n, γ) captures are balanced by (γ, n) photodisintegrations leading to a kind of statistical equilibrium modified by (relatively slow) β-decays. Eventually (e.g. after a SN explosion), the neutron supply is switched off and the products then undergo a further series of β-decays ending up at the nearest stable isobar, which will be on the neutron-rich side of the β-stability valley (Fig. 1.5). Some species have contributions from both r- and s-processes, while others are pure r (if by-passed by the s-process) or pure s (if shielded by a more neutron-rich stable isobar); the actinides (U, Th etc.) are pure r-process products. A few species on the proton-rich side of the stability valley are not produced by neutron captures at all and are attributed to what B^{2}FH called the 'p-process'; these all have very low abundances and are believed to arise predominantly from (γ,n) photodisintegrations of neighbouring nuclei.

Neutron capture processes give rise to the so-called magic-number peaks in the abundance curve, corresponding to closed shells with 50, 82 or 126 neutrons (see Chapter 2). In the case of the s-process, the closed shells lead to low neutron-capture cross-sections and hence to abundance peaks in the neighbourhood of Sr, Ba and Pb (cf. Fig. 1.4), since such nuclei will predominate after exposure to a chain of neutron captures. In the r-process, radioactive progenitors with closed shells are more stable and hence more abundant than their neighbours and their subsequent decay leads to the peaks around Ge, Xe and Pt on the low-A side of the corresponding s-process peak.

1.4   Brief outline of stellar evolution

(1) An interstellar cloud collapses forming a generation of stars. The masses of the stars are spread over a range from maybe 0.01 $M_\odot$ or less to maybe $100M_\odot$ or so. The distribution function of stellar masses at birth, known as the 'initial mass function' (IMF), has more small stars than big ones.

(2) The young stars, which appear as variable stars with emission lines known as T Tauri stars and related classes, initially derive their energy from gravitational contraction, which leads to a steady increase in their internal temperature (see Chapter 5). Eventually the central temperature becomes high enough ($\sim 10^7$ K or 1 keV) to switch on hydrogen burning and the star lies on the 'zero-age main sequence' (ZAMS) of the Hertzsprung–Russell (HR) diagram in which luminosity is plotted against surface temperature (Fig. 1.6). Stars spend most of their lives (about 80 per cent) in a main-sequence band stretching slightly upwards from the ZAMS; the corresponding time is short (a few $\times 10^6$ years) for the most massive and luminous stars and very long ($> 10^{10}$ years) for stars smaller then the Sun, because the luminosity varies as a high power of the mass and so bigger stars use up their nuclear fuel supplies faster.

(3) When hydrogen in a central core occupying about 10 per cent of the total mass is exhausted, there is an energy crisis. The core, now consisting of helium, contracts gravitationally, heating a surrounding hydrogen shell, which consequently ignites to form helium and gradually eats its way outwards (speaking in terms of the mass coordinate). At the same time, the envelope expands, making the star a red giant in the upper right part of the diagram.

(4) Eventually the contracting core becomes hot enough to ignite helium ($3\alpha \rightarrow {}^{12}C$; ${}^{12}C + {}^4He \rightarrow {}^{16}O$) and the core contraction is halted.

(5) In sufficiently big stars ($>\sim 10M_\odot$) this process repeats; successive stages of gravitational contraction and heating permit the ashes of the previous burning stage to be ignited leading to C, Ne, O and Si burning in the centre with less advanced burning stages in surrounding shells leading to an onion-like structure with hydrogen-rich material on the outside. Silicon burning leads to a core rich in iron-group elements and with a temperature of the order of 10^9 K, i.e. about 100 keV.

(6) The next stage of contraction is catastrophic, partly because all nuclear energy supplies have been used up when the iron group is reached, and partly because the core, having reached nearly the Chandrasekhar limiting mass for a white

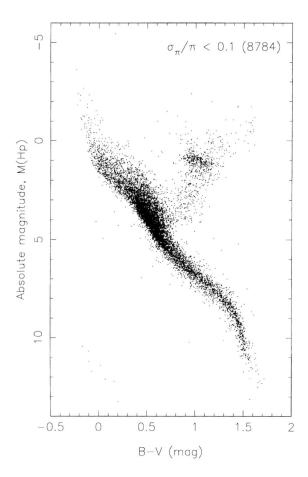

Fig. 1.6. HR diagram for nearby stars with well-known parallaxes (i.e. distances) from the HIPPARCOS astrometric satellite mission, after M.A.C. Perryman *et al.*, *Astr. Astrophys.*, **304**, 69 (1995). The abscissa, B (blue) − V (visual) magnitude is a measure of 'redness' or inverse surface temperature ranging from about 2×10^4 K on the left to about 3000 K on the right, while the ordinate measures luminosity in the visual band expressed by absolute magnitude M_{Hp} in the HIPPARCOS photometric system. Luminosity decreases by a factor of 100 for every 5 units increase in M and the Sun's $M_V \simeq M_{\mathrm{Hp}}$ is 4.8. The main sequence forms a band going from top left to bottom right in the diagram, with the ZAMS forming a lower boundary. Subgiants and red giants go upward and to the right from the MS and a few white dwarfs can be seen at the lower left. Courtesy Michael Perryman.

dwarf supported by electron degeneracy pressure, is close to instability and also suffers loss of pressure due to neutronization by inverse β-decays. Further contraction leads to photodisintegration, which absorbs energy, and this leads to dynamical collapse of the core which continues until it reaches nuclear density and forms a neutron star. (If the mass of collapsing material is too large, then

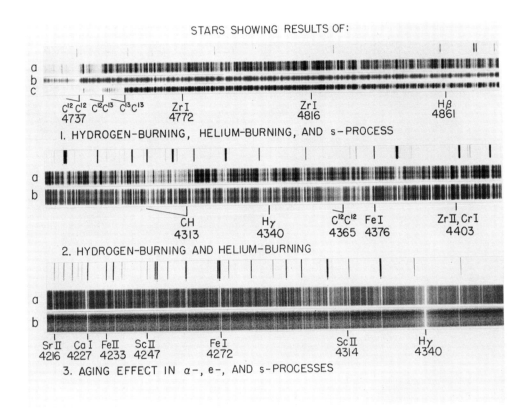

Fig. 1.7. Photographic (negative) spectra of stars showing various aspects of nucleosynthesis. Top: (a) carbon star X Cancri with $^{12}C/^{13}C \simeq 4$ from H-burning by the CNO cycle and a suggestion of enhanced Zr; (b) peculiar carbon star HD 137613 without ^{13}C bands in which hydrogen is apparently weak (H-deficient carbon star); (c) carbon star HD 52432, with $^{12}C/^{13}C$ ~ 4. Middle: (a) Normal carbon star HD 156074, showing the CH band and Hγ; (b) Peculiar (H-poor) carbon star HD 182040 showing C$_2$ but weak Hγ and CH. Bottom: (a) normal F-type star ζ Pegasi (slightly hotter than the Sun); (b) old Galactic halo population star HD 19445 with similar temperature (shown by Hγ) but very low metal abundance (about 1/100 solar). After E.M. and G.R. Burbidge, W.A. Fowler & F. Hoyle 1957, *Rev. Mod. Phys.*, **29**, 547 (B^{2}FH). Courtesy Margaret and Geoffrey Burbidge.

a black hole will probably form instead.) The stiff equation of state of nuclear matter leads to a bounce which sends a shock out into the surrounding layers. This heats them momentarily to high temperatures, maybe 5×10^9 K in the silicon layer, leading to explosive nucleosynthesis of iron-peak elements, mainly ^{56}Ni (which later decays by electron capture and β^+ to ^{56}Fe); more external layers are heated to lower temperatures resulting in milder changes. Assisted by high-energy neutrinos, the shock expels the outer layers in a supernova explosion (Type II and related classes); the ejecta eventually feed the products into the

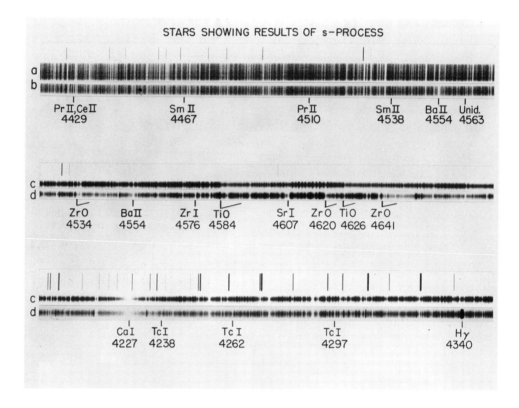

Fig. 1.8. Spectra showing effects of s-process. Top: (a) normal G-type giant κ Gem (similar temperature to the Sun); (b) Ba II star HD 46407, probably not an AGB star but affected by a companion which was. Middle: (c) M-type giant 56 Leo, showing TiO bands; (d) S-type AGB star R Andromedae showing ZrO bands, partly due to enhanced Zr abundance. Bottom: another spectral region of the same two stars showing Tc features in R And which indicate s-processing and dredge-up within a few half-lives of technetium (2×10^5 yrs) before the present. After B^{2}FH (1957). Courtesy Margaret and Geoffrey Burbidge.

ISM which thus becomes enriched in 'metals' in course of time. This scenario was first put forward in essentials by Hoyle (1946), and modern versions give a fairly good fit to the local abundances of elements from oxygen to calcium. The iron yield is uncertain because it depends on the mass cut between expelled and infalling material in the silicon layer, but can be parameterized to fit observational data, e.g. for SN 1987A in the Large Magellanic Cloud (LMC). The upshot is that iron-group elements are probably underproduced relative to local abundances, but the deficit is plausibly made up by contributions from supernovae of Type Ia. A subset of Type II supernovae may also be the site of the r-process (see Chapter 6).

(7) For intermediate mass stars (IMS), $\sim 1M_\odot \leq M \leq \sim 8M_\odot$, stages (i) to (iii) are much as before, but these never reach the stage of carbon burning because

the carbon–oxygen core becomes degenerate first by virtue of high density, and later evolution is limited by extensive mass loss from the surface. After core helium exhaustion, these stars re-ascend the giant branch along the so-called asymptotic giant branch (AGB) track (see Fig. 5.14) with a double shell source: helium-burning outside the CO core and hydrogen-burning outside the He core. This is an unstable situation giving rise to thermal pulses or 'helium shell flashes' in which the two sources alternately switch on and off driving inner and outer convection zones (in which mixing takes place) during their active phases (see Chapter 5). The helium-burning shell generates ^{12}C and neutrons, either from ^{22}Ne$(\alpha, n)^{25}$Mg or from ^{13}C$(\alpha, n)^{16}$O, leading to s-processing, and the products are subsequently brought up to the surface in what is known as the third dredge-up process. This process leads to observable abundance anomalies in the spectra of AGB stars, carbon and S stars; see Figs. 1.7, 1.8). The products are then ejected into the ISM by mass loss in the form of stellar winds and planetary nebulae (PN), leaving a white dwarf as the final remnant. If the white dwarf is a member of a close binary system, it can occasionally be 'rejuvenated' by accreting material from its companion, giving rise to cataclysmic variables, novae and supernovae of Type Ia (cf. Chapter 5).

2 Thermonuclear reactions

2.1 General properties of nuclei

Nuclear matter has an approximately constant density, the nuclear radius being about $R_0 A^{1/3}$ where $R_0 \simeq 1.2$ fm (1 fm, Fermi or femtometre $= 10^{-13}$ cm) and the density 2.3×10^{14} gm cm^{-3} or 0.14 amu fm^{-3} (the atomic mass unit, amu, is defined as $1/12$ of the mass of a neutral ^{12}C atom). R_0 is of the order of the Compton wavelength of the π meson, $\hbar/m_\pi c = 1.4$ fm. Some properties of nuclei can be explained on the basis of Niels Bohr's 'liquid drop' model, but others depend on the very complicated many-body effects of interactions among the constituent nucleons. These, in turn, are partially explained by the shell model, in which each nucleon is assumed to move in the mean field caused by all the others and has quantum numbers analogous to those of electrons in an atom. Gross features of the beta-stability valley (Fig. 1.2) and the binding-energy curve (Fig. 1.3) are summarised in the semi-empirical mass formula originally due to C.F. von Weizsäcker

$$M(A,Z) = Zm(^1\text{H}) + (A - Z)m_n - \text{B.E.}/c^2, \tag{2.1}$$

where M is the mass of the neutral atom, B.E. is the total binding energy, $c^2 = 931.5$ MeV/amu, and

$$\text{B.E.} \simeq a_v A - a_s A^{2/3} - a_c Z(Z-1)A^{-1/3} - a_{sym}\frac{(A-2Z)^2}{A}$$
$$+ \frac{1+(-1)^A}{2}(-1)^Z a_p A^{-3/4}. \tag{2.2}$$

Here $a_v \simeq 15.5$ MeV represents a constant term in B.E. per nucleon, $a_s \simeq 16.8$ MeV provides a surface term allowing for a reduced contribution to B.E. from nucleons at the surface and $a_c \simeq 0.72$ MeV $\simeq 0.6e^2/R_0$ allows for the electrostatic repulsion of the protons. $a_{sym} \simeq 23$ MeV gives a term favouring equal numbers of protons and neutrons, especially at low A, which arises in part from the Pauli Exclusion Principle and in part from symmetry effects in the interaction between nucleons. The last term, with $a_p \simeq 34$ MeV, comes from a 'pairing force' due to the propensity of identical particles to form pairs with opposite spins. This last term permits several stable isobars to exist for even A.

To the extent that the nuclear potential is central, each nucleon has an orbital angular momentum ℓ as well as a spin $\mathbf{s}$, where $s = 1/2$, and total angular momentum $\mathbf{j}$. These

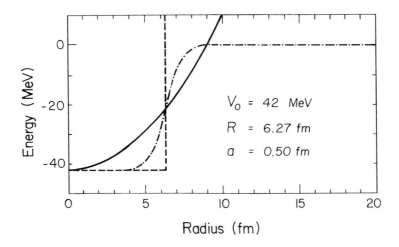

Fig. 2.1. Approximate potentials for the nuclear shell model. The solid curve represents the 3-dimensional harmonic oscillator potential, the dashed curve the infinite square well and the dot-dashed curve a more nearly realistic Woods–Saxon potential, $V(r) = -V_0/[1+\exp\{(r-R)/a\}]$ (Woods & Saxon 1954). Adapted from Cowley (1995).

combine by $j-j$ coupling to form the total nuclear 'spin' **I**. Nuclear states are specified by their energy (relative to the ground state), and their spin and parity I^π, e.g. 0^+, where $\pi = (-1)^{\sum \ell}$. Because the individual j's of nucleons are half-integral, I is always an integer for even A and half-integral for odd A. The nuclear force is almost charge-independent, but depends on the relative orientation of particle spins and has a small non-central or tensor component. The force is of short range, overcoming the Coulomb repulsion of protons at distances $\sim R_0$, but becomes highly repulsive at shorter distances < 0.5 fm maintaining a nearly constant nuclear density.

Many significant properties of nuclear states are explained by the shell model, for which Maria G. Mayer and Hans Jensen shared the 1963 Nobel prize. This is patterned after the shell model of the atom, in which electrons fill consecutively the lowest lying available bound states and it is assumed that each nucleon effectively moves in a spherically symmetrical mean potential $V(r)$, giving it a definite orbital angular momentum quantum number ℓ and a quantum number n related to the number of nodes in the radial wave function. For the Coulomb potential, which does not apply here, the energy depends uniquely on the principal quantum number $N = n + \ell$. In the more general case, the energy depends on both n and ℓ, and n is simply used as a label to indicate successively higher energy states (starting with $n = 1$) for given ℓ, so we can have (using atomic notation) $1s, 2s \ldots, 1p, 2p \ldots$ etc.

Some simple models for $V(r)$ are shown in Fig. 2.1. Two crude approximations, the infinite square well (ISW) and the 3-dimensional harmonic oscillator (3DHO), have the advantage of leading to analytical solutions of the Schrödinger equation which lead to the following energy levels:

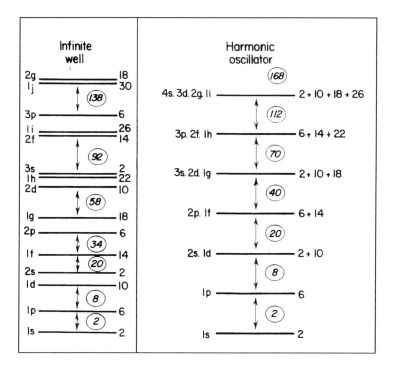

Fig. 2.2. Energy levels for ISW and 3DHO potentials. Each shows major gaps corresponding to closed shells and the numbers in circles give the cumulative number of protons or neutrons allowed by the Pauli principle. In a more realistic potential, the levels for given (n, ℓ) are intermediate between these extremes, in which the lower magic numbers 2, 8 and 20 are already apparent. Adapted from Krane (1988).

For ISW ($V = -V_0$ for $r < R$; $V = \infty$ for $r > R$), Clayton (1968) has given the approximation

$$E(n, \ell) \simeq -V_0 + \frac{\hbar^2}{2mR^2}\left[\pi^2\left(n + \frac{\ell}{2}\right)^2 - \ell(\ell + 1)\right], \tag{2.3}$$

whereas for 3DHO ($V = -V_0 + \frac{1}{2}m\omega^2 r^2$),

$$E(n, \ell) = -V_0 + \hbar\omega\left(2n + \ell - \frac{1}{2}\right) \tag{2.4}$$

and the levels are highly degenerate. Fig. 2.2 shows the resulting energy levels for the two cases, in which distinct energy gaps appear corresponding to closed shells of either protons or neutrons, each of which separately obey the Pauli principle. A more realistic form for $V(r)$ leads to energy levels intermediate between the two, shown on the left side of Fig. 2.3, but the higher cumulative occupation numbers above 20 do not correspond to reality. The solution to this problem came from the inclusion of a spin-orbit potential, analogous to the spin-orbit interaction that causes fine structure in atomic spectra, but in this case not of electromagnetic origin. Thus each (n, ℓ) level is split into two, with $j = \ell + s$ and $j = \ell - s$ respectively, and the splitting is proportional to $\mathbf{l.s}$, for which the expectation value can be calculated by an elementary argument:

Since

$$\mathbf{j}^2 = (\mathbf{l} + \mathbf{s})^2 = \mathbf{l}^2 + 2\mathbf{l.s} + \mathbf{s}^2, \tag{2.5}$$

$$\mathbf{l.s} = \frac{1}{2}(\mathbf{j}^2 - \mathbf{l}^2 - \mathbf{s}^2) \tag{2.6}$$

and its expectation value (in units of $\hbar^2$) is

$$<\mathbf{l.s}> = \frac{1}{2}[j(j+1) - \ell(\ell+1) - s(s+1)], \tag{2.7}$$

whence

$$<\mathbf{l.s}>_{\ell+1/2} - <\mathbf{l.s}>_{\ell-1/2} = \frac{1}{2}(2\ell + 1) \tag{2.8}$$

so that the splitting increases with ℓ. Taking the higher j value to correspond to the lower energy (i.e. a negative coefficient of $\mathbf{l.s}$ in the energy term, specifically $-13A^{-2/3}$ MeV) gives the correct magic numbers for shell closures shown on the right side of Fig. 2.3.

These shell closures have a profound influence on nuclear properties, in particular the binding energy (adding terms not accounted for in Eq. 2.2), particle separation energies and neutron capture cross-sections. The shell model also forms a basis for predicting the properties of nuclear energy levels, especially the ground states. For even-A, even-Z nuclei, the pairing effect expressed by the a_{sym} term in Eq. (2.2) combines with the j-dependence of the energy levels to make all their ground states 0^+. For even-A, odd-Z nuclei, on the other hand, the extra proton and neutron have an overall symmetrical wave function in their respective coordinates (leading to a more compact wave function and hence lower energy than an antisymmetrical one), with parallel spins in many cases. Thus among the few stable even-A, odd-Z nuclei (^{2}H, ^{6}Li, ^{10}B, ^{14}N), three have 1^+ ground states, while ^{10}B has 3^+.

In the case of odd-A nuclei, I^π in the ground state is fixed by the single 'valency' nucleon in the lowest vacant level shown in Fig. 2.3. In ^{15}O, for example, 8 protons fill closed shells up to $1p_{1/2}$, 6 neutrons fill closed shells up to $1p_{3/2}$ and the last neutron occupies $1p_{1/2}$ making the state $\frac{1}{2}^-$. In ^{17}O, on the other hand, all states up to $1p_{1/2}$ are filled by both protons and neutrons and the extra neutron occupies $1d_{5/2}$, making the state $\frac{5}{2}^+$. These rules are generally valid for $A < 150$ and $190 < A < 220$; outside these ranges, cooperative effects of many single-particle shell-model states may combine to produce a permanent nuclear deformation, and even within them the single-particle shell model does not account for all excited states. A more general classification uses the concept of isospin which treats protons and neutrons as states of a single nucleon with $+1/2$ and $-1/2$ components of a spin-like vector along the z-axis in an abstract space.

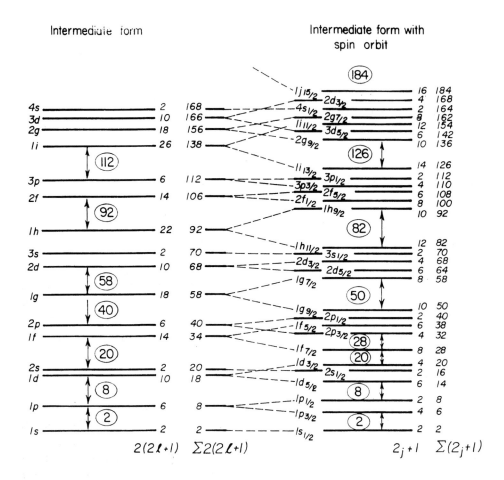

Fig. 2.3. At left, energy levels for a Woods–Saxon potential with $V_0 \simeq 50$ MeV, $R = 1.25A^{1/3}$ fm and $a = 0.524$ fm, neglecting spin-orbit interaction. At right, the same with spin-orbit term included. Adapted from Krane (1988).

2.2   Nuclear reaction physics

Typical reactions relevant to astrophysics are

- H-burning in the Big Bang ($T \simeq 10^9$ K, $kT \simeq 0.1$ MeV) and in stars (10^7 K $< T < 10^8$ K, 1 keV $< kT < 10$ keV).
- He-burning in stars ($T \simeq 10^8$ K, $kT \simeq 10$ keV).
- C, Ne, O burning in stars (10^8 K $< T < 10^9$ K, 10 keV $< kT < 0.1$ MeV).
- Si burning, hydrostatically near 10^9 K (0.1 MeV) and explosively at several times 10^9 K.
- Neutron capture in the Big Bang, s-process (a few $\times$ 10^8 K) and r-process (a few $\times$ 10^9 K).

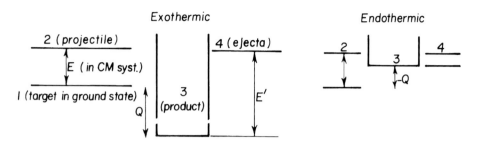

Fig. 2.4. Schematic illustration of a generic nuclear reaction.

- Spallation reactions, especially those involving cosmic rays in the ISM (non-thermal, with MeV to GeV energies). Spallation reactions are those in which one or a few nucleons are split off from a nucleus (see Chapter 9).

Fig. 2.4 illustrates two kinds of generic nuclear reaction. A target labelled '1', which is in its ground state, is impacted by a projectile "2" with initial kinetic energy E in the centre-of-mass (CM) coordinate system, leading to a product '3' and ejecta '4' with energy E'. The reaction is summarised in compact notation as 1(2,4)3, e.g. $^{15}N(p, \alpha)^{12}C$. The energy released, $E' - E$, is usually called Q, and

$$Q = c^2(M_1 + M_2 - M_3 - M_4), \tag{2.9}$$

where $c^2 = 931.5$ MeV/amu and the reaction is said to be exothermic or endothermic according to whether Q is positive or negative. The value of Q can be deduced from tables of nuclear masses, and the CM kinetic energy is $\frac{1}{2}mv^2$, where m is the reduced mass $M_1M_2/(M_1 + M_2)$ and v the relative velocity, since relativistic effects are negligible for nuclear particles at the relevant energies. Nuclear reactions are subject to the standard conservation laws in addition to that of mass-energy, i.e. conservation of linear and angular momentum, parity, baryon and lepton numbers and isospin, except that parity and isospin are not always conserved in weak interactions.

The probability of a particular reaction (or scattering) taking place is measured by the relevant cross-section, which can be thought of in classical terms as the cross-sectional area of a sphere that has the same probability of being hit. The corresponding nuclear dimensions would be of order

$$\begin{aligned} \pi(R_1 + R_2)^2 &= \pi[1.2 \times 10^{-13}(A_1^{1/3} + A_2^{1/3})\,\text{cm}]^2 \\ &= 4.5 \times 10^{-26}(A_1^{1/3} + A_2^{1/3})^2\,\text{cm}^2 \\ &\sim 10^{-24}\,\text{cm}^2 \equiv 1\,\text{barn}, \end{aligned} \tag{2.10}$$

but at relevant energies the cross-section is in fact related to the de Broglie wave-length and mutual orbital angular momentum of the particles. The incoming particle is represented quantum-mechanically as a plane wave which in turn can be decomposed into a series of spherical waves characterised by their angular momentum quantum number ℓ, corresponding to an impact parameter r_ℓ where

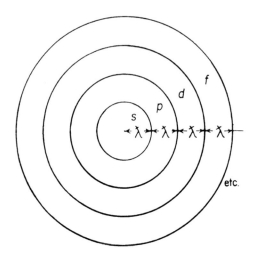

Fig. 2.5. Dependence of total reaction cross-section on the angular momentum quantum number.

$$mv\, r_\ell = \ell\hbar, \tag{2.11}$$

i.e. the impact parameters are quantised in units of the reduced de Broglie wave-length $\bar{\lambda} = \hbar/mv$. A semi-classical interpretation is that the plane of impact parameters is divided into zones of differing values of ℓ as shown in Fig. 2.5. The area of the ℓth zone, which is the total cross-section for given ℓ, is then

$$\pi(\hbar/mv)^2\left[(\ell+1)^2 - \ell^2\right] = \pi(\hbar/mv)^2\,(2\ell+1). \tag{2.12}$$

Thus for an s-wave ($\ell = 0$), which corresponds to hitting the bull's eye and so usually dominates unless forbidden by selection rules or upstaged by a resonance, the total cross-section (of which the major component is normally elastic scattering) is given by

$$\sigma_{1,2\,tot} = (1+\delta_{12})\pi\bar{\lambda}^2 = (1+\delta_{12})\frac{65.7}{AE}\ \text{barn}, \tag{2.13}$$

where A is the reduced mass in amu and E in keV. The factor $(1+\delta_{12})$ comes from the fact that, when two identical particles interact, the cross-section is doubled because either of them can be identified as the projectile or as the target. Actual cross-sections vary widely according to the nature of the interaction involved, e.g. for $E = 2$ MeV some characteristic cross-sections are as follows:

- for $^{15}\text{N}(p,\alpha)^{12}\text{C}$, $\sigma = 0.5$ b (strong nuclear force);
- for $^{3}\text{He}(\alpha,\gamma)^{7}\text{Be}$, $\sigma \simeq 10^{-6}$ b (electromagnetic interaction);
- for $p(p,e^{+}v)d$, $\sigma \simeq 10^{-20}$ b (weak force, based on theoretical calculations, since the rate is too small to be measured).

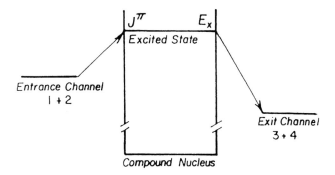

Fig. 2.6. Schematic of a compound-nucleus reaction.

2.3   **Non-resonant reactions**

Reactions may be either 'direct', where an energetic particle has such a small wave-length that it only 'sees' one nucleon of the target, or 'compound nucleus' reactions, where the projectile's energy is shared among many nucleons in successive collisions within a compound nucleus which can then decay into one of a number of 'exit channels' (Fig. 2.6). The first case is the more common one in in reactions between light nuclei, whereas the second one dominates for heavier ones.

A compound-nucleus reaction can be described as

$$1 + 2 \rightarrow C \rightarrow 3 + 4 + Q \tag{2.14}$$

and the cross-section results from the joint probability that the compound nucleus is formed from 1+2 and that it then decays into the channel 3+4, i.e.

$$\sigma = \sigma_{tot}\,\omega\,|< 3,4\,|\,H_{II}\,|\,C >< C\,|\,H_I\,|\,1,2 >|^2, \tag{2.15}$$

where $\omega = (2I_3+1)(2I_4+1)/[(2I_1+1)(2I_2+1)]$ is the statistical-weight factor (obtained by averaging over initial states and summing over final states). The matrix elements in angle brackets contain nuclear factors and (in the case of charged particles) the Coulomb barrier penetration probabilities or Gamow factors, originally calculated in the theory of α-decay, which can be roughly estimated as follows (Fig. 2.7).

A particle with energy E moving one-dimensionally along the negative x-axis in a potential $V(x)$ is described quantum-mechanically by a plane wave

$$\Psi = A\,e^{i(\omega t + kx)} = A\,e^{\frac{i}{\hbar}(Et + \sqrt{[2m(E-V)]}x)}. \tag{2.16}$$

The probability current is conserved for $E > V$. When penetrating the barrier, $E - V$ becomes negative and $\Psi^*\Psi$ decays as $e^{-(2/\hbar)\sqrt{[2m(V(x)-E)]}\delta x}$ as the particle moves from position $x + \delta x$ to position x. The total probability of barrier penetration (neglecting angular momentum which supplies an additional 'centrifugal' barrier) then includes a

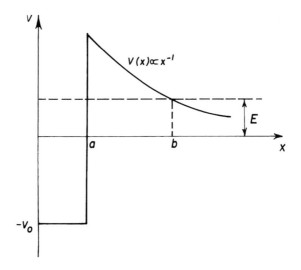

Fig. 2.7. Coulomb barrier penetration by a charged particle. a is the range of the nuclear force and b the classical turning point.

factor e^{-G} where

$$G = \frac{2}{\hbar} \int_a^b \sqrt{[2m(V(x) - E)]}\, dx = \frac{2(2mE)^{1/2}}{\hbar} \int_a^b \sqrt{\left(\frac{b}{x} - 1\right)}\, dx \qquad (2.17)$$

and

$$b = \frac{Z_1 Z_2 e^2}{E} = \frac{2Z_1 Z_2 e^2}{mv^2} \qquad (2.18)$$

is the classical turning-point. The integral can be evaluated by an elementary trigonometric substitution, giving

$$G = \frac{4Z_1 Z_2 e^2}{\hbar v} [\arccos \sqrt{(a/b)} - \sqrt{(a/b)}\sqrt{(1 - a/b)}]. \qquad (2.19)$$

Since $a \ll b$ (typically by a factor of order 100), the factor in brackets is nearly $\pi/2$ and is quite insensitive to the value of a. Numerically, the barrier penetration probability for s-waves is thus

$$P \propto e^{-2\pi Z_1 Z_2 e^2 /(\hbar v)} \equiv e^{-2\pi \eta} = e^{-BE^{-1/2}} \qquad (2.20)$$

where

$$B = (2m)^{\frac{1}{2}} \pi Z_1 Z_2 e^2 /\hbar = 31.3\, Z_1 Z_2 A^{\frac{1}{2}}; \qquad (2.21)$$

A is the reduced mass in amu and E is in keV. E.g. for 1 keV protons, $P \sim 10^{-9}$, but the penetration probability is enhanced at high densities by electron screening effects. At very high densities, so-called pycnonuclear reactions occur, even at low temperatures, owing to quantum tunnelling and kinetic energy from vibrations in a solid lattice.

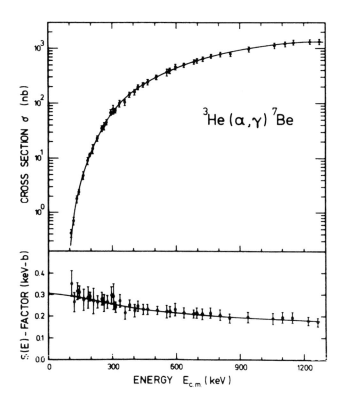

Fig. 2.8. Energy dependence of the cross-section and of $S(E)$ for a typical nuclear reaction. The extrapolation of $S(E)$ to zero energy is carried out with the aid of theory. After C.E. Rolfs & W.S. Rodney (1988), *Cauldrons in the Cosmos*, University of Chicago Press, Fig. 4.4, p. 157. ©1988 by the University of Chicago. Courtesy Claus Rolfs.

Pycnonuclear reactions set in quite suddenly above a certain critical density, which (in gm cm^{-3}) is of the order of 10^6 for hydrogen, 10^9 for He and 10^{10} for C; such densities can be reached in accreting white dwarfs and advanced burning stages of massive stars.

The steep dependence of P on energy leads to the use of the 'astrophysical S-factor' defined by

$$\sigma(E) = \frac{1}{E} e^{-2\pi\eta} S(E), \qquad (2.22)$$

which contains the purely nuclear part of the cross-section, plus factors of a few in the barrier penetration probability, and varies only slowly with E in the absence of resonances. Laboratory measurements of cross-sections, which usually have to be carried out at higher energies than those mostly relevant in stars, are converted into $S(E)$ to facilitate extrapolation to these lower energies (Fig. 2.8).

A more exact expression for the barrier penetration factor, derived for example by Clayton (1968), is

$$P_\ell(E) = \left(\frac{E_c}{E}\right)^{1/2} \exp\left[-BE^{-1/2} + 4\left(\frac{E_c}{\hbar^2/2mR^2}\right)^{1/2} - 2\left(\ell + \frac{1}{2}\right)^2 \left(\frac{\hbar^2/2mR^2}{E_c}\right)^{1/2}\right] \quad (2.23)$$

or numerically

$$\begin{aligned}
P_\ell(E) &= \left(\frac{E_c}{E}\right)^{1/2} \exp[-BE^{-1/2} + 1.05(ARZ_1Z_2)^{1/2} \\
&\quad - 7.62\left(\ell + \frac{1}{2}\right)^2 (ARZ_1Z_2)^{-1/2}]
\end{aligned} \quad (2.24)$$

where

$$E_c = \frac{Z_1Z_2 e^2}{R} = 1.0\frac{Z_1Z_2}{A_1^{1/3} + A_2^{1/3}} \text{ MeV} \quad (2.25)$$

is the height of the Coulomb barrier and R is in fm.

2.4   Sketch of statistical mechanics

Many situations in both nuclear and other parts of astrophysics can be treated to a greater or lesser degree of approximation using concepts of thermodynamic equilibrium. An obvious example is the velocity distribution in classical plasmas, which is Maxwellian to a high degree of approximation under a very wide range of conditions because of the high frequency of elastic collisions, but there are many others. In nuclear reactions at very high temperatures, many reverse reactions proceed with almost the same frequency as forward reactions, leading to a situation of quasi-equilibrium between abundances of the relevant nuclei. Excitation and ionisation of atoms, and the dissociation of molecules, follow very precisely thermal equilibrium relations in stellar interiors, and these relations also provide a more or less usable first approximation in stellar atmospheres which is known as local thermodynamic equilibrium. This section therefore gives a brief sketch of relevant parts of statistical thermodynamics, which will have many applications in what follows.

Consider N fermions in a box with fixed energy levels $\tilde{E}_i$ (including rest-mass which itself includes internal excitation energy). To each $\tilde{E}_i$ there correspond ω_i distinct (degenerate) states, made up of g_i internal states and $4\pi V p^2 dp/h^3$ kinetic degrees of freedom, where p is momentum and V is the volume.

Because fermions are indistinguishable and satisfy the Pauli Exclusion principle, the number of different ways in which each set of ω_i states with energy $\tilde{E}_i$ can have N_i of them occupied is

$$W(N_i, \tilde{E}_i) = \prod_i C_{N_i}^{\omega_i} = \prod_i \frac{\omega_i!}{N_i!(\omega_i - N_i)!}. \quad (2.26)$$

Stirling's formula for huge numbers is

$$n! \simeq n^n e^{-n} \tag{2.27}$$

so

$$\ln W = \sum_i [\omega_i \ln \omega_i - N_i \ln N_i - (\omega_i - N_i) \ln(\omega_i - N_i)]. \tag{2.28}$$

Thermal equilibrium implies that, under fixed conditions, W takes on its maximum possible value, i.e.

$$0 = \delta \ln W = \sum_i \delta N_i \ln \left(\frac{\omega_i}{N_i} - 1 \right) \tag{2.29}$$

subject to overall conditions

$$\sum_i N_i = N, \tag{2.30}$$

the total number (which becomes irrelevant in cases where particles are easily created and destroyed), and

$$\sum_i N_i \tilde{E}_i = \tilde{E}, \tag{2.31}$$

the total energy.

These conditions are incorporated into the stationary condition (2.29) using Lagrange multipliers

$$\sum_i \delta N_i \left[\ln \left(\frac{\omega_i}{N_i} - 1 \right) - \alpha - \beta \tilde{E}_i \right] = 0 \quad \forall \, \delta N_i, \tag{2.32}$$

whence

$$N_i = \frac{\omega_i}{e^{\alpha + \beta \tilde{E}_i} + 1}. \tag{2.33}$$

We now appeal to Boltzmann's entropy postulate

$$S = k \ln W \tag{2.34}$$

and let a small reversible change take place. Specifically, we allow some heat and particles to enter a cylinder containing perfect gas at a fixed temperature and pressure. This will lead to a variation in the ω_is by virtue of the volume change, and

$$\begin{aligned}
\delta S &= k \sum_i \delta N_i \ln \left(\frac{\omega_i}{N_i} - 1 \right) - k \sum_i \delta \omega_i \ln \left(1 - \frac{N_i}{\omega_i} \right) \\
&= k(\alpha \delta N + \beta \delta \tilde{E} + \frac{N}{V} \delta V - \ldots).
\end{aligned} \tag{2.35}$$

Compare the equation from thermodynamics

$$T \, dS = d\tilde{E} + P \, dV - \sum_j \mu_j dN_j. \tag{2.36}$$

μ_j is the *chemical potential* of particle species j which comes in as a sort of driving force (analogous to population pressure) when numbers of particles are allowed to change.[1] Carrying out some divisions in Eq. (2.36),

$$\left(\frac{\partial S}{\partial \tilde{E}}\right)_{V,N_j} = k\beta = \frac{1}{T}; \quad \left(\frac{\partial S}{\partial N_j}\right)_{\tilde{E},V} = k\alpha = -\frac{\mu_j}{T}. \tag{2.37}$$

Therefore

$$\beta = \frac{1}{kT}, \quad \alpha = -\frac{\mu}{kT} \tag{2.38}$$

for a particular species, and the Fermi–Dirac distribution becomes (per unit volume)

$$n_i = \frac{g_i}{h^3} \frac{4\pi p^2 dp}{e^{(\tilde{E}-\mu)/kT} + 1} \tag{2.39}$$

$$= \frac{4\pi g_i}{(hc)^3} \frac{\tilde{E}(\tilde{E}^2 - m^2c^4)^{1/2} d\tilde{E}}{e^{(\tilde{E}-\mu)/kT} + 1}, \tag{2.40}$$

where we have used the exact relativistic expression $\tilde{E}^2 = m^2c^4 + p^2c^2$. $g_i = 2$ for free electrons etc. and in general g_i is the statistical weight of some internal energy level for nuclei, atoms and molecules. A similar expression can be derived for bosons (particles with integral spin), but with -1 in the denominator instead of $+1$; for photons, $\mu = m = 0$.

Consider a non-degenerate ($\mu \ll mc^2$), non-relativistic ($\tilde{E} = mc^2 + p^2/2m$) gas. The total number density is

$$n = \frac{4\pi u}{h^3} e^{(\mu-mc^2)/kT} \int_0^\infty p^2 e^{-p^2/2mkT} dp \tag{2.41}$$

$$= \left(\frac{mkT}{2\pi\hbar^2}\right)^{3/2} u\, e^{(\mu-mc^2)/kT} \tag{2.42}$$

where u is the internal partition function $\sum_i g_i e^{-(\tilde{E}_i - \tilde{E}_1)/kT}$, or

$$\mu = kT \ln\left[n/\left\{u\, e^{-mc^2/kT} \left(\frac{mkT}{2\pi\hbar^2}\right)^{3/2}\right\}\right] \equiv mc^2 + kT \ln[n/(un_Q)]. \tag{2.43}$$

$n_Q \equiv (mkT/2\pi\hbar^2)^{3/2}$ is called the *quantum concentration*. Replacing g_i by u in Eq. (2.39) and dividing by Eq. (2.42) immediately gives the Maxwellian velocity distribution function

$$\frac{n(v)dv}{n} = \left(\frac{m}{2\pi kT}\right)^{3/2} 4\pi v^2 e^{-\frac{1}{2}mv^2/kT} dv. \tag{2.44}$$

For two particles having masses m_1, m_2, the distribution function of relative velocity is obtained by substituting for m the reduced mass $m_1 m_2/(m_1 + m_2)$.

---

[1] This case is more rigorously treated in the theory of the 'Grand Canonical Ensemble', which consists of a number of identical systems that are able to exchange both heat and particles with a common thermal bath.

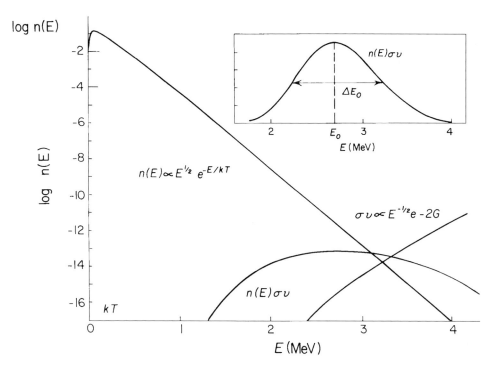

Fig. 2.9. Dependence of particle number and barrier penetration probability on energy for ^{12}C + ^{12}C at a temperature $kT = 100$ keV. In this case, the energy at the Gamow peak, shown on a linear scale in the inset, is well within the range of typical laboratory energies. Adapted from Krane (1988).

In the application to chemical equilibrium, we consider a generic reaction of the form

$$AB \rightleftharpoons A + B \tag{2.45}$$

(nuclear disintegration, atomic ionization, molecular dissociation). This is in equilibrium when the entropy is a maximum with respect to the numbers on each side, i.e. the total chemical potentials are the same on both sides:

$$T\delta S = -\mu_A \delta N_A - \mu_B \delta N_B - \mu_{AB}\delta N_{AB} = \delta N_A(\mu_{AB} - \mu_A - \mu_B) = 0 \quad \forall \, \delta N_A. \tag{2.46}$$

In combination with expression (2.43) for μ, this leads to Saha's equation in the form

$$\frac{n_A n_B}{n_{AB}} = \frac{u_A u_B}{u_{AB}} \left(\frac{m_A m_B}{m_{AB}} \frac{kT}{2\pi\hbar^2} \right)^{3/2} e^{(m_{AB} - m_A - m_B)c^2/kT}. \tag{2.47}$$

2.5 Thermonuclear reaction rates

In a hot plasma, the reaction rate per unit volume between particles of types i and j is

$$r_{ij} = n_i n_j (1 + \delta_{ij})^{-1} <\sigma v>_{ij} \tag{2.48}$$

where n_i, n_j are the respective number densities and $<\sigma v>$ is the velocity-averaged product of cross-section and relative velocity, also known as the rate coefficient. The factor $(1 + \delta_{ij})^{-1}$ comes from the fact that, if i and j are identical, the number density of particle pairs is $n(n-1)/2 \simeq n^2/2$. Nucleus i has a corresponding mean life

$$\tau_i = \left(\sum_j n_j <\sigma v>_{ij} \right)^{-1} \tag{2.49}$$

without the $(1 + \delta_{ij})$ factor. Number densities are related to the total mass density ρ by

$$n_i = 6.02 \times 10^{23} \rho X_i / A_i \equiv 6.02 \times 10^{23} \rho Y_i, \tag{2.50}$$

where ρ is in gm cm^{-3} and X_i is the abundance by mass fraction of nucleus i with atomic mass number A_i. The net energy production per unit mass is

$$\epsilon = (r_{ij} - r_{ji})Q/\rho - \text{neutrino losses.} \tag{2.51}$$

Averaging over the Maxwell distribution,

$$<\sigma v> = (8/\pi m)^{1/2} (kT)^{-3/2} \int_0^\infty E\sigma(E) e^{-E/kT} \, dE. \tag{2.52}$$

For non-resonant reactions, the S-factor defined in Eq. (2.22) is a slowly varying quantity which can be replaced by an appropriate mean value, i.e.

$$<\sigma v> = (8/\pi m)^{1/2} (kT)^{-3/2} <S(E)> \int_0^\infty e^{-\frac{E}{kT} - 2\pi\eta} \, dE. \tag{2.53}$$

(For endothermic reactions, the lower limit of integration is $-Q$.) Now from Eq. (2.20) it is easily derived that the integrand of Eq. (2.52), expressed as

$$f(E) = e^{-(\frac{E}{kT} + BE^{-1/2})} \tag{2.54}$$

(where B is given by Eq. 2.21) has a maximum for an energy

$$E_0 = (BkT/2)^{2/3} \tag{2.55}$$

where its value is

$$f(E_0) = e^{-(B^2/kT)^{1/3}(2^{1/3} + 2^{-2/3})} = e^{-3E_0/kT} \equiv e^{-\tau}, \tag{2.56}$$

and

$$\tau \equiv 3 (B/2kT)^{2/3}. \tag{2.57}$$

This is known as the 'Gamow peak'. Owing to the two contrary exponential dependences on energy in Eq. (2.53), reaction rates are significant only for particles in a more or less narrow range of energy around E_0, which is much larger than kT (see Fig. 2.9), but still usually small compared to energies at which cross-sections can be measured in the laboratory. The integral can be approximated by the 'method of steepest descents' in which $f(E)$ is approximated by a Gaussian with the same peak value and the same second derivative at that peak, i.e.

$$\int_0^\infty f(E)\,dE \simeq f(E_0)\sqrt{[2\pi f(E_0)/-f''(E_0)]} = \frac{2}{3}\pi^{1/2}kT\,\tau^{1/2}\,e^{-\tau}. \tag{2.58}$$

Owing to the asymmetry in $f(E)$, the best value to take for $<S(E)>$ turns out to be $S(E_0' \simeq E_0 + \frac{5}{6}kT)$ rather than $S(E_0)$. Making all the substitutions in Eq. (2.53) one finds

$$<\sigma v> \simeq \frac{8}{81}\frac{\hbar}{\pi Z_1 Z_2 e^2 m}\tau^2 e^{-\tau}S(E_0') = \frac{7.2\times10^{-19}}{AZ_1Z_2}\tau^2 e^{-\tau}S(E_0')\ \text{cm}^3\ \text{s}^{-1}, \tag{2.59}$$

where $S(E_0')$ is in keV barns, and

$$\tau = 3E_0/kT = 19.7\,(Z_1^2 Z_2^2 A/T_7)^{1/3} \tag{2.60}$$

where T_7 is in units of 10^7 K.

From Eqs. (2.59), (2.60), one can deduce an approximate power-law dependence of specific reaction rates on temperature, T^ν, since

$$\nu \equiv \frac{\partial\log<\sigma v>}{\partial\log T} = \frac{\tau-2}{3}. \tag{2.61}$$

E.g. in the interior of the Sun ($T_7 = 1.5$; $kT = 1.3$ keV), one has the following:

Reaction	E_0(keV)	τ	ν
$p+p$	5.9	13.7	3.9
$p+^{14}$N	27	63	20

These temperature dependences are illustrated in Fig. 5.4.

2.6 Resonant reactions

When the compound nucleus has an excited level coinciding in energy with that of the projectile in the CM system, i.e. the projectile energy E is close to E_P where

$$E_P + Q = E_R, \tag{2.62}$$

the energy of the level above the ground state (see Fig. 2.10), the reaction cross-section is greatly enhanced by resonance effects, subject to the standard selection (or conservation) rules. In these cases, Q is often referred to as the threshold. The excited level is broadened owing to its finite lifetime τ and thus has (in the system of the compound nucleus) a probability distribution

$$P(E)dE = \frac{\Gamma/2\pi}{(E-E_R)^2+(\Gamma/2)^2}\,dE \tag{2.63}$$

where

$$\Gamma = \hbar/\tau = \Gamma_p + \Gamma_n + \Gamma_\alpha + \Gamma_\beta + \Gamma_\gamma + \dots \tag{2.64}$$

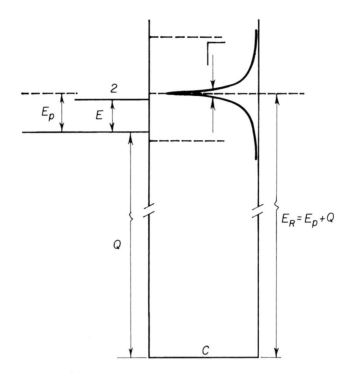

Fig. 2.10.   Schematic of input channel for a resonant reaction.

is the total width, and is the sum of partial widths for the various possible decay channels. The rate at which the excited compound nucleus decays into any channel is proportional to its partial width. The rate at which the excited state is formed by incoming particles with energy E is proportional to $P(E)$. If the reaction is to be initiated by a particle of type a, say, the cross-section must also be proportional to Γ_a. A reaction occurs if the excited state breaks up into any other channel, i.e. with a probability proportional to $\Gamma - \Gamma_a$, and the probability of ejecting particles of type b, say, is proportional to Γ_b. Putting all these factors together and normalising to the maximum cross-section at $E = E_P$, one obtains the Breit–Wigner formula for a single-level resonance

$$\sigma(E) = \pi \bar{\lambda}^2(E_P)(1 + \delta_{12})\,\omega\,\frac{\Gamma_a \Gamma_b}{(E - E_P)^2 + (\Gamma/2)^2} \tag{2.65}$$

with

$$\omega = \frac{2I + 1}{(2I_1 + 1)(2I_2 + 1)}, \tag{2.66}$$

where I is the angular momentum of the excited state in the compound nucleus: $\mathbf{I} = \mathbf{I_1} + \mathbf{I_2} + \ell$. Here Γ_a is the width for elastic resonant scattering and the reaction cross-section is a maximum for $\Gamma_a = \Gamma_b = \Gamma/2$, i.e. any reaction is accompanied

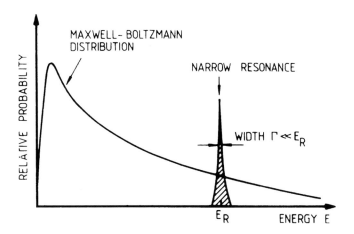

Fig. 2.11. A narrow resonance near or not too far above the Gamow peak. After C.E. Rolfs & W.S. Rodney (1988), *Cauldrons in the Cosmos*, University of Chicago Press, Fig. 4.12, p. 174. ©1988 by the University of Chicago. Courtesy Claus Rolfs.

by elastic resonant scattering. The Γs for charged particle interactions are generally energy-dependent, owing to the penetration factor, whereas Γ_γ is constant and small, of the order of 1 eV or less, depending on the energy and order (electric or magnetic dipole, quadrupole etc.) of the transition. A Γ of 1 eV corresponds to a mean lifetime of 6.6×10^{-16} s.

A narrow resonance not too far above the Gamow peak may supplant the latter (see Fig. 2.11). The reaction rate coefficient is then given in analogy to Eq. (2.52) by

$$<\sigma v> = (8/\pi m)^{1/2}(kT)^{-3/2}E_P\, e^{-E_P/kT} \int_0^\infty \sigma(E)dE, \tag{2.67}$$

where

$$\int \sigma(E)dE = 2\pi^2\, \bar{\lambda}_P^2\, \omega\, \frac{\Gamma_a\Gamma_b}{\Gamma} \equiv 2\pi^2\, \bar{\lambda}_P^2\, \omega\, \gamma. \tag{2.68}$$

$\gamma \equiv \Gamma_a\Gamma_b/\Gamma$ is referred to as the strength of the resonance and the reaction rate coefficient finally becomes

$$<\sigma v> = \left(\frac{2\pi}{mkT}\right)^{3/2} \hbar^2(\omega\gamma)_P f\, e^{-E_P/kT} \tag{2.69}$$

$$= 2.6 \times 10^{-13}(AT_7)^{-3/2}f\omega\gamma\, e^{-1.16E_P/T_7} \text{ cm}^3\, \text{s}^{-1} \tag{2.70}$$

where f is a correction factor ≥ 1 for electron screening and E_P and γ are in keV. For several narrow resonances that are well separated, the rate coeffients are simply summed; in heavier nuclei these become numerous and can be treated by statistical methods. When resonances overlap, however, interference effects can occur.

The dependence of Γ on energy for charged-particle decay can be found as follows. The decay rate is the probability density flux integrated over surface area at infinity, i.e.

$$\Gamma/\hbar = \lim_{r\to\infty} v \int_\Omega \mid \psi(r,\theta,\phi) \mid^2 r^2 d\Omega. \tag{2.71}$$

For a spherical potential, the wave function has the form

$$\psi(r,\theta,\phi) = \frac{\chi_\ell(r)}{r} Y_{\ell m}(\theta,\phi), \tag{2.72}$$

where the spherical harmonics $Y_{\ell m}$ have the normalisation

$$\int_\Omega \mid Y_{\ell m}(\theta,\phi) \mid^2 d\Omega = 1, \tag{2.73}$$

so that

$$\Gamma/\hbar = v \mid \chi_\ell(\infty) \mid^2 = v P_\ell(v) \mid \chi_\ell(R) \mid^2, \tag{2.74}$$

where v is the speed at infinity, R the nuclear radius and $P_\ell(v)$ the barrier penetration probability given by Eq. (2.24). $\mid \chi_\ell(R) \mid^2$ is the probability per unit dr of finding the particle within dr of the nuclear surface, which for a uniform nuclear density is given by

$$\mid \chi_\ell(R) \mid^2 = 3/R, \tag{2.75}$$

so that one may write

$$\Gamma_\ell = \frac{3\hbar v}{R} P_\ell(v) \theta_\ell^{\,2}, \tag{2.76}$$

where the dimensionless quantity $\theta_\ell^{\,2}$, called the reduced width, is a constant between 0 and 1 depending on details of the nucleus. This is especially useful when $\theta_\ell^{\,2}$ can be estimated from laboratory measurements around E_P and the resonance is broad enough to affect the S-factor at much lower energies. Allowance for these effects is often straightforward, but the case of $^{12}C(\alpha,\gamma)^{16}O$ presents some difficulties (Fig. 2.12) because the main ^{16}O resonance at 2.5 MeV above threshold is accompanied by two subthreshold resonances (see Fig. 5.7).

When there are only two possible decay channels for the excited state of the compound nucleus, i.e. $\Gamma = \Gamma_a + \Gamma_b$, two simple limiting cases arise:

(1) $\Gamma_b \gg \Gamma_a$; $\gamma \simeq \Gamma_a$. This could be the case for a (p,α) reaction, or for (p,γ) or (α,γ) when the energy of the incident particle is so low that Γ_a falls below 1 eV or so. In this case, $\gamma \simeq \Gamma_a$ and is independent of Γ_b. The reaction occurs virtually every time the resonant state is formed.

(2) $\Gamma_a \gg \Gamma_b$; $\gamma \simeq \Gamma_b$. This could be the case for resonances of energy sufficiently high that the state decays predominantly by re-emission of the incident particle and the reaction rate coefficient is given by Eq. (2.68) with γ replaced by Γ_b. If the number density of excited nuclei is n^*, say, then we now have for the

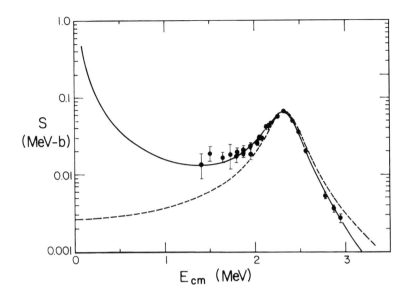

Fig. 2.12. *S*-factor for $^{12}C(\alpha,\gamma)^{16}O$. The dashed curve ignores the subthreshold resonances, while the continuous curve allows for them, but the uncertainties are still significant at low energies. Reproduced from *Nucl. Phys. A*, **220**, S.E. Koonin, T.A. Tombrello & G. Fox, 'A "Hybrid" R-Matrix-Optical Model Parametrization of the $^{12}C(\alpha,\gamma)^{16}O$ Cross Section', pp 221-232, ©1974, with kind permission from Elsevier Science - NL, Sara Burgerhartstraat 25, 1055 KV Amsterdam, The Netherlands. Courtesy S.E. Koonin.

reaction rate

$$r_{12} = n_1 n_2 <\sigma v> = n^* \Gamma_b / \hbar, \qquad (2.77)$$

whence

$$\frac{n^*}{n_1 n_2} = \left(\frac{2\pi\hbar^2}{mkT} \right)^{3/2} \omega f e^{-E_P/kT}, \qquad (2.78)$$

equivalent to the Saha equation (2.47) apart from the factor f which represents modification of the laboratory dissociation energy by the same plasma effects as give rise to electron screening. This is an example of statistical equilibrium brought about by the balancing of forward and backward reactions, and it applies in particular to helium burning by the 3α reaction. In such cases, many reaction cross-sections can be replaced in calculations by simpler statistical properties.

2.7 Neutron capture reactions

Neutron captures have no Coulomb barrier (there is a centrifugal barrrier for $\ell > 0$) and generally take place through a series of broad and overlapping resonances, the

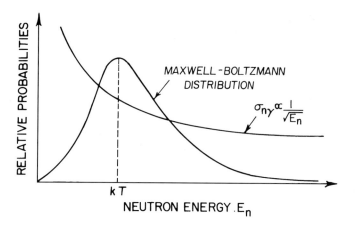

Fig. 2.13. Schematic superposition of the Maxwell energy distribution and neutron capture cross-section. The most probable energy for the capture process in stars is near kT. After C.E. Rolfs & W.S. Rodney (1988), *Cauldrons in the Cosmos*, University of Chicago Press, Fig. 9.2, p. 452. ©1988 by the University of Chicago. Courtesy Claus Rolfs.

breadth of the levels reflecting the ease with which a neutron can escape from an excited nucleus. For magic-number nuclei with closed neutron shells, the resonances have much higher energies and the neutron capture cross-sections are then particularly small (cf. Fig. 6.2).

The capture rate is dominated at thermal energies around 30 keV by s-waves for which the Breit–Wigner formula gives

$$\sigma_n(E_n) \propto \bar{\lambda}^2 \, \Gamma_n(E_n) \, \Gamma_X(E_n + Q). \tag{2.79}$$

The last factor is essentially $\Gamma_X(Q)$, since $Q \gg E_n$, and is virtually independent of energy; X may refer to γ, which is the normal exit channel in the s- and r-processes, or to p etc. Γ_n, on the other hand, is proportional to v by virtue of Eq. (2.76). This combines with the v^{-2} dependence of $\bar{\lambda}^2$ to make σ_n approximately proportional to $1/v$, i.e.

$$\sigma_n v \simeq \text{const.} \simeq <\sigma_n v>, \tag{2.80}$$

so the reaction rate follows directly from Eq. (2.48).

2.8   **Inverse reactions**

At low stellar temperatures, nuclear reactions occur predominantly in the direction leading to positive values of Q, but at higher temperatures the inverse reactions become increasingly significant. In Eq. (2.15), owing to time-reversal invariance, the matrix elements are the same for both forward and reverse reactions, so that the ratio of the two cross-sections is

$$\frac{\sigma_{12}}{\sigma_{34}} = \frac{m_3 m_4 E_{34}(2I_3 + 1)(2I_4 + 1)(1 + \delta_{12})}{m_1 m_2 E_{12}(2I_1 + 1)(2I_2 + 1)(1 + \delta_{34})}, \tag{2.81}$$

where $\bar{\lambda}_{ik}^2$ has been replaced by the equivalent quantity $\hbar^2/2mE$. This means that, by measuring a reaction cross-section in one direction, one can infer the cross-section in the opposite direction provided that the same specific nuclear states are involved. This is particularly useful in predicting photodisintegration rates, which are important at high temperatures in the Big Bang and in advanced stages of stellar evolution, and are not easy to measure.

When four particles are involved, e.g. for $^{15}N(p, \alpha)^{12}C$ and its reverse, averaging over the Maxwell distribution as in Eq. (2.52) and using Eq. (2.81) above, one finds

$$\frac{<\sigma v>_{34}}{<\sigma v>_{12}} = \frac{(2I_1 + 1)(2I_2 + 1)(1 + \delta_{34})}{(2I_3 + 1)(2I_4 + 1)(1 + \delta_{12})} \left(\frac{m_{12}}{m_{34}}\right)^{3/2} e^{-Q/kT} \tag{2.82}$$

provided only single states of each nucleus are involved; the same result can be found from Saha's equation (2.47) using the principle of detailed balancing. For photodisintegration, e.g. $^{14}N(\gamma, p)^{13}C$, an analogous expression can be derived by calculating the lifetime τ_3 from Saha's equation:

$$\frac{n_3}{n_1 n_2} = \tau_3 <\sigma v>_{12} \tag{2.83}$$

$$= \frac{2I_3 + 1}{(2I_1 + 1)(2I_2 + 1)} \left(\frac{2\pi\hbar^2}{m_{12}kT}\right)^{3/2} e^{Q/kT}, \tag{2.84}$$

whence

$$n_\gamma <\sigma v>_\gamma = \tau_3^{-1} = <\sigma v>_{12} \left(\frac{m_{12}kT}{2\pi\hbar^2}\right)^{3/2} \frac{(2I_1 + 1)(2I_2 + 1)}{2I_3 + 1} e^{-Q/kT}. \tag{2.85}$$

At high temperatures, the rates have to be summed over relevant nuclear states and the $(2I + 1)$-factors in Saha's equation replaced by partition functions.

2.9 α-decay and fission

From the binding-energy curve, Fig. 1.3, it appears that it is energetically favourable for many of the heavier nuclei to split into lighter ones. Considering the nucleus as a vibrating liquid drop, this could occur as a result of its changing shape from a nearly spherical to an elongated one from which a piece could break off, resulting in a gain in binding energy which would initially appear as kinetic energy of the nucleons. However, escape of the resulting fragment from the nucleus is strongly inhibited by the height of the Coulomb barrier.

One of the easiest ways for a heavy nucleus to split is by α-decay, because α-particles retain some of their identity within the nucleus, they are highly favoured energetically compared to any lighter particles (protons, neutrons or deuterons) and the Coulomb barrier is still relatively low: for an energy gain $\Delta Q = 1$ MeV, $2\pi\eta \simeq 3Z$ from Eq. (2.21). (For ^{238}U $\to^{234}$Th $+\alpha$, $\Delta Q \simeq 4$ MeV.) Much higher values of ΔQ are available from fission into two nearly equal fragments, but this occurs at the expense of much higher Coulomb barriers, which explains why fission is less common than α-decay.

From the semi-empirical mass formula Eq. (2.2), neglecting the last two terms, the gain in B.E. for a nucleus splitting into two equal fragments is

$$\Delta Q = a_s A^{2/3} + a_c Z^2 A^{-1/3} - 2a_s \left(\frac{A}{2}\right)^{2/3} - 2a_c \left(\frac{Z}{2}\right)^2 \left(\frac{A}{2}\right)^{-1/3} \tag{2.86}$$

$$= a_c Z^2 A^{-1/3}(1 - 2^{-2/3}) - a_s A^{2/3}(2^{1/3} - 1) \simeq 250 \text{ MeV} \tag{2.87}$$

if $Z \simeq 100$ and $A \simeq 250$. The first term in Eq. (2.87) is a gain due to reduced Coulomb repulsion and the second is a loss due to increased surface area. In practice, the fission fragments are not usually equal in mass, but the above calculation is adequate to get a general idea.

The Coulomb barrier is roughly given by

$$E_{\text{Coul}} \simeq \left(\frac{Ze}{2}\right)^2 / 2R_0 \left(\frac{A}{2}\right)^{1/3} \simeq 0.28 a_c Z^2 A^{-1/3} \simeq 320 \text{ MeV} \tag{2.88}$$

for the same parameters, leaving a substantial barrier or threshold to spontaneous fission; $2\pi\eta \simeq 1200$. The barrier can be overcome or penetrated if the nucleus is excited by capturing a particle, say a neutron, leading to induced fission. Spontaneous fission will only occur if ΔQ is comparable to, or greater than, E_{Coul}, i.e. if

$$\frac{Z^2}{A} \geq \sim 3\frac{a_s}{a_c} \simeq 67 \tag{2.89}$$

according to the constants adopted in Eq. (2.2), but this number is quite inaccurate because no account has been taken of ellipticity. A better value for the fission parameter Z^2/A would be in the neighbourhood of 45. One consequence of the form of this parameter is that a nucleus can become liable to spontaneous fission as a result of β-decay, an effect known as β-delayed fission. All three kinds of fission play a role in the r-process, where successive neutron captures lead to very heavy unstable nuclei (see Chapter 6). Each fission can be followed by more neutron captures leading to a cyclic process in which the fission fragments and heavier nuclei are doubled in abundance every time the cycle operates.

2.10 Weak interactions

Unstable nuclei decay to an isobar in the β stability valley by β^- or β^+ emission or electron capture:

$$n \to p + e^- + \bar{\nu}_e; \quad p \to n + e^+ + \nu_e; \quad p + e^- \to n + \nu_e. \tag{2.90}$$

The (electron) neutrino ν_e and anti-neutrino $\bar{\nu}_e$ ensure conservation of energy, momentum, angular momentum and (electron) lepton number. (The lepton numbers are

+1 for electrons and v_e, and -1 for their anti-particles.) In the Standard Model of particle physics, there are three families each consisting of two quarks and two leptons, together with their antiparticles. Ordinary matter is made up from the first family: up and down quarks constitute protons and neutrons and the associated leptons are $e^{\pm}$ and $v_e, \bar{v}_e$. The second family has strange and charm quarks and its leptons are the $\mu^{\pm}$ and the muon neutrinos $v_{\mu}, \bar{v}_{\mu}$; the third has beauty and top quarks and its leptons are the $\tau^{\pm}$ and $v_{\tau}, \bar{v}_{\tau}$. The muon and tauon are massive and unstable, and their decay is accompanied by the emission of the corresponding neutrino or anti-neutrino, which play an analogous role to electron neutrinos in preserving lepton number separately for each family. The idea of conserved lepton numbers means that emission of a neutrino can be treated as absorption of an anti-neutrino and *vice-versa*.

v_{μ} and v_{τ} and their anti-neutrinos are produced along with $v_e, \bar{v}_e$ in the early universe and in collapsing cores of massive stars at the end of their evolution. No rest mass has been detected for any kind of neutrino; current upper limits are of the order of 5 eV, 160 keV and 24 MeV for v_e, v_{μ} and v_{τ} respectively. Neutrinos are fermions with spin 1/2, but they differ from protons, neutrons and electrons (apart from those emitted in β-decay) in having a fixed helicity; the spin vector of neutrinos points in the opposite direction to their velocity vector, while that of anti-neutrinos points in the same direction. Thus they are sometimes referred to as 'left-handed'. Right-handed neutrinos, if they exist at all, do not take part in the regular weak interaction and are referred to as 'sterile'. This effect is an example of the non-conservation of parity in weak interactions.

Unlike α-decay (and the related nuclear processes involving strong and electromagnetic interactions), β-decay involves creation of new particles and relativistic mechanics. Basic understanding of the process was provided by Enrico Fermi, who postulated the existence of a weak interaction, analogous to the electromagnetic interaction responsible for radiation. Time-dependent perturbation theory (see Appendix 3) leads to the Golden Rule:[1]

$$\lambda(E_f) = \frac{2\pi}{\hbar} |V_{fi}|^2 \rho(E_f), \tag{2.91}$$

where $\lambda \equiv \ln 2/t$ is the decay constant, and

$$V_{fi} = g \int [\psi_f^* \phi_e^* \phi_v^*] O_X \psi_i \, dv \tag{2.92}$$

is a matrix element involving a coupling constant g and an overlap integral between the perturbing potential assumed to act like some operator O_X on the initial nuclear wave function ψ_i and the final wave functions of the daughter nucleus ψ_f, the electron ϕ_e and the neutrino ϕ_v. (Strictly one should speak of an electron and anti-neutrino, or positron and neutrino, but the distinction is not important for the present purpose. The wave functions are actually 4-component spinors and the operator a 4×4 matrix.)

---

[1] Sometimes referred to as Fermi's Golden Rule, although it was originally derived by Dirac. Mandl (1992) has remarked that 'Fermi is not in need of "borrowed feathers".'

The operator O_X can in principle behave under coordinate transformations in one of five different ways all consistent with Special Relativity: like a (polar) 4-vector (as in electric dipole radiation), like an axial vector, like a scalar, pseudo-scalar or tensor; in practice only the first two are found in nature and they are referred to as Fermi and Gamow–Teller decays respectively. Finally, the phase-space factor $\rho(E_f)$ is the density of states (i.e. the number of states per unit energy interval) available to the outgoing leptons within the range of final energies E_f allowed by the Uncertainty Principle as a result of the finite lifetime of the initial state. For an electron momentum p to $p + dp$ and a neutrino momentum q to $q + dq$ in a volume V, the number of states (bearing in mind the fixed helicity) is

$$d^2n = (4\pi)^2 \, V^2 p^2 dp \, q^2 dq / h^6. \tag{2.93}$$

For typical lepton energies of a few MeV, the de Broglie wave-length is of order 100 times the nuclear radius and when orbital angular momentum is zero, one can use the 'allowed' approximation for their wave functions

$$\phi_{e,\nu} = \text{const.} = V^{-1/2}. \tag{2.94}$$

The partial decay constant then becomes

$$d\lambda = \frac{2\pi}{\hbar} \, g^2 \, |M_{fi}|^2 \, (4\pi)^2 \, \frac{p^2 dp \, q^2}{h^6} \, \frac{dq}{dE_f}, \tag{2.95}$$

where $M_{fi} \equiv \int \psi_f^* O_X \psi_i \, dv$ is the nuclear matrix element. It then readily follows that the kinetic-energy spectrum of the electron (neglecting the small amount of energy taken up by recoil of the nucleus) is given by

$$N(T_e) \, dT_e = N(p) \, dp \propto p^2 dp \, (Q - T_e)^2, \tag{2.96}$$

a function which vanishes at $T_e = 0, Q$ and has a single maximum. [1]

The actual curve, however, is somewhat modified by Coulomb interaction between the electron or positron and the nucleus. This is allowed for by multiplication with a dimensionless function $F(Z', p)$, which leads to a correction factor f for the total decay rate, and it is the product ft that is used for purposes of comparing measured lifetimes with theory. The most rapid decays, with $ft = 10^3$ to 10^4 s, are known as 'superallowed'. These include 0^+ to 0^+ decays having $|M_{fi}|^2 = 2$ and ft is found experimentally to be close to 3000s, giving the coupling constant for the Fermi interaction

$$g_F = 0.88 \times 10^{-4} \text{ MeV fm}^3 \ = 1.4 \times 10^{-49} \text{ erg cm}^3 \tag{2.97}$$

and its dimensionless equivalent (analogous to the fine structure constant $\alpha \sim 10^{-2}$)

$$G_F \equiv \frac{g_F}{m_p c^2 \, (\hbar / m_p c)^3} = g_F \, m_p^2 c / \hbar^3 = 1.02 \times 10^{-5}. \tag{2.98}$$

---

[1] For β^+ decay, since Q is calculated from atomic masses, the last factor has $Q - 2m_e c^2$ instead of Q.

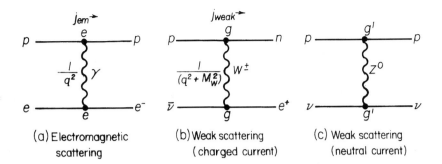

Fig. 2.14. Electromagnetic and weak interactions mediated by virtual boson exchange. q is the 4-momentum transferred in the interaction. Adapted from Perkins (1982).

Allowed decays are those in which the leptons can be treated as originating at the origin in the nucleus, so that they cannot carry orbital angular momentum or a change in parity. Their spins may be anti-parallel, leading to a Fermi decay, or parallel, leading to a Gamow–Teller decay. In the first case, the nuclear spin is unchanged; in the second, it changes by one vector unit, i.e. $\Delta I = 0$ or ± 1, except that $I = 0 \rightarrow I = 0$ is forbidden. Thus $0^+ \rightarrow 0^+$ transitions are purely of the Fermi type, whereas free neutron decay $(\frac{1}{2}^+ \rightarrow \frac{1}{2}^+)$ has contributions from both types, involving different coupling constants and matrix elements. In cases where the nuclear parity changes, the leptons carry orbital angular momentum; such transitions are 'forbidden' and accordingly slower.

The modern theory of weak interactions treats them as mediated by virtual weak-interaction quanta (known as intermediate vector bosons) acting between fermion currents, the fermions being endowed with a 'weak charge' g in analogy with electric charge (Fig. 2.14). Experiments show that, in addition to charge-changing (called 'charged-current') interactions mediated by $W^\pm$ bosons, there are also 'neutral-current' interactions mediated by a neutral particle, the Z^0. The electroweak theory of S. Weinberg, A. Salam and S. Glashow, which unifies electromagnetic and weak interactions at high energies, predicted masses of order 80 and 90 GeV for the $W^\pm$ and Z^0 respectively, which were confirmed experimentally at CERN in 1983, and subsequent measurements of the width (i.e. lifetime) of the Z^0 resonance show that there cannot be any additional neutrinos (above the three pairs in the Standard Model) coupling to the Z^0 and having a mass below $m_Z/2$. In accordance with the hypothesis of a 'universal Fermi interaction', supported by many experiments, the ν_μ and ν_τ, as well as the ν_e, interact with matter through the neutral current mechanism, which can give rise to elastic and inelastic scattering events, but at typical stellar energies (below many MeV) only the ν_e and $\bar{\nu}_e$ have charged-current interactions (corresponding to absorption), since these require creation of the corresponding charged lepton.

Inverse β-decays, e.g.

$$p + e^- \quad \rightarrow \quad n + \nu_e, \tag{2.99}$$

$$\bar{\nu}_e + p \quad \rightarrow \quad n + e^+, \tag{2.100}$$

$$\nu_e + {}^{37}\text{Cl} \quad \rightarrow \quad {}^{37}\text{Ar} + e^-, \tag{2.101}$$

are important in primordial nucleosynthesis and advanced stages of stellar evolution, as well as in the experimental detection of neutrinos (from the Sun and Supernova 1987A) and antineutrinos (from reactors and also from Supernova 1987A). Experimental detection is difficult because the interaction with matter is so weak. The cross-section for neutrino interactions can be estimated from Eqs. (2.91), (2.92) and (2.94) with a fixed neutrino momentum q:

$$n_\nu \sigma c \equiv \lambda = \frac{2\pi}{\hbar} g^2 \frac{|M_{fi}|^2}{V^2} \frac{4\pi p^2}{h^3} \frac{dp}{dE} V, \tag{2.102}$$

where $n_\nu = 1/V$ is the number density of neutrinos, σ the cross-section and $E = T_e = pc$ for relativistic electrons, whence

$$\sigma = \frac{2\pi}{\hbar c} g^2 |M_{fi}|^2 \frac{4\pi pE}{c^2 h^3} = 1.6 \times 10^{-20} |M_{fi}|^2 \left(\frac{E}{1\,\text{MeV}}\right)^2 \text{barn}. \tag{2.103}$$

Thus the mean free path of a 10 MeV neutrino (not counting scattering processes) is of the order of 10^{42} nucleons cm^{-2} or 10^{18} gm cm^{-2}, around 1 light-year through water density. ν_μ and ν_τ interact still more weakly. In very dense matter, however, as in the collapsing core of a supernova, high-energy neutrinos are effectively trapped, mostly by scattering processes, ν_μ and ν_τ at somewhat deeper and hotter levels than ν_e.

Notes to Chapter 2

More information about the topics discussed in this chapter can be found in the following textbooks:

D.D. Clayton, *Principles of Stellar Evolution and Nucleosynthesis*, McGraw-Hill 1968, University of Chicago Press 1984. This classic text is particularly well written and gives more accurate and rigorous arguments for results derived here often in a somewhat heuristic manner. A more up-to-date (though in some ways less complete) description of nuclear astrophysics is given in

David Arnett, *Supernovae and Nucleosynthesis*, Princeton University Press 1996.

K.S. Krane, *Introductory Nuclear Physics*, John Wiley & Sons 1988, is an excellent account of nuclear physics at a fairly simple level.

D.H. Perkins, *Introduction to High Energy Physics*, Addison-Wesley Publishing Co., 1987, is also strongly recommended.

C.E. Rolfs & W.S. Rodney, *Cauldrons in the Cosmos*, University of Chicago Press 1988. This book gives a good overview of modern astrophysics as well as a detailed account of how reaction rates are measured in the laboratory.

G.W.C. Kaye & T.H. Laby, *Tables of Physical and Chemical Constants*, Longman 1995, gives a table of properties of the nuclides including isotopic abundance or half-life, decay modes, mass excess, neutron capture cross-section and ground-state spin and parity.

Detailed formulae and tables for nuclear reaction rates are given by
Fowler, Caughlan & Zimmerman (1967, 1975), with updates by Harris *et al.* (1983) and Caughlan & Fowler (1988), for light nuclei. Rates for unstable light nuclei are given by Malaney & Fowler (1988, 1989a). Rates for medium and heavy nuclei which exhibit a high density of excited states at capture energies, calculated using the Hauser–Feshbach statistical model, are given by Woosley *et al.* (1978). Experimental neutron capture cross-sections are summarised by Bao & Käppeler (1987). Neutron capture cross-sections and other properties of heavy unstable nuclei are discussed by Cowan, Thielemann & Truran (1991a).

An introduction to chemical thermodynamics (which is entirely outside the scope of this book) as well as many relevant aspects of elementary nuclear physics can be found in:

C.R. Cowley, *An Introduction to Cosmochemistry*, Cambridge University Press, 1995.

Problems

1. Show that the zero-order energies E_0 of nuclear energy states with quantum numbers ℓ, s, where $s = 1/2$, are shifted respectively by the spin-orbit interaction $-a\ell.s$ to

$$E_1 = E_0 - \frac{a}{2}\ell; \qquad j = \ell + \frac{1}{2}; \tag{2.104}$$

$$E_2 = E_0 + \frac{a}{2}(\ell + 1); \quad j = \ell - \frac{1}{2}; \tag{2.105}$$

and that the average energy is still E_0.

2. Use the shell model (Fig. 2.3) to find the ground-state spins and parities of the following nuclei: ^{3}He, ^{7}Li, ^{9}Be, ^{11}B, ^{13}C, ^{15}N.

3. Given that the tabulated masses refer to the neutral atom, verify that,

- for β^--decay, $M(A, Z) = M(A, Z + 1) + Q$;
- for β^+-decay, $M(A, Z) = M(A, Z - 1) + Q + 2m_e$;
- for K-electron capture, $M(A, Z) = M(A, Z - 1) + Q$,

where $m_e = 5.5 \times 10^{-4}$ amu is the electron rest-mass.

4. Given the atomic masses tabulated below, find the Q-values and the changes in B.E. per nucleon (both in MeV) for the following reactions:

- $4\,^1\text{H} \rightarrow {}^4\text{He}$;
- $3\,^4\text{He} \rightarrow {}^{12}\text{C}$;
- $^{12}\text{C} + {}^4\text{He} \rightarrow {}^{16}\text{O}$;
- $^{13}\text{C}(\alpha, n)^{16}\text{O}$;
- $^{116}\text{Sn}(n, \gamma)^{117}\text{Sn}$.

Neutron	1.008665	^{13}C	13.003355
^{1}H	1.007825	^{16}O	15.994915
^{4}He	4.002603	^{116}Sn	115.901747
^{12}C	12.000000	^{117}Sn	116.902956

(Remember that two electrons are annihilated in the first reaction above.)

5. Verify that the kinetic energy of two (non-relativistic) particles with masses m_1, m_2 and velocities $\mathbf{v_1}$, $\mathbf{v_2}$ can be written

$$E = \frac{1}{2}(m_1 + m_2)V^2 + \frac{1}{2}mv^2 \qquad (2.106)$$

where

$$\mathbf{V} = \frac{m_1\mathbf{v_1} + m_2\mathbf{v_2}}{m_1 + m_2}, \qquad (2.107)$$

$$m = \frac{m_1 m_2}{m_1 + m_2} \qquad (2.108)$$

(the reduced mass) and

$$\mathbf{v} = \mathbf{v_2} - \mathbf{v_1}. \qquad (2.109)$$

The first term in E is the K.E. of the centre of mass, which conserves its momentum and hence its velocity (almost, as the change in mass from Q is relatively very small), and the second term is the K.E. in the CM system which is available for penetrating the Coulomb barrier.

What are the momenta of each of the two particles in the CM system?

6. Prove the assertion in the text that the relative velocity of two sets of particles having individual Maxwellian velocity distribution functions also has a Maxwellian distribution with the masses replaced by the reduced mass.

Hint: This is most easily done by projecting each distribution into one dimension, where it becomes a gaussian, and using the theorem from statistics that the variance of the difference (or sum) of two random variables is the sum of the individual variances.

What is the distribution function for the velocity of the centre of gravity?

7. Use Lagrange's method of multipliers to derive the law of refraction of light from Fermat's principle of least time between two fixed points.

8. Use the expression (2.28) with the approximations $\omega_i \gg N_i$, $\omega_i/N_i = \exp[(\tilde{E}_i - \mu)/kT]$, $\tilde{E} = mc^2 + p^2/2m$ to derive the Sackur–Tetrode equation for the entropy of an ideal gas

$$S = kN \left[\frac{5}{2} + \ln \left(\frac{un_Q}{n} \right) \right] \qquad (2.110)$$

and verify that this is consistent with the Second Law Eq. (2.36) and the chemical potential from Eq. (2.43).

9. Find the number density of positrons resulting from pair production by γ-rays in thermal equilibrium in oxygen at a temperature of 10^9 K and a density of 1000 gm cm^{-3}, using the twin conditions that the gas is electrically neutral and that the chemical potentials of positrons and electrons are equal and opposite. (At this temperature, the electrons can be taken as non-relativistic.) The quantum concentration for positrons and electrons is $8.1 \times 10^{28} T_9^{3/2}$ cm^{-3}, the electron mass is 511 keV and $kT = 86.2\, T_9$ keV.

10. Verify Eq. (2.61) for the power-law dependence of reaction rate on temperature.

11. At what centre-of-mass energy should one study the reaction ^{28}Si $(\alpha, \gamma)^{32}$S to duplicate stellar conditions at a temperature of the order of 10^9 K? If α-particles are incident on a Si target at rest in the laboratory, what energy should be chosen?

12. Show that the integration procedure over the Gamow peak implies for that peak a $1/e$ full width of $\Delta = (4/\sqrt{3})\sqrt{(E_0 kT)}$, i.e. essentially twice the geometric mean of E_0 and kT.

Evaluate E_0 and Δ for ^{12}C$(p, \gamma)^{13}$N as functions of T_7.

13. Find the range of neutrino energies resulting from the reaction

$$p + p \rightarrow d + e^+ + v, \tag{2.111}$$

given the following mass excesses:
H 7.289 MeV
D 13.136 MeV.
(The electron mass is 0.507 MeV).
   Which of the following reactions might be used to detect such neutrinos from the Sun?
^{37}Cl$(v, e^-)^{37}$Ar $(Q = -0.81$ MeV$)$; ^{71}Ga$(v, e^-)^{71}$Ge $(Q = -0.23$ MeV$)$; ^{115}In$(v, e^-)^{115}$Sn $(Q = 0.49$ MeV$)$.
   Estimate the kinetic energy of the D nucleus when all the available energy is taken up (a) by the positron, (b) by the neutrino.

14. Show that the slope of the electron energy spectrum for allowed β-decays is zero near $T_e = Q$ if the neutrino has zero rest-mass, but becomes infinite if it has a finite rest-mass.

3 Cosmic abundances of elements and isotopes

3.1 Introduction: data sources

The data sources can be classified into 'tangible', for which samples can be directly handled and analysed by chemical and other laboratory methods, and 'intangible' requiring spectroscopy or considerations from theoretical astrophysics. Table 3.1 lists some typical objects and methods relating to each class.

Our discussion will concentrate on astrophysical spectroscopy.

3.2 Analysis of absorption lines

The bulk of stellar radiation comes from the surface layers or 'atmosphere' of a star, more particularly the 'photosphere', which is defined as the region having optical depths for continuum radiation between about 0.01 and a few. The optical depth τ_λ is measured inwards from the surface and represents the number of mean free paths of radiation travelling vertically outwards before it escapes from the star. It is related to the geometrical height z above some arbitrary layer by

$$\tau_\lambda(z) = \int_z^\infty \kappa_\lambda(z)\rho(z)dz, \qquad (3.1)$$

where ρ is the mass density and κ the mass absorption coefficient or opacity, measured in, e.g., cm^2 gm^{-1}. One of the more important parameters governing the structure of the photosphere is the effective temperature $T_{\rm eff}$, defined as the temperature of a black body having the same emissivity (integrated over all wavelengths) as the stellar surface. $T_{\rm eff}$ is equal to the actual (kinetic) temperature at $\bar{\tau} \simeq 2/3$. In hotter stars ($T_{\rm eff} >\sim 10^4$ K), the atmospheric opacity comes mainly from photo-ionization of hydrogen atoms in different excitation levels

$$H^* + h\nu \rightarrow H^+ + e^-. \qquad (3.2)$$

This gives rise to absorption in the Lyman ($\lambda \leq 912$ Å), Balmer ($\lambda \leq 3646$ Å), Paschen ($\lambda \leq 8208$ Å) etc. continua, together with free–free transitions of electrons passing by protons, which dominate in the infra-red. In cooler stars like the Sun ($T_{\rm eff} = 5800$ K), the main opacity source in the optical region is photo-detachment of negative hydrogen

Table 3.1. *Data sources and methods for cosmic abundances*

Data sources		
Tangible:-		Intangible:-
	crust	Astrophysical objects:-
Earth:	atmosphere	
	oceans	Sun, planets, stars
		PN, SNR, H II regions,
Moon rocks (Apollo)		cold ISM,
Meteorites		galaxies,
		intra-cluster gas (X-r),
	Cosmic rays	QSOs, intervening gas
Space plasmas:		
	Solar wind	

Methods		
Chem. analysis		
Mass spectrometry		Spectrum analysis
Electron microscope		(abs. or em. lines)
n-activation,	for	
X-r fluorescence spectr.,	trace	Deductions from
Ion + electron microprobe	species	stellar structure
		and evolution
	nucl. em./plastic	
Space-borne		
	ion counters	

ions (for $\lambda < 1.66\,\mu\text{m}$),

$$\text{H}^- + h\nu \; (\geq 0.75\,\text{eV}) \rightarrow \text{H} + \text{e}^-, \tag{3.3}$$

again dominated in the infra-red by free–free transitions, this time in the fields of neutral hydrogen atoms which now dominate the population. Emission is due to the same processes in reverse and in many cases there is a good approximation to local thermodynamic equilibrium (LTE). In LTE, the relative populations of protons, neutral hydrogen atoms and H^- ions are as predicted by the Boltzmann and Saha equations (cf. Eq. 2.47):

$$\frac{n_i}{n_j} = \frac{g_i}{g_j} e^{-(E_i - E_j)/kT} = \frac{g_i}{g_j} 10^{-\theta \chi_{ij}} \tag{3.4}$$

where g_i, g_j are the statistical weights $(2J + 1)$ of two atomic levels, χ_{ij} is the difference in excitation potential in volts and

$$\theta \equiv \frac{5040}{T}; \tag{3.5}$$

Table 3.2. *Logarithm of opacity $\kappa_{0.5\mu m}$ in stellar photospheres as a function of θ and* $\log P_e$

Spec type(I)	θ	−2	−1	0	1	2	3	θ	Spec type(V)
				(log P_e)					
	0.1			Thomson scat.			−0.1	0.1	O5
B0	0.2			$\log \sigma =$		−0.3	0.4	0.2	B0
B3	0.3			$\log x_H - 0.5$		0.0	0.9	0.3	B3
A0	0.5				−0.2	0.6	1.2	0.5	A0
A5	0.7			−0.8	−0.8	0.0	0.9	0.7	A5
F2	0.8			−1.4	−0.9	0.1		0.8	F8
F8	0.9			−2.4	−1.7	−0.7		0.9	G2
G3	1.0		Rayl.	−2.4	−1.5	−0.5		1.0	G8
G8	1.2		−3.3	−2.1	−1.2			1.2	K3
K5	1.5		−2.6	−1.6	−0.6			1.5	M0
$\log P_e$:		−2	−1	0	1	2	3		

Letters OBAFGKM define spectral types in a sequence of decreasing effective temperature, with decimal subdivisions, while I to V indicate declining luminosity, corresponding to increasing gravity, main-sequence stars being of luminosity class V and normal giants class III; the Sun is classified as G2V. x_H is the fraction of hydrogen ionized to H^+. The solid lines represent boundaries between different opacity regimes; Rayl. stands for Rayleigh scattering which varies as λ^{-4}.

and

$$\frac{n_+ n_e}{n_0} = 2 \frac{u_+}{u_0} \left(\frac{m_e k T}{2\pi \hbar^2} \right)^{3/2} e^{-E_I/kT} \qquad (3.6)$$

or

$$\log \frac{n_+}{n_0} = \log \left(\frac{u_+}{u_0} \right) + 9.08 - 2.5 \log \theta - \theta I - \log P_e, \qquad (3.7)$$

where I is the ionization potential in volts to detach an electron from a species labelled with 0 and the us are partition functions defined in Section 2.4. Broadly speaking, the temperature is fixed as a function of optical depth by considerations of radiative transfer, and the pressure P or density by hydrostatic equilibrium at a mass column density governed by the opacity which itself depends mainly on the electron pressure P_e:

$$dP = -g\rho dz \quad \text{and} \quad d\tau_\lambda = -\kappa_\lambda \rho dz, \qquad (3.8)$$

whence

$$P(\tau_\lambda) = g \int_0^{\tau_\lambda} \frac{d\tau'_\lambda}{\kappa_\lambda(T, P_e[P, T])}. \qquad (3.9)$$

At solar-like temperatures, most of the free electrons come from easily ionized metals (Na, Mg, Al, Si, Ca, Fe) which for solar chemical composition total about 10^{-4} of

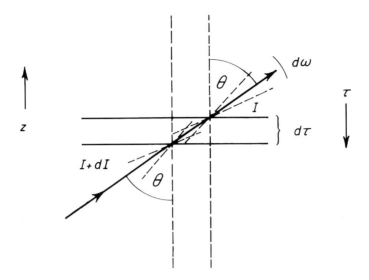

Fig. 3.1. Derivation of the equation of radiative transfer.

the hydrogen number density leading to an electron pressure that is about 10^{-4} times the gas pressure. The processes of absorption and emission in LTE, which lead to an emissivity governed by Planck's law for specific intensity,

$$I_v dv = B_v(T)dv \equiv \frac{2hv^3}{c^2} \frac{dv}{\exp(hv/kT) - 1}, \qquad (3.10)$$

are often referred to as 'pure absorption'. Large departures from LTE occur in stellar chromospheres (rarefied hotter layers above the photosphere which dominate the continuum in the far IR and UV, but contribute little at optical and near IR wavelengths) and at extreme temperatures and low densities. The result of these processes is that stars emit a black-body-like continuum modulated by the dependence of κ_λ on wavelength. Neutral hydrogen in hotter stars causes a discontinuous decrease in intensity as one moves shortward in wavelength across the Paschen, Balmer and especially the Lyman limit (cf. Fig. 3.18), whereas H^- opacity varies smoothly with wavelength and does not lead to very notable departures from a black-body-like continuum in the visible and near IR. At very high temperatures, electron scattering dominates the opacity and leads again to a more black-body-like spectrum because it is independent of wavelength. Table 3.2 gives a rough overview of stellar atmospheric opacities, $\log(\kappa_{0.5\mu m} + \sigma)$, in cm^2 gm^{-1}, as a function of θ and $\log P_e$, with a rough indication of corresponding spectral types where these parameters are typical of photospheric conditions.

Some insight into the structure of stellar atmospheres can be obtained by considering the simple case of a plane parallel grey atmosphere ($\kappa + \sigma$ independent of wavelength) in radiative equilibrium.

Fig. 3.1 shows a ray with specific intensity $I(\tau, \theta)$ (integrated over all wavelengths) travelling outwards at an angle θ to the normal. On passing through a layer with vertical thickness dz, it loses $I\,d\tau/\mu$ from absorption (where $\mu \equiv \cos\theta$) and gains $j\,dz/\mu \equiv S\,d\tau/\mu$ from emission, where $j(\tau)$ is the emission per unit solid angle per unit volume and $S(\tau) \equiv j/\kappa\rho$ is known as the source function. The transfer equation then reads

$$\mu\frac{\partial I}{\partial \tau} = I - S. \tag{3.11}$$

Integrating Eq. (3.11) over solid angles and dividing by 4π, one obtains

$$\frac{dH}{d\tau} = J - S, \tag{3.12}$$

where $J(\tau) \equiv \int I\,d\omega/4\pi \equiv \frac{1}{2}\int_{-1}^{1} I\,d\mu$ is the (angle-averaged) mean intensity and $H \equiv \int I\mu\,d\omega/4\pi \equiv \frac{1}{2}\int_{-1}^{1} I\mu\,d\mu$ is known as the 'Eddington flux'; the actual flux is $\sigma T_{\text{eff}}^4 \equiv \pi F \equiv 4\pi H$. In radiative equilibrium, the flux is constant, so

$$S = J, \tag{3.13}$$

and in LTE,

$$S = B(T) = \sigma T^4/\pi \tag{3.14}$$

(Stefan's law). Multiplying Eq. (3.11) by μ and integrating over solid angles again, one has

$$\frac{dK}{d\tau} = H \tag{3.15}$$

where $K \equiv \frac{1}{2}\int_{-1}^{1} I\mu^2\,d\mu$ is $(c/4\pi)\times$ the radiation pressure. Eddington's approximation consists in setting

$$K = \frac{1}{3}J \tag{3.16}$$

(which is exact if $I(\tau,\mu)$ can be expanded in odd powers of μ, plus a constant), so from Eqs. (3.13) to (3.16)

$$J = S = B = 3H\tau + \text{const.} \tag{3.17}$$

A simple boundary condition at the surface is obtained by assuming $I(\tau = 0, \mu > 0) = \text{const.}$ (i.e. limb-darkening is neglected), $I(\tau = 0, \mu < 0) = 0$. Then $J(\tau = 0) = 2H$ and we have

$$S = J = H(2 + 3\tau) \tag{3.18}$$

and in LTE using $\pi F = 4\pi H = \sigma T_{\text{eff}}^4$

$$T^4 = T_{\text{eff}}^4 \left(\frac{1}{2} + \frac{3}{4}\tau\right). \tag{3.19}$$

Eq. (3.19) gives a first approximation to the temperature structure of an atmosphere in radiative equilibrium, and departures from greyness can also be treated approximately by defining a suitable mean absorption coefficient (see Chapter 5). The emergent

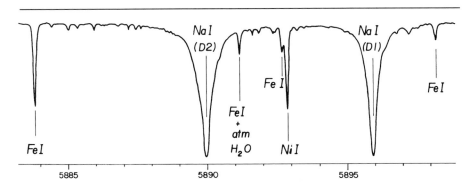

Fig. 3.2. Portion of the solar spectrum in the neighbourhood of the sodium D-lines. Adapted from L. Delbouille, G. Roland & L. Neven, *Spectrophotometric Atlas of the Solar Spectrum,* Liège, 1973.

monochromatic intensity at an angle θ to the normal (relevant to some point on the solar disk) is also found by integrating the equation of transfer (3.11):

$$I_\lambda(0,\mu) = \int_0^\infty B_\lambda[T(\tau_\lambda)]e^{-\tau_\lambda/\mu}d\tau_\lambda/\mu \tag{3.20}$$

and the flux (relevant to radiation from an unresolved stellar surface) from $F_\lambda(0) = 2\int_0^1 I_\lambda(0,\mu)\mu d\mu$. In the Eddington–Barbier approximation, $B_\lambda(\tau_\lambda)$ is treated as a linear function and then one has simply

$$I_\lambda(0,\mu) \simeq B_\lambda(\tau_\lambda = \mu) \tag{3.21}$$

and

$$F_\lambda(0) \simeq B_\lambda(\tau_\lambda = 2/3). \tag{3.22}$$

Atoms, ions and molecules present in the stars provide additional opacity at wavelengths corresponding to specific atomic transitions; these give rise to comparatively narrow absorption lines (see Fig. 3.2) with intensities related to the abundances of the relevant elements (and much else). Despite the name, processes other than pure absorption (e.g. scattering and fluorescence) are involved in the production of these lines and, while they are often treated in LTE, this is now only a simplifying approximation which often works fairly well, but needs to be checked by more detailed calculations for each particular case. (In some cases, there are even emission lines or emission components, e.g. the solar Ca^+ H and K lines in the near UV which are so strong that the chromosphere affects their central parts.)

Accurate abundance analysis of stellar absorption lines is an elaborate physical and numerical exercise involving the following steps (see Fig. 3.3):-

- Calculate a model atmosphere, normally based on principles of radiative and (where necessary) convective equilibrium, together with hydrostatic equilibrium.

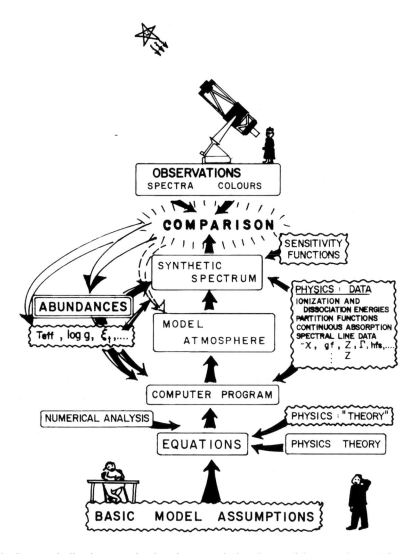

Fig. 3.3. Cartoon indicating steps in abundance analysis using model atmospheres. After Gustafsson (1980).

This gives the run of physical parameters (temperature T, P_g, P_e, κ_λ) with optical depth $\tau_{\lambda 1}$ at some chosen wavelength, using as input parameters $T_{\rm eff}$, the surface gravity g and relevant parameters relating to chemical composition, e.g. He/H and m/H, where m represents easily ionized metals. Effects of line opacity ('line blanketing') on the structure of the atmosphere need to be allowed for, raising a major computational problem. A further parameter affecting line profiles and intensities is the so-called micro-turbulent velocity, which has to be found empirically for each star.

- Deduce atomic level populations at each depth point, using LTE or preferably something better.
- Deduce the selective absorption coefficients $\kappa_l(\Delta v, \tau_\lambda)$ as functions of depth and distance from the line centre.
- Integrate the equation of radiative transfer to produce a synthetic spectrum as a function of assumed abundances of each element of interest.
- Convolve the synthetic spectrum with the surface velocity field (rotation, granulation) and with the instrumental profile.
- Compare the convolved synthetic spectrum with the observed spectrum, line profiles or equivalent widths.

The precision that can be achieved depends on having an accurate model with the right T_{eff}, good control over non-LTE effects and the availability of suitable lines with known atomic parameters, especially oscillator strengths (defined in Appendix 3). Various consistency checks can be made, in particular the model should fit the continuum energy distribution and hydrogen line profiles, and in some cases (e.g. CNO in the Sun), one may check that the same set of abundances fits a variety of atomic and molecular lines (CI, [CI], NI, [NI], OI, [OI], CO, CH, CN, OH). Some solar abundance determinations are based on a semi-empirical model (Holweger & Müller 1974) which takes into account observable centre-limb effects, rather than on a purely theoretical model, and scaled versions of this or other solar models are sometimes applied to stars having similar characteristics to the Sun. Special difficulties arise when the star is either very hot ($T_{eff} > 30\,000$ K) or very cool ($T_{eff} < 4000$ K) or has a very low gravity ($\log g < 0$; cf. $\log g_\odot = 4.4$ in cgs units). In favourable cases, an accuracy of perhaps ± 0.1 dex can be achieved, although abundances are often quoted to two decimal places in the logarithm.

3.2.1 Equivalent widths and curves of growth

We now consider in somewhat more detail a simplified approach based on the 'curve of growth'. For this, we ignore fine details of the observed line profile and use the equivalent width (EW) defined in Fig. 3.4, $W_\lambda \equiv \int R d\lambda$ or $W_v \equiv \int R dv$, where $R(\Delta\lambda)$ or $R(\Delta v)$ is the relative depression below the continuum at some part of the line. The curve of growth is a relationship between the equivalent width of a line and some measure of the effective number of absorbing atoms. Equivalent widths are useful (also in model-atmosphere analyses) as long as the line is isolated and has a well-defined continuum; this is not always the case, especially for rare elements represented by few lines, but when it is, the curve of growth is still a very useful tool for the analysis of many lines of one element, especially Fe I, and its theory offers good insights into the factors involved in the production of absorption lines. Under ideal conditions, the equivalent width is independent of instrumental broadening and velocity fields on the stellar surface.

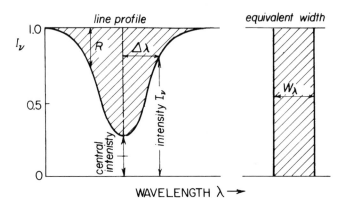

Fig. 3.4. Definition of equivalent width.

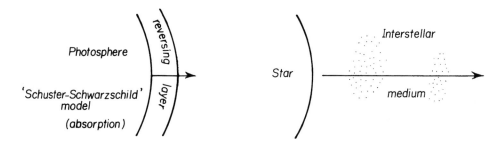

Fig. 3.5. Path of radiation through a stellar atmosphere (left) and through the interstellar medium (right).

3.2.1.1 *The exponential curve of growth*

The exponential curve of growth is derived by caricaturing line formation as the result of an incident smooth continuum passing through a uniform gaseous column that absorbs only in the lines (Fig. 3.5). When applied to stars, it is based on an old and crude picture in which there is a gaseous 'reversing layer' that imprints absorption lines on continuous radiation coming from the photosphere below,[1] but it gives an exact picture of the formation of interstellar lines.

Consider continuous radiation with specific intensity I_1 incident normally on a uniform slab with a source function $S \equiv B_\nu(T_{ex}) \ll I_1$ (Fig. 3.6). S is the ratio of emission per unit volume per unit solid angle to the volume absorption coefficient $\kappa\rho$ and is equal to the Planck function B_ν of an 'excitation temperature' T_{ex} obtained by force-fitting the ratio of upper to lower state atomic level populations to the Boltzmann

---

[1] The reversing-layer picture is sometimes referred to as the 'Schuster–Schwarzschild model', but Arthur Schuster and Karl Schwarzschild considered more sophisticated models of radiative transfer (including scattering) than the one used here.

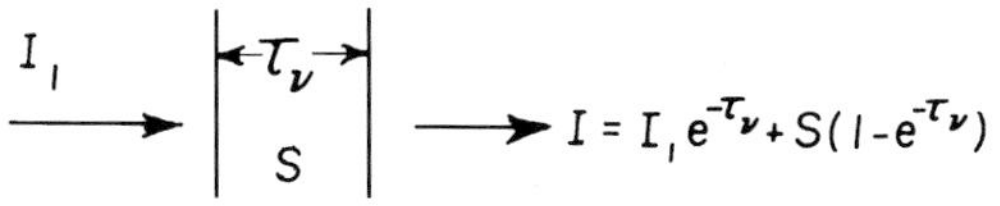

Fig. 3.6. The slab model.

formula Eq. (3.4). For the interstellar medium at optical and UV wavelengths, effectively $S = 0$.

The intensity emerging from the slab is

$$I = I_1 e^{-\tau_v} + S(1 - e^{-\tau_v}) \tag{3.23}$$

where τ_v is the optical thickness of the slab at the appropriate distance from the line centre. Hence

$$R = 1 - \frac{I}{I_1} = \left(1 - \frac{S}{I_1}\right)(1 - e^{-\tau_v}) = R_\infty(1 - e^{-\tau_v}) \tag{3.24}$$

and

$$\frac{W_\lambda}{\lambda} = \frac{W_v}{v} = R_\infty \int_0^\infty (1 - e^{-\tau_v})dv/v. \tag{3.25}$$

When the absorbing slab is effectively very cold compared to the brightness temperature of I, notably in the interstellar case, $R_\infty = 1$.

The optical depth τ_v is given — apart from a stimulated emission factor $(1 - e^{-hv/kT_{ex}})$ explained in Appendix 3 — by a product of the column density N of absorbing atoms in the appropriate state and the atomic absorption cross-section α_v. The latter depends on atomic constants and on a convolution of two line-broadening mechanisms: Doppler effect from a gaussian distribution of velocities in the line of sight and damping from finite lifetimes of the atomic energy states which latter usually leads to a Lorentzian distribution similar to Eq. (2.63); the effect of these two mechanisms is illustrated in Fig. 3.7.

$$\begin{aligned} \tau_v &= N\frac{\pi e^2 f}{m_e c}\frac{\lambda e^{-v^2}}{b\sqrt{\pi}} \otimes \frac{\gamma}{\Delta\omega^2 + (\gamma/2)^2} \tag{3.26} \\ &= N\alpha_0 H(a, v) \equiv \tau_0 H(a, v). \tag{3.27} \end{aligned}$$

τ_0 is the (nominal) optical depth at the line centre, given by

$$\tau_0 = \frac{\pi^{1/2}e^2 f\lambda}{m_e cb}N \simeq 10^{-11}f\left(\frac{1\,\text{km s}^{-1}}{b}\right)\left(\frac{\lambda}{0.5\,\mu\text{m}}\right)N_{\text{cm}^{-2}}. \tag{3.28}$$

Here

- $\pi e^2 f/(m_e c) = \int \alpha(v)dv$ (in cgs units) from classical electromagnetic theory modified by quantum mechanics (e.g. time-dependent perturbation theory; see Appendix 3) which introduces the dimensionless oscillator strength f. f may

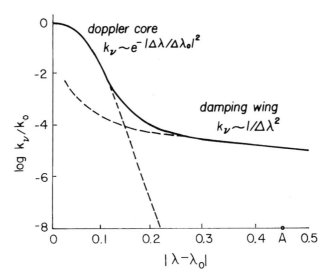

Fig. 3.7. Dependence of the line absorption coefficient on distance from the line centre, for the particular case of the Na I D-lines in the outer solar atmosphere. Reproduced with permission after A. Unsöld, *The New Cosmos*, p. 171, Fig. 19.3, ©1977 by Springer-Verlag New York Inc.

be calculated theoretically or measured experimentally. Under certain circumstances it can also be cancelled out in a differential abundance analysis of one star relative to a standard (often the Sun) giving directly a logarithmic abundance change usually symbolized by [M/H] where M stands for some element. This method has become less important in recent years owing to advances in knowledge of f-values.

- $\pm b$ is the 'typical' line-of-sight velocity in a gaussian distribution, specifically $\sqrt{2} \times$ the rms velocity, so that the probability of a velocity between V and $V + dV$ is $\exp(-V^2/b^2)dV/b\sqrt{\pi} \equiv \exp(-v^2)dv/\sqrt{\pi}$. The dimensionless velocity variable v has the equivalent definitions

$$v \equiv \frac{V}{b} \equiv \frac{\Delta\lambda}{\Delta\lambda_D} \equiv \frac{\Delta v}{\Delta v_D} \equiv \frac{\Delta\omega}{\Delta\omega_D}. \tag{3.29}$$

- $b^2 = b_{\text{th}}^2 + \xi^2$, where

$$b_{\text{th}} = \sqrt{\left(\frac{2kT}{Am_{\text{H}}}\right)} = 12.9 \left(\frac{T_4}{A}\right)^{1/2} \text{ km s}^{-1} \tag{3.30}$$

(with A the atomic mass number and T_4 the temperature in units of 10^4 K) is the thermal velocity, and ξ is an empirical parameter independent of A known as 'microturbulence' which is needed to account for the Doppler width of the absorption coefficient and amounts in most cases to between about 0.5 and a few km s^{-1}.

- γ is the damping parameter or inverse mean lifetime of an excited level involved in the transition, made up of three components:

$$\gamma = \gamma_{\text{nat}} + \gamma_{\text{Stark}} + \gamma_{\text{coll}}. \tag{3.31}$$

Natural damping dominates in the interstellar medium (ISM). It is given for resonance lines (those involving the ground state) by $\gamma_{\text{nat},i} = \sum_j A_{ij}$, where A_{ij} is the Einstein spontaneous transition probability from an excited state i to any lower state j. On the other hand, pressure broadening from $\gamma_{\text{Stark}} + \gamma_{\text{coll}}$ dominates in main-sequence stars. (The Lorentzian broadening profile in Eq. (3.26) does not apply to stellar hydrogen lines, which have a linear Stark effect; it does apply, however, to interstellar lines because they just have natural damping.) An approximation to γ_{nat} for intrinsically strong lines is the classical damping constant

$$\gamma_{\text{cl}} = \frac{2e^2\omega^2}{3m_e c^3} = 8.9 \times 10^7 \left(\frac{0.5\,\mu\text{m}}{\lambda}\right)^2 \text{s}^{-1}. \tag{3.32}$$

- $\otimes$ denotes a convolution of the Doppler broadening due to a gaussian velocity field with the Lorentzian from damping, expressed in the Voigt function

$$\frac{\alpha(v)}{\alpha_0} = H(a,v) \equiv \frac{a}{\pi} \int_{-\infty}^{\infty} \frac{e^{-x^2}dx}{a^2 + (v-x)^2} \tag{3.33}$$

where

$$a \equiv \frac{\gamma}{2\Delta\omega_D} = \frac{\gamma\lambda}{4\pi b} = 3.53 \times 10^{-3} \left(\frac{\gamma}{\gamma_{\text{cl}}}\right)\left(\frac{0.5\,\mu\text{m}}{\lambda}\right)\left(\frac{1\,\text{km s}^{-1}}{b}\right) \tag{3.34}$$

is the dimensionless damping parameter.

To develop the curve of growth, we shall roughly approximate the convolution integral in Eq. (3.33) by treating the two broadenings separately, i.e. we take

$$\alpha(v) \simeq \alpha_0 e^{-v^2} \tag{3.35}$$

for $v < 3$, and

$$\alpha(v) \simeq \frac{\pi e^2 f}{mc} \frac{\gamma}{\Delta\omega^2 + (\gamma/2)^2} = \alpha_0 \frac{a}{\sqrt{\pi(a^2 + v^2)}} \simeq \frac{\alpha_0 a}{\sqrt{\pi v^2}} \tag{3.36}$$

for $v > 3$.

At moderate optical depths, only the gaussian part of the profile is significant:

$$\frac{W_\lambda}{\lambda R_\infty} = \int_{-\infty}^{\infty} [1 - \exp(-\tau_0 e^{-v^2})] \frac{b}{c} dv. \tag{3.37}$$

For $\tau_0 \ll 1$, this tends to

$$\frac{b}{c}\tau_0\sqrt{\pi} = \frac{\pi e^2}{m_e c^2} Nf\lambda \propto N \tag{3.38}$$

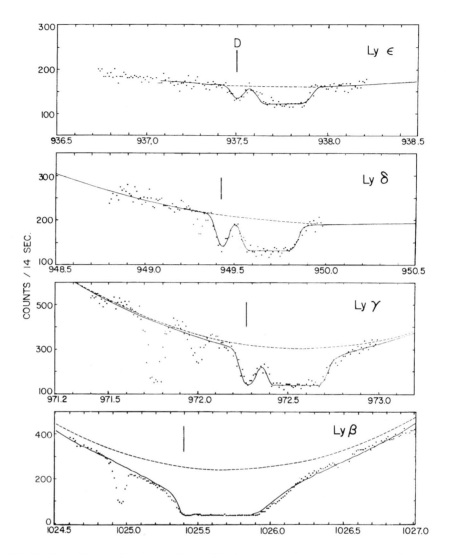

Fig. 3.8. Profiles of interstellar Lyman lines of hydrogen and deuterium towards γ Cas, observed with the *Copernicus* satellite. Solid curves represent a synthetic spectrum fitted to data, assuming a continuum (modified by instrument sensitivity) shown by the dashed lines. The hydrogen lines have flat bottoms as predicted by Eq. (3.24) with damping wings becoming noticeable in the case of the strong line Lyman-β, whereas the deuterium lines, displaced 81 km s^{-1} to the violet and with 1 to 2 $\times 10^{-5}$ of the H I column density of 1.0×10^{20} cm^{-2}, show only Doppler-broadened profiles. The hydrogen Doppler width b^H is 14 km s^{-1}. Adapted from Vidal-Madjar *et al.* (1977).

and we have the linear regime in which the EW is proportional to Nf and independent of line broadening. For $\tau_0 \gg 1$, the factor in brackets in Eq. (3.37) is effectively 1 for $v^2 < \ln \tau_0$ and 0 for $v^2 > \ln \tau_0$, leading to a flat-bottomed profile (cf. Fig. 3.8) and the

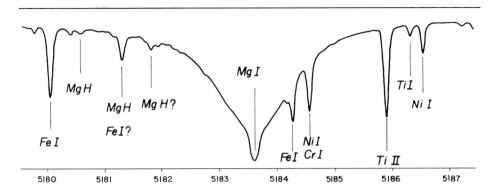

Fig. 3.9. A very strong absorption line of Mg I in the solar spectrum, dominated by damping wings. Adapted from the Liège Atlas (Delbouille *et al.* 1973).

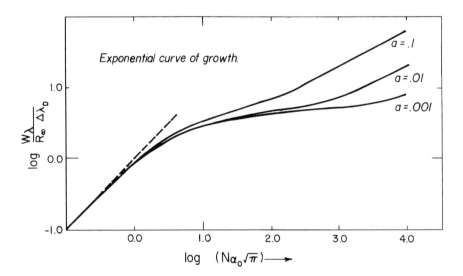

Fig. 3.10. Generic exponential curves of growth, normalized so that abscissa and ordinate are the same in the linear regime.

EW is given by

$$\frac{W_\lambda}{\lambda R_\infty} \simeq 2\frac{b}{c}\sqrt{\ln \tau_0} \propto b; \qquad (3.39)$$

this is the saturated regime in which the EW depends mainly on b and increases only very slowly with the column density, leading to a flat portion of the curve of growth (Fig. 3.10). The flat portion can also be raised by line broadening due to hyperfine structure, when present.

For still larger column densities, the damping wings completely overcome the contribution of the Doppler core to the EW (see Fig. 3.9) and we can take for the whole

Table 3.3. *Exponential curves of growth*

$\log N\alpha_0\sqrt{\pi}$	$\log(W_\lambda/R_\infty\Delta\lambda_D)$		
	$a = 0.001$	$a = 0.01$	$a = 0.1$
-0.75	-0.76	-0.76	-0.76
-0.50	-0.53	-0.52	-0.52
-0.25	-0.30	-0.29	-0.29
0.00	-0.08	-0.08	-0.07
0.50	0.27	0.27	0.30
1.00	0.47	0.47	0.54
1.50	0.56	0.58	0.69
2.00	0.65	0.68	0.87
2.50	0.70	0.74	1.07
3.00	0.73	0.88	1.31
3.50	0.79	1.07	1.55
4.00	0.92	1.31	1.80

After Powell (1969).

line profile $\Delta\omega \gg \Delta\omega_D \gg \gamma$. Then

$$\frac{1}{R_\infty}\frac{W_\lambda}{\lambda} = \int_{-\infty}^{\infty}\left\{1 - \exp\left(-\frac{N\pi e^2 f\gamma}{m_e c}\frac{1}{\Delta\omega^2}\right)\right\} d\,\Delta\omega/(2\pi v) \tag{3.40}$$

$$= \frac{\lambda}{c}\left(\frac{e^2}{m_e c}Nf\gamma\right)^{1/2} = \frac{2\pi e^2}{m_e c^2}\left(\frac{2}{3}Nf\frac{\gamma}{\gamma_{\rm cl}}\right)^{1/2} \tag{3.41}$$

$$= 1.45 \times 10^{-12}\left(N_{\rm cm^{-2}}f\frac{\gamma}{\gamma_{\rm cl}}\right)^{1/2}, \tag{3.42}$$

the square-root law. In this case, the EW is once again independent of b. A generic set of exponential curves of growth is shown in Fig. 3.10 and tabulated in Table 3.3, while Fig. 3.11 shows a pair of curves fitted to interstellar lines for two different values of the Doppler width b. By the use of two or more lines from the same multiplet or ground state with different f-values, it is possible to sort out both the column density and the Doppler width, provided that at least one sufficiently weak line is available, e.g. in the doublet ratio method applied to the sodium D-lines. When several lines from the same lower atomic state or term are available, one can plot an empirical curve of growth, $\log(W/\lambda)$ against $\log f\lambda$ or $\log gf\lambda$ as appropriate and slide it over the theoretical one to deduce N and $\Delta\lambda_D$. However, complications can arise if there is more than one cloud in the line of sight.

3.2.1.2 *The MEMMU curve of growth*
A form of the curve of growth more relevant to stellar (as opposed to interstellar) absorption lines is derived from work by E.A. Milne, A.S. Eddington, M. Minnaert, D.H. Menzel and A. Unsöld. In the 'Milne–Eddington' model of a stellar photosphere,

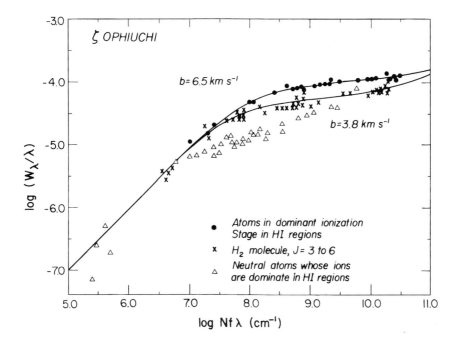

Fig. 3.11. Curves of growth for interstellar lines towards ζ Oph, observed with the *Copernicus* satellite. This method of plotting shows the dependence on *b* along the flat portion and its disappearance as the square-root portion is approached (provided all lines have the same value of $\gamma\lambda^2$). After L. Spitzer & E.B. Jenkins (1975), with permission, from the *Annual Review of Astronomy and Astrophysics*, Vol. **13**, ©1975 by Annual Reviews, Inc.

the continuum source function (equated to the Planck function in the LTE approximation) increases linearly with continuum optical depth τ_λ and there is a selective absorption $\eta\kappa_\lambda$ in the line, where $\eta(\Delta v)$, the ratio of selective to continuous absorption, is a constant independent of depth given by

$$\eta(\Delta v) = \left(\frac{M}{H}\right)\left(\frac{M_{ij}}{M}\right)\left(\frac{\alpha(\Delta v)}{\alpha_\lambda}\right). \tag{3.43}$$

Here M/H is the atomic abundance, M_{ij}/M is the proportion of M atoms in the appropriate state of excitation and ionization to absorb the line, $\alpha(\Delta v)$ is the selective absorption cross-section implied by Eq. (3.27) and α_λ the continuous absorption cross-section per hydrogen atom; this latter is related to κ_λ (e.g. as in Table 3.2) by the number of H atoms per gram.

Expressing the source function by

$$B(\tau) = A\left(1 + \frac{3}{2}\beta\tau\right) \tag{3.44}$$

(where β is an increasing function of $h\nu/kT$ by virtue of Planck's law and a fixed $T(\tau)$ relation as in Eq. (3.19)), the continuum flux obtained by solving the equation of radiative transfer for a typical angle of emergence of a ray, $\arccos 2/3$, is

$$F_1 \simeq A(1 + \beta), \tag{3.45}$$

while the line flux is

$$F \simeq A\left(1 + \frac{\beta}{1 + \eta}\right) \tag{3.46}$$

which basically expresses the fact that one sees less far down at the line frequency (cf. the Eddington–Barbier approximation Eq. 3.22). Hence

$$R \equiv \frac{F_1 - F}{F_1} = \frac{\beta}{1 + \beta}\frac{\eta}{1 + \eta} = R_\infty \frac{\eta}{1 + \eta}, \tag{3.47}$$

or

$$\frac{1}{R} = \frac{1}{R_\infty} + \frac{1}{x(\Delta\nu)}, \tag{3.48}$$

where

$$x = \eta R_\infty = \frac{\eta\beta}{1 + \beta}. \tag{3.49}$$

This is the Menzel–Minnaert–Unsöld interpolation formula (often used assuming $R_\infty = 1$). It gives a better approximation to stellar absorption-line profiles (which are definitely not flat-bottomed) than does the exponential formula; the shape of the corresponding curve of growth is much the same, but its use leads to a b-parameter that is about 25 per cent higher for the same observational data.

Calling the central value of η η_0, the Doppler part of the curve is given by

$$\frac{1}{R_\infty}\frac{W_\lambda}{\Delta\lambda_D} = \int_{-\infty}^{\infty} \frac{\eta d\nu}{1 + \eta} = \eta_0 \int_{-\infty}^{\infty} \frac{\exp(-\nu^2)d\nu}{1 + \eta_0 \exp(-\nu^2)}. \tag{3.50}$$

As before, there are three regimes:

For $\eta_0 \ll 1$,

$$\frac{W_\lambda}{\lambda} = R_\infty \frac{\eta_0 b}{c}\sqrt{\pi} \tag{3.51}$$

independent of b. Here η_0 plays the same role as does $N\alpha_0$ in the exponential curve of growth.

For a mild degree of saturation ($\eta_0 < 1$), the denominator of Eq. (3.50) can be expanded as a power series

$$\frac{1}{R_\infty}\frac{W_\lambda}{\Delta\lambda_D} = \eta_0 \int_{-\infty}^{\infty} e^{-\nu^2}(1 - \eta_0 e^{-\nu^2} + \eta_0^2 e^{-2\nu^2} - \ldots)d\nu \tag{3.52}$$

$$= \eta_0\sqrt{\pi}\left(1 - \frac{\eta_0}{\sqrt{2}} + \frac{\eta_0^2}{\sqrt{3}} - \ldots\right), \tag{3.53}$$

where the expression in brackets represents the saturation factor.

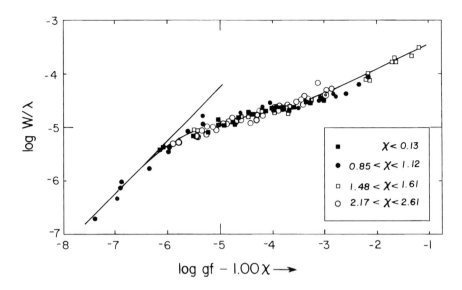

Fig. 3.12. Simple (exponential) curve of growth for low-excitation Fe I lines with wavelengths between 4000 and 8700 Å at the centre of the solar disk, with $\theta_{ex} = 1.00, b = 1$ km s^{-1} (assuming $R_\infty = 1$), $a = 0.02$. Equivalent widths are from Moore, Minnaert & Houtgast (1966). gf-values are from furnace measurements by the Oxford group (Blackwell *et al.* 1986 and references therein).

On the damping branch ($\eta_0 a \gg 1$), we have

$$\frac{1}{R_\infty} \frac{W_\lambda}{\Delta\lambda_D} = \eta_0 \int_{-\infty}^{\infty} \frac{dv}{\eta_0 + \frac{\sqrt{\pi}}{a}v^2} = \sqrt{(\pi^{3/2}a\eta_0)} \tag{3.54}$$

or

$$\frac{1}{R_\infty} \frac{W_\lambda}{\lambda} = \frac{1}{2c}\sqrt{(\pi^{1/2}\eta_0 b\gamma\lambda)}, \tag{3.55}$$

again independent of b.

These approximations relate to model-atmosphere analysis like a one-point integration formula. $\eta_0 bR_\infty$ can be replaced by an exact integral through the atmosphere representing the equivalent width in the weak-line (unsaturated) limit, providing the abscissa for a curve of growth which then simply takes on the role of a saturation curve. This is sometimes known as the 'method of weighting functions', after a certain way of formulating the integral.

3.2.1.3 *Curve-of-growth fitting procedures*

In a simple-minded application known as 'coarse analysis', one simply assumes the lines to be formed in in a single layer in quasi-LTE with a fixed excitation temperature θ_{ex}, a fixed ionization temperature θ_{ion} and a fixed electron pressure P_e. The logarithmic equivalent column density is then $\log gf - \theta_{ex}\chi + \text{const.}$, using gf rather than $gf\lambda$

mainly because the H^- opacity is more or less proportional to λ throughout the optical region and θ_{ex} is chosen to give the best fit between lines with differing lower excitation potential (see Fig. 3.12). θ_{ion} and $\log P_e$ are best determined from model-atmosphere considerations based on effective temperature (e.g. for the Sun, $\theta_{ion} \simeq 0.9$ for $\theta_{eff} = 0.87$) and gravity; $\log P_e$ can be taken from Table 3.2 for the relevant spectral type and luminosity class. Deducing $\eta_0 b/gf$ for ground-state lines by comparing the linear portion of the curve with Eq. (3.51), we then have from Eq. (3.28)

$$Agf = \eta_0 \frac{b}{c}\sqrt{\pi} = \frac{\pi e^2}{m_e c^2}\frac{M_1/g_1}{H}\frac{g_1 f\lambda}{\alpha_\lambda(H)} \tag{3.56}$$

where $\log A$ is the constant from the curve-of-growth shift, M_1/H is the abundance of the ground-state neutral atom relative to hydrogen and $\alpha_\lambda(H)$ the absorption cross-section per H atom, related to $\kappa_{0.5\mu m}$ given in Table 3.2 by

$$\alpha_{0.5\mu m} = 1.4\kappa_{0.5\mu m}m_H. \tag{3.57}$$

(The factor 1.4 comes from the assumption that the He/H ratio is 0.1.) We thus have

$$\frac{M_1/g_1}{H} = A\frac{1.4 m_H\, \kappa_{0.5\mu m}}{\pi e^2\lambda/m_e c^2} = 5.2\times 10^{-8}A\,\kappa_{0.5\mu m}. \tag{3.58}$$

Correcting $g_1^{-1}M_1/H$ to the total abundance (mostly the first ion) using Saha's equation Eq. (3.7), one can obtain an estimate of a metallic abundance M/H good to about 0.2 dex, provided that the linear part of the curve of growth is well defined and the value of θ_{ion} judiciously chosen.

In a more sophisticated application, one calculates an abscissa $\log X$, which is a theoretical value of $\log W/\lambda$ taking into account all atmospheric effects except saturation, as a function of the desired abundance ratio M/H; $\log X = \log(M/H) + \log gf + \log \Gamma$, where Γ is calculated for given excitation and ionization potentials, ionic partition functions and the model atmosphere. The abundance is then chosen to give the optimal fit for weak lines. The same curve can also be used (with due precautions) to predict saturation effects for elements represented by only a few lines that may not be weak.

3.2.1.4 *Differential curve-of-growth analysis*

Differential curve-of-growth analysis is a method of comparing stars with a standard (often the Sun) that was pioneered by L.H. Aller and J.L. Greenstein in the 1950s and developed further by G. Wallerstein, R. Cayrel, J. Jugaku, B.E.J. Pagel and others in the 1960s. The method can be quite accurate and it is described here in some detail both because it gives a good insight into the issues arising in abundance analysis and because it has provided some graphic illustrations of how abundance differences translate into visible curve-of-growth shifts. The idea is to find $\log X_\odot$ for a given solar line from an assumed normalised solar curve of growth and predict the corresponding $\log X$ for a stellar spectrum, as a function of differential abundance $[M/H] \equiv \log(M/H) - \log(M/H)_\odot$ for some element M using either detailed calculations

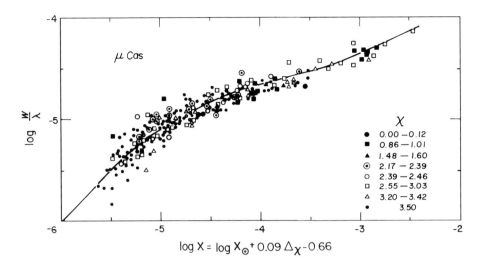

$$\log X = \log X_\odot + 0.09\,\Delta\chi - 0.66$$

Fig. 3.13. Differential curve of growth for Fe I in μ Cas relative to the Sun, after Catchpole, Pagel & Powell (1967).

of Γ from model atmospheres (e.g. Cayrel & Jugaku 1963) or simple considerations of the effect of temperature differences $\Delta\theta$ and electron pressure differences $[P_e] \equiv \Delta\log P_e$ between the star and the standard; these differences are a better approximation to the effects of stratification in the stellar atmosphere than is the use of single values of θ, $\log P_e$ in the simple curve-of-growth analysis. A further advantage, less important now than it was then, is that unknown oscillator strengths cancel out (or one can use 'solar oscillator strengths' designed to give accepted solar abundances from the given line in the solar spectrum), and it is still customary and useful to express stellar abundances in the relative [M/H] form.

A particularly simple case is that of neutral metals in F to early K-type stars where these metals are predominantly singly ionized. Applying Saha's equation Eq. (3.7) to such metals and also to H^-, which is itself predominantly 'ionized' to neutral H, one finds

$$[X] = [M/H] + [\Gamma] \simeq [\eta_0 b] \simeq [M^+/H] + \Delta\theta(I - \chi - 0.75). \tag{3.59}$$

$\Delta\chi \equiv I - \chi$ is the ionization potential from the lower state of the line and 0.75 eV is the electron detachment potential of H^-. $[M^+/H] = [M/H] + [x]$, where x is the degree of ionization which changes negligibly while it is close to one, and the electron pressure cancels out. $\Delta\theta$ can be identified with $\Delta\theta^1_{ex}$ obtained by optimally fitting neutral lines with different excitation potentials to one curve of growth (see Fig. 3.13), or deduced from red-infra-red colours. As a refinement, a small term $[\theta]$ should be added to the rhs of Eq. (3.59) to allow for an increase of the weighting function integral towards lower effective temperatures.

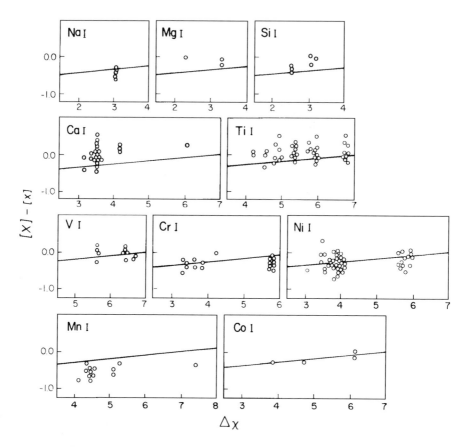

Fig. 3.14. Curve-of-growth shifts for individual lines of individual neutral elements in μ Cas, plotted against $\Delta\chi$. The solid line shows the corresponding relation for Fe I, so that the offsets give directly the differential abundance changes [M/Fe]. Adapted from Catchpole, Pagel & Powell (1967).

Fig. 3.14 shows horizontal curve-of-growth offsets for different elements in the mild 'subdwarf' star μ Cassiopeiae, relative to those for Fe I at the same $I - \chi$. Several elements maintain a nearly constant abundance ratio relative to iron, so that [Fe/H], often known as the 'metallicity', is sometimes taken as representing the abundance change in the heavy-element mass fraction Z, which is made up of all the elements from carbon upwards; these are also often loosely referred to as 'metals' although they include nitrogen, oxygen etc. However, Fig. 3.14 shows some abundance differences characteristic of stars with significantly lower metallicity than the Sun: elements with α-particle nuclei (Mg, Si, Ca) and Ti are slightly overabundant compared to iron, and manganese slightly underabundant, although in this latter case the interpretation is complicated by hyperfine structure effects on the saturated portion of the curve of growth.

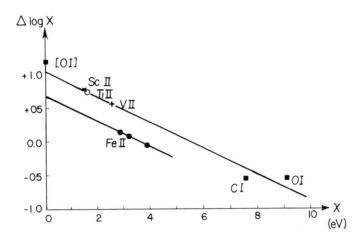

Fig. 3.15. Curve-of-growth shifts (relative to the Sun) for forbidden [O I], permitted O I and C I and singly ionized metals in the spectrum of Arcturus, after R.E.M. Gasson & B.E.J. Pagel, *Observatory*, **86**, 196 (1966).

The curve-of-growth shift for singly ionized lines (or for neutral lines of elements with large ionization potentials like C, N, O) is given by

$$[X]^{\mathrm{II}} = [M^+/H] - \Delta\theta_{\mathrm{ex}}^{\mathrm{II}}(\chi + 0.75) - \frac{3}{2}[\theta] - [P_e]. \tag{3.60}$$

$\Delta\theta_{\mathrm{ex}}^{\mathrm{II}}$ is generally less than $\Delta\theta$ because excited lines of dominant species tend to be formed deeper in cooler atmospheres, and the $[P_e]$ term introduces further depth dependence, but accurate offsets between dominant species can be obtained by comparing plots of $[X]$ against χ among themselves. Fig. 3.15 gives such a plot for the red giant Arcturus ($[Fe/H] \simeq -0.5$), which shows that oxygen shares the relative overabundance of the α-elements, while carbon does not. (The position of Fe II in this plot is ambiguous, because it is not the dominant species throughout the atmosphere of Arcturus, whereas it is so in the Sun, but the other metals are predominantly ionized in both stars.)

Complications arise in differential curve-of-growth analysis when the star to be analysed is extremely metal-deficient, making it necessary to use strong lines on the damping branch of the curve of growth for the Sun which reappear as weak lines in the star, and of high luminosity (low gravity) leading to near or complete absence of pressure broadening. In such cases one has to be cautious in using solar X or gf values, especially for neutral lines (although a reasonable result can still be obtained from ionized metal lines; cf. Fig. 3.16) and in modern work, direct analyses using oscillator strengths measured in the laboratory are preferred.

3.2.1.5 *Integrated spectra of stellar populations*

While it is possible to obtain colours and spectra for the most luminous stars in the nearest galaxies — the Magellanic Clouds and some dwarf spheroidal galaxies

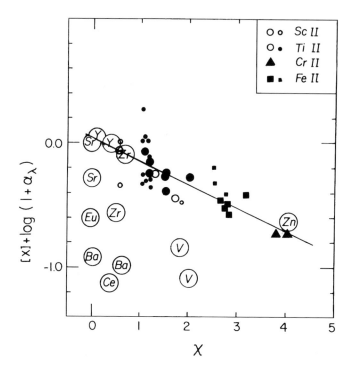

Fig. 3.16. Curve-of-growth shifts (corrected by a factor $(1 + \alpha_\lambda)$ for Rayleigh scattering opacity) for singly ionized metals in the giant HD 122563 ([Fe/H] $= -2.6$), against excitation potential. The line joining plotted points for Ti II, Fe II and Cr II defines a locus from which one can judge the presence or absence of additional deficiencies in the heavier elements. Adapted from Pagel (1965).

of the Local Group — this cannot be done for more distant galaxies. In gas-poor systems such as elliptical galaxies and the bulges of spirals one can only measure the integrated spectra and colours of an entire stellar population. Several line features in the integrated spectrum are sensitive to metallicity, but their interpretation requires assumptions about the age of the stars (and its uniformity) and about the spread in metallicity which must be either assumed to be negligible or predicted from a theoretical population synthesis model. One feature that has proved especially useful in such studies is the Mg *b* triplet, accompanied by MgH molecular bands, which can be measured using interference filters centred on the region and continuum side bands (see Fig. 3.17). Assuming coeval, single-metallicity populations, and allowing for age differences on the basis of theoretical models, the feature can be calibrated using Galactic globular clusters for which the metallicity is known from individual stars. Following earlier work by Spinrad & Taylor (1969), Faber, Burstein & Dressler (1977) have defined a so-called Mg$_2$ index, measured in stellar magnitudes, for which Bender,

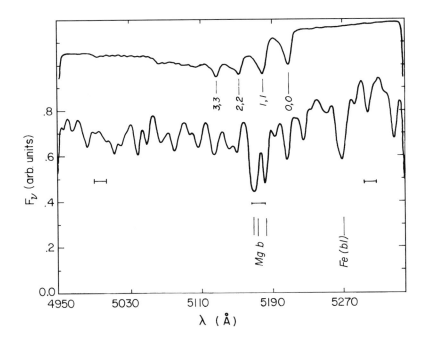

Fig. 3.17. Synthetic spectrum of a red giant, $T_{\mathrm{eff}} = 4500$ K, $\log g = 2.25$ in the region of the strong Mg I *b* lines (cf. Fig. 3.9). The upper spectrum is the same with atomic lines 'switched off' and shows molecular bands of MgH. Adapted from Mould (1978).

Burstein & Faber (1993) have given the calibration

$$\mathrm{Mg}_2 \simeq 0.1 \left[\frac{Z}{Z_\odot} t(\mathrm{Gyr}) \right]^{0.41} \tag{3.61}$$

where $\log(Z/Z_\odot)$ is more or less equivalent to [Mg/H]. This corresponds to a highly non-linear dependence of Mg_2 on $\log Z$, with a slope that increases strongly towards high metallicities. A problem with this method (and others) is the lack of well-observed globular clusters with solar metallicity and above, but it undoubtedly gives a good ranking. Another problem is that the Mg/Fe ratio is not universal, so that it is important to measure other features as well (cf. Chapter 11).

3.3 Photometric methods

Photoelectric or CCD photometry through colour filters is widely used for 'quantitative classification', i.e. to measure major spectral features with low wavelength resolution but rapidly and precisely to obtain major properties of large numbers of stars, such as

- T_{eff}, from continuum slope, Balmer jump in hotter stars and hydrogen-line strengths in F–G stars.

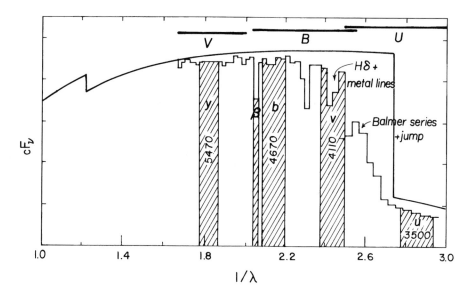

Fig. 3.18. Idealised continuum and actual fluxes measured in 50-Å-wide bands of A7 stars in the Hyades open cluster with $T_{eff} = 8000$ K, plotted against inverse wavelength in μm^{-1}. Horizontal lines above the spectrum show the locations of the Johnson UBV pass bands and the vertical boxes show schematically the corresponding properties of of the Strömgren system with central wavelengths in Å. (In that system, there are actually two Hβ pass bands, one narrow and one broad, so that comparison of the two gives a measure of the strength of the line.) Some prominent spectral features are marked.

- Luminosity or gravity from H-line strengths in hotter stars, Balmer jump and molecular features in F – K stars.

- Reddening of the continuum by interstellar dust (which leads to excess redness, known as colour excess, relative to the spectral type from H or other line features).

- Metallicity, from integrated strength of metal lines, mostly due to Fe I, giving essentially [Fe/H].

These methods are a development from older photographic spectral methods developed by D. Barbier, D. Chalonge and others in the 1930s and 40s.

Some of the most widely used systems are

- The Johnson UBV(RIJKLMNO) broad-band system, $\lambda/\Delta\lambda \simeq 5$ (cf. Figs. 3.18, 3.19).

- The Strömgren $uvby\beta$ intermediate-band system, $\lambda/\Delta\lambda \simeq 40$ (cf. Fig. 3.18) using interference filters.

- The Geneva $UBB_1B_2VV_1G$ broad-band system, $\lambda/\Delta\lambda \simeq 10$.

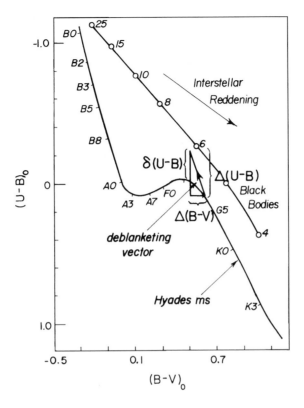

Fig. 3.19. Two-colour plot of U−B against B−V. The curve shows the main sequence for stars with the metallicity of the Hyades, loci of black bodies (with temperatures marked in units of 1000 K) and a deblanketing vector illustrating schematically the effect of metal deficiency in F and G stars. Adapted from A. Unsöld, *The New Cosmos*, Springer-Verlag 1977.

- The DDO (Toronto) (38)(41)(45)(48) intermediate-band system, designed for the study of cooler stars, mainly giants, on the basis of metallic and molecular features.

The Balmer jump causes strong non-linearity in the (U−B), (B−V) relation (see Fig. 3.19), which is useful in sorting out reddening effects for the hotter stars (in which metal lines are weak) and metallicity effects in F − G stars, where lowered metallicity causes a bluer colour for given T_{eff} due to reduced blocking of radiation by metal lines, which increase in number and strength towards shorter wavelengths. (There is also a reduction in the 'blanketing' or 'backwarming' effect, which is a raising of photospheric temperatures by radiation thrown back by the lines and works in the opposite direction.) This leads to an ultra-violet excess $\delta(U-B)$ relative to the relation for the Hyades, which are among the most metal-rich stars in the solar neighbourhood, having solar or slightly higher metallicity, and are commonly used as a standard. For metallicities that are not too low,

$$\delta(U - B) \simeq -0.2\,[\text{Fe/H}], \tag{3.62}$$

where $\delta(U - B)$ is measured in stellar magnitudes. Weaknesses in the UBV system for measuring metallicities arise from the dependence of $U-B$ on luminosity among F stars and from a loss of sensitivity when metallicities are very low.

The Strömgren system is a very elegant and precise method of quantitative classification (cf. Fig. 3.18), in which

- $b - y$ and β together indicate T_{eff} and reddening; differential blanketing effects on $b - y$ are rather small for stars warmer than the Sun;
- $m_1 \equiv (v - b) - (b - y)$ measures line blanketing: $[\text{Fe/H}] \simeq -14\Delta m_1$ for given (intrinsic) $b - y$;
- $c_1 \equiv (u - v) - (v - b)$ measures the Balmer jump, a measure of gravity in F and G stars from which luminosities and ages can be deduced.

Because the continuum bands are equally spaced in $1/\lambda$, these double indices are relatively insensitive to T_{eff} and reddening, for which small corrections are applied.

The Geneva system makes use of some of the ideas of both broad and intermediate-band systems, being similar to UBV but with the B and V each divided into two non-overlapping bands. The DDO system measures metallic, CH and CN features for luminosity and metallicity classification of red giants.

These methods, using well-defined band passes, are capable of a better internal precision than high-resolution spectroscopy (as well as being vastly quicker), but they give (in general) only one abundance parameter and they need high-resolution spectroscopy for calibration. Other quick methods, useful in particular contexts (e.g. discovery of extremely metal-deficient stars), include comparison of hydrogen and Ca^+ K-line (λ 3933) intensities (Beers, Preston & Shectman 1992) and numerical comparisons of digital spectra with low signal:noise ratio (Carney *et al.* 1987).

3.4   Emission lines from nebulae

In gas clouds containing one or more hot stars ($T_{\text{eff}} > 30\,000$ K), hydrogen atoms are ionized by the stellar UV radiation in the Lyman continuum and recombine to excited levels; their decay gives rise to observable emission lines such as the Balmer series. Examples are planetary nebulae (PN), which are envelopes of evolved intermediate-mass stars in process of ejection and are ionized by the hot, exposed stellar core, and H II regions or 'diffuse nebulae' such as the Orion Nebula, which are portions of molecular clouds in which new star formation has taken place within the last few million years and the most massive stars (spectral type O, mass $> \sim$ $15 M_\odot$) are still in the hot main-sequence stage.   Giant H II regions, thousands of times larger than Orion, can be studied in distant galaxies and give important information about abundances in those systems, while Galactic H II regions like

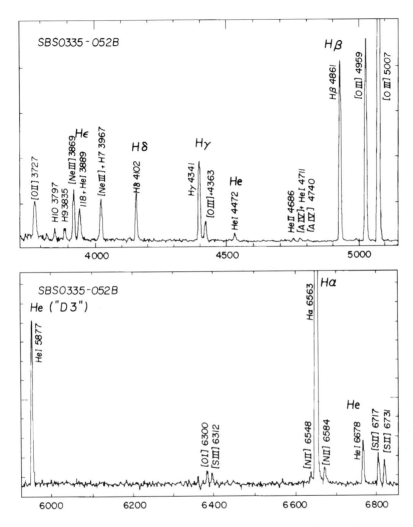

Fig. 3.20. Spectrum of an extragalactic H II region with low abundances, SBS 0335-0520 from the Second Byurakan Survey. Adapted from Melnick *et al.* (1992).

Orion provide local and more remote Galactic data on light elements (He, Ne, Ar) that do not appear in the solar photospheric spectrum, as well as others, notably N, O and S. Somewhat similar data are provided by PN, but with some modifications (particularly among the CNO elements) due to previous internal nuclear evolution. Other emission nebulae giving interesting abundance information are Wolf–Rayet and nova shells, in which the cause of ionization is similar, supernova remnants (SNR, excited by radiative shocks) and active galactic nuclei (AGNs, excited by some combination of shocks, non-thermal radiation and thermal radiation from very hot stars).

3.4.1   Sketch of H II region physics

We consider a single hot star or compact cluster containing hot stars surrounded by a cloud of hydrogen and minority elements. If there is enough hydrogen to soak up all the UV photons, the region is said to be 'ionization bounded'; this is usually the case for H II regions, but not necessarily for PN (which can be ionization bounded in some directions but not in others). The opposite case is called 'matter bounded'. In the ionization-bounded case, all Lyman photons (continuum and lines) are eventually degraded into Lyman-α and recombination lines or continua of higher series which eventually escape, or may be absorbed by dust if the optical depth is large; this is often the case for Lyman-α. Long ago, J.G. Baker & D.H. Menzel identified two extreme cases of optical depth: Case A in which all optical depths are small and all photons escape; and Case B in which Lyman line and continuum photons are absorbed 'on the spot'. In practice, Case B is almost always a better approximation, and in this approximation one can treat the hydrogen atom as though its ground state did not exist, since every emission of a Lyman photon (apart from Ly-α which is multiply scattered) is cancelled by an absorption. In particular, the total recombination rate of protons and electrons is effectively the sum of the rates to the second and higher quantum levels.

The size of a spherically symmetrical ionization-bounded nebula (known as a 'Strömgren sphere') can be found by equating the total number of recombinations in Case B to the total emission rate of ionizing photons from the central star(s):

$$\frac{4}{3}\pi R^3 \epsilon \, n_e n_p \alpha_B(T_e) = Q(N, T_{\text{eff}}, L) \tag{3.63}$$

where we suppose a fraction ϵ of the volume (the 'filling factor') to have uniform electron density n_e and proton density $n_p \simeq n_e$. α_B is the Case B total recombination coefficient, which is a slowly varying function of the electron temperature T_e and the rate Q of ionizing photon production by the embedded star(s) is a function of their number N, effective temperature and luminosity L. The radius of the Orion nebula is a few pc (1 pc or parsec is 3.3 light years), while that of a giant H II region like 30 Doradus in the Large Magellanic Cloud is of the order of 100 pc; still larger giant H II regions are seen in gas-rich spiral galaxies (especially Scd in the Hubble classification) and in star-forming dwarf galaxies, usually classified as 'blue compact' or 'H II' galaxies or both, according to the method of discovery.

The excess energy ($h\nu - 13.6$ eV) of the ionizing photons supplies heat to the ionized gas, which is cooled chiefly by the emission of collisionally excited lines from ions such as O^{++}, O^+, N^+ etc. This leads to a thermal balance at an electron temperature of the order of 10^4 K, depending on the stellar temperatures, the chemical composition (T_e increases with a lowering of abundances because there are then fewer coolants) and the ionization parameter $u = Q/(4\pi R^2 n_e c)$. Most of these collisionally excited lines are from 'forbidden' transitions, i.e. transitions within the ground configuration of the atom or ion having no electric dipole ($\Delta \ell = 0$), and hence a very low transition probability from magnetic dipole or electric quadrupole interaction, so that they are not seen in the laboratory and have to be identified from spectroscopic term analysis. But at the low densities of nebulae (typically 10 to 10^4 cm^{-3}), collisional de-excitation is so slow that a photon is eventually emitted after a time of the order of seconds. The green [O III] lines (forbidden lines of O^{++}) $\lambda\lambda$ 4959, 5007 are sometimes the strongest in the entire optical spectrum (and were once attributed to an unknown element, 'nebulium', before they were identified by Ira S. Bowen in 1928).

3.4.2 Nebular spectrum analysis

Helium and hydrogen lines are formed by similar recombination processes, so that their intensity ratios, with effective recombination coefficients calculated from atomic physics, give the helium abundance subject to corrections for unseen neutral helium and some other effects (see Chapter 4). (The effective recombination coefficient for a given line, e.g. α_{42} for Hβ, is the sum of all direct recombination coefficients to its upper level and all higher levels, multiplied by the probability that a hydrogen atom in each of these levels will eventually emit an Hβ photon.) To obtain intrinsic line ratios one has to correct the measured ratios for interstellar reddening, which is usually done by comparing measured ratios of Balmer lines (the 'Balmer decrement', going down from Hα in the red to higher series members) with theoretical values for Case B, using an assumed law for reddening as a function of wavelength.

Intensities of collisionally excited lines relative to hydrogen lines depend on the ionic abundance and on the balance between excitation by electron collisions and de-excitation by both electron collisions and radiation. The emission rates per unit volume are given respectively by:

$$j(H_{nn'}) = n_e n_p \alpha_{nn'} h\nu_{nn'} \tag{3.64}$$

and

$$j(4959 + 5007) = n_{O^{++}}(N_2/N_{O^{++}})A_{21}h\nu(4995), \tag{3.65}$$

where N_2 represents the population in the appropriate excited state. In the low-density limit (collisional de-excitation negligible), one 4959 or 5007 photon is emitted for each collisional excitation (see Fig. 3.21), so

$$j(4959 + 5007) = n_{O^{++}}n_e h\nu [(g_2/g_1)q_{21}(T_e)e^{-E/kT_e}], \tag{3.66}$$

where g_2/g_1 is a statistical weight factor, E the excitation energy ($E = h\nu$ in this case) and q_{21} is the collisional de-excitation rate coefficient, given by

$$g_2 q_{21} = 8.63 \times 10^{-8} t^{-1/2}\Omega_{12} \text{ cm}^3\text{s}^{-1}; \tag{3.67}$$

t is the electron temperature in units of 10^4 K and Ω a dimensionless constant, often of order 1, calculable from quantum mechanics and known as the 'target area'.[1] Hence

---

[1] In more refined calculations, Ω is replaced by its average over the Maxwellian velocity distribution and is a slowly varying function of t.

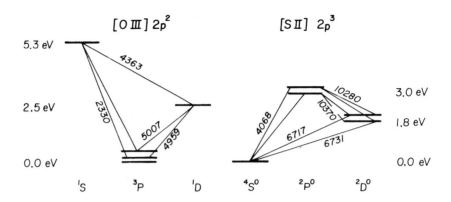

Fig. 3.21. Energy levels of O^{++} and S^+ relevant to the determination of t and n_e.

$$\frac{n_{O^{++}}}{n_p} = \frac{I(5007)}{I(H\beta)} \times f(T_e) \tag{3.68}$$

where $f(T_e)$ is a strong function driven by the exponential in Eq. (3.66), and similar relations exist for other ions. The total oxygen abundance can usually be found by adding those of O^{++} and O^+ which produces a strong line at $\lambda\,3727$; in some planetary nebulae, novae and AGNs higher ionization states of oxygen (detectable in the UV) may need to be considered. For most other elements, only one ionization state is visible in the optical and ionization correction factors need to be estimated, or found from observations in the UV and/or IR.

At higher densities, the population factor in Eq. (3.65) ceases to be proportional to the collisional excitation rate, but is rather given (in a two-level approximation) by

$$\frac{N_2}{N_1} = \frac{q_{12}n_e}{q_{21}n_e + A_{21}} = \frac{g_2}{g_1}\frac{e^{-E/kT_e}}{1 + A_{21}/(q_{21}n_e)}. \tag{3.69}$$

Thus, above a certain critical density

$$n_{\mathrm{crit}} = A_{21}/q_{21}, \tag{3.70}$$

the relative population of the excited state ceases to increase with n_e and the forbidden line is 'suppressed'; when the density approaches within an order of magnitude of n_{crit}, collisional de-excitation has to be allowed for.

3.4.3   Determination of electron temperatures and densities

The left panel of Fig. 3.21 shows the energy levels of the ground configuration $2p^2$ of O^{++} and transitions between them. Direct transitions from the first excited level to the ground term, which in this case produce the famous 'nebulium' lines, are referred to as 'nebular', while the transition at $\lambda\,4363$ is called 'auroral', by analogy with the similar transition of [O I] $\lambda\,5577$ that is prominent in aurorae and the night sky in general.

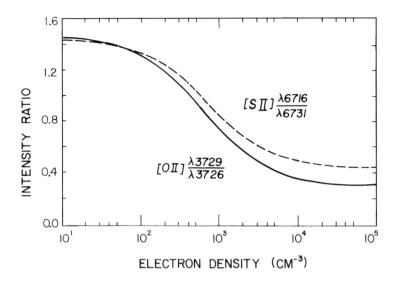

Fig. 3.22. Dependence of [O II] and [S II] doublet ratios on electron density when $t = 1$. Reproduced with permission from D.E. Osterbrock (1988, *Publ. Astr. Soc. Pacific*, **100**, 412). ©1988, Astronomical Society of the Pacific.

The third transition, from the upper state of λ 4363 to the ground state, occurs (in this case) in the UV and is called 'transauroral'. The ratio of the auroral line λ 4363 to the nebular lines $\lambda\lambda$ 4959, 5007 is a sensitive measure of electron temperature by virtue of an exponential factor analogous to that in Eq. (3.66). Similar configurations occur for N^+ and S^{++} and these can also be used in favourable cases, mainly PN and H II regions where the heavy-element abundance is below solar; in high-abundance H II regions t becomes so low that auroral lines become unmeasurably weak, but the situation is better for PN because they can have very hot ionizing stars and high surface brightness.

The right panel of Fig. 3.21 shows energy levels of the $2p^3$ electron configuration, exemplified by O^+ and S^+. The nebular lines form a close doublet ([O II] $\lambda\lambda$ 3729, 3726; [S II] $\lambda\lambda$ 6717, 6731) with an intensity ratio that depends on n_e because at low density it is just the ratio of collisional excitation rates from the ground state, while at high density the upper states are relatively near thermal equilibrium and the line intensities are governed by statistical weights and radiative transition probabilities. The critical density is around 10^3 cm^{-3} (see Fig. 3.22), whereas for [O III] it is much higher. In practice, [S II] is used more commonly than [O II] because it does not demand such a high spectral resolution.

Two limitations of the use of auroral:nebular line ratios to obtain t are (i) the possibility that there may be temperature fluctuations or gradients in the nebula, which would cause a bias towards overestimating t and therefore underestimating abundances (the extent of this effect in H II regions is still not well known, but it

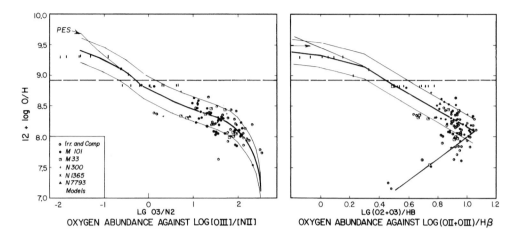

Fig. 3.23. Trends of nebular line strengths in H II regions with oxygen abundance. Dots show observational data, short vertical tick marks indicate model predictions and the curves indicate calibration relations that can be used to estimate the oxygen abundance. Solar abundance is indicated by the horizontal broken lines. Adapted from Edmunds & Pagel (1984).

could be of the order of 0.1 dex); and (ii) the disappearance of auroral lines when $t < 0.75$ or so, depending on the surface brightness. To overcome the latter problem, so-called 'empirical' methods have been developed that can give a rough idea of t and abundances from nebular lines alone, inspired partly by photo-ionization models and partly by observational trends of line strengths with galactocentric distance in gas-rich spirals which are believed to be due to a radial abundance gradient with abundances decreasing outwards (Fig. 3.23). As abundances of cooling ions such as O^{++} decrease from near-solar values (shown by the horizontal line), the intensity of the [O III] nebular lines relative to hydrogen lines paradoxically increases because of the rise in t, which itself is largely controlled by emission of [O III] fine-structure transitions in the far infra-red. This goes on up to a certain point where the nebular lines take over a significant cooling role themselves and reach a maximum around $12 + \log(O/H) = 8.1$; for still lower abundances, the cooling is dominated by hydrogen free–free emission, and the lines revert to the more intuitively natural behaviour of decreasing with diminishing abundance. The resulting ambiguity is resolved by reference to the line ratio [O III]/[N II], which varies very much more than the abundance ratio N/O (Alloin *et al.* 1979).

3.4.3.1 *Other kinds of emission nebulae*
Apart from H II regions and PNs, interesting abundance data can be derived from a number of other nebular-type objects:

- Novae and symbiotic stars have shells which are excited by extremely hot stars (several $\times 10^5$ K) like some PNs, but denser, and display overabundances of

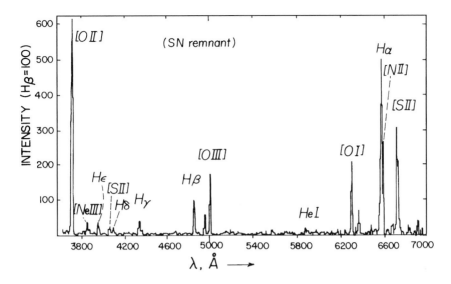

Fig. 3.24. Spectrum of a supernova remnant in the Scd spiral M33. Adapted from Dopita, D'Odorico & Benvenuti (1980).

elements up to Ne due to thermonuclear processes affecting matter accreted from a companion by a white dwarf.

- Supernova remnants (SNR) in early stages display results of advanced (explosive) nucleosynthesis, but in later stages light up the interstellar medium by shock excitation and give information about the ISM in external galaxies complementary to that derived from H II regions.

- AGNs range from their most extreme form in quasi-stellar objects (QSOs) and radio galaxies to milder forms in Seyfert galaxies, low-ionization emission line regions (LINERs) and starburst galaxies. The latter are essentially giant H II regions, usually rich in dust, in nuclei of Sc galaxies, and perhaps should not be classified as 'active' at all, but they can have very high luminosities in the infra-red and may be relevant to luminous distant galaxies discovered by IRAS[1], which are among the most luminous objects known.

QSOs[2] have high-excitation spectra with very broad lines, while Seyfert galaxies are early-type spirals (Sa to Sc) with nuclear emission-line spectra displaying differing proportions of a broad-line component similar to a QSO (Sy 1) and a narrow-line component (Sy 2) somewhat like the spectrum of a PN, but with very strong forbidden lines. The broad and narrow-line components appear to come respectively from an inner, high-density region and an outer relatively low-density region, surrounding a 'central engine' (usually believed to

[1] Infra-red Astronomy Satellite, launched in 1983.
[2] The term 'QSOs' is used to cover both radio-loud quasi-stellar objects, commonly known as 'quasars', and radio-quiet objects that otherwise have similar characteristics.

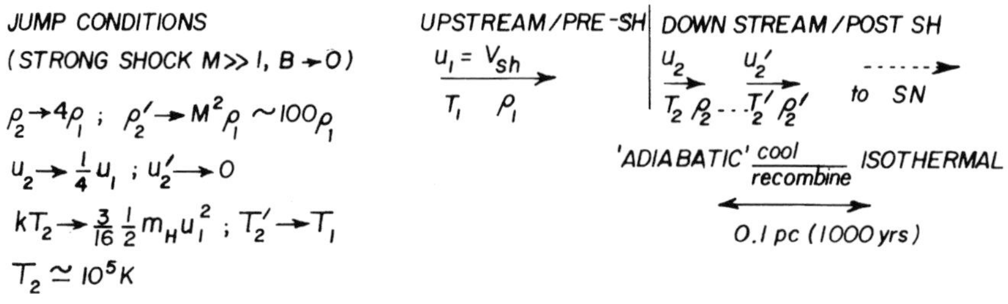

JUMP CONDITIONS

(STRONG SHOCK M ≫ 1, B → 0)

$\rho_2 \to 4\rho_1$; $\rho_2' \to M^2 \rho_1 \sim 100 \rho_1$

$u_2 \to \frac{1}{4} u_1$; $u_2' \to 0$

$kT_2 \to \frac{3}{16} \frac{1}{2} m_H u_1^2$; $T_2' \to T_1$

$T_2 \simeq 10^5 K$

UPSTREAM/PRE-SH

$u_1 = V_{sh}$ →

T_1 ρ_1

DOWN STREAM/POST SH

u_2 u_2' - - - - - →

$T_2 \, \rho_2$ - - $T_2' \, \rho_2'$ to SN

'ADIABATIC' $\frac{cool}{recombine}$ ISOTHERMAL

← 0.1 pc (1000 yrs) →

Fig. 3.25. Schematic picture of conditions surrounding a supernova shock. *M* is the Mach number, *B* the magnetic field and velocities are relative to the shock front.

be a black hole swallowing material from an accretion disk). The 'engine' emits a non-thermal spectrum with a component given by a power law, $F_\nu \propto \nu^{-n}$ where $n \sim 1$; and there is a more black-body-like component from the accretion disk. There may also be contributions to the ionizing continuum from radiative shocks and from very hot stars that have evolved by extensive mass loss to an extreme Wolf–Rayet (WR) stage, since fresh star formation is usually observed around Seyfert nuclei.

LINERs are weaker emission-line regions in certain elliptical and early-type spiral galaxies (e.g. M 51 and M 81) showing relatively strong lines of [O I], [N II] and [S II], similar to SNR. It is not clear whether they are excited by shocks like SNR or by a very dilute (i.e. low *u*) non-thermal spectrum.

The analysis of SNR spectra is more complicated than that of ordinary H II regions because there are strong gradients in ionization and electron temperature in the shocked gas which may also photo-ionize the pre-shock gas. Thus a typical SNR, like the Cygnus loop or N159 in the Large Magellanic Cloud (LMC), is an X-ray source and shows a wide range of excitation conditions in its optical spectrum going from Mg I to [Ne V]. Electron temperatures range from $t \simeq 2$ for [O III] to $t \simeq 1.2$ for [N II] and $t \simeq 0.9$ for [S II].

The evolution of a supernova remnant is believed to involve three main stages:

(1) Free expansion of the ejecta until they have swept up a mass of ISM comparable to their own mass, $\sim 10 M_\odot$; this would be the mass within a radius ~ 1 pc that would be covered in a time of the order of 50 years. However, the SN is likely to be surrounded by one or more lumpy circumstellar shells due to previous mass ejection, which would make the time much shorter; the impact of ejecta from SN 1987A in the LMC on such a shell is currently awaited.

(2) The Sedov adiabatic expansion phase, driven by thermal pressure. In this phase, which covers a distance ~ 15 pc over a period $\sim 2 \times 10^4$ yr, the swept material cools down and accumulates in a shell behind the shock front while inner material forms a very hot bubble emitting X-rays.

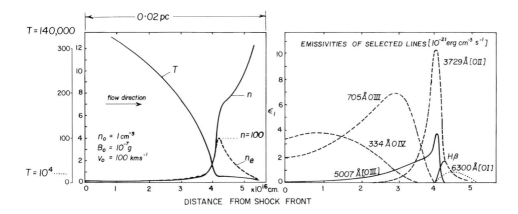

Fig. 3.26. Physical parameters and line emissivities in the post-shock gas according to a model of the Cygnus Loop SNR. Adapted from Cox (1972).

(3) The snowplough phase in which the swept-up material cools radiatively for up to about 10^5 yr giving rise to emission lines and ionizes the ISM ahead of the shock. In a steady state, the shock velocity $\simeq 100$ km s^{-1} and the pre-shock number density $\simeq 1$ cm^{-3}. The jump conditions are crudely illustrated in Fig. 3.25 and results of a specific model calculation in Fig. 3.26.

Ionization by a power-law continuum also leads to a much greater range in t and ionization than exists in H II regions ionized by O-stars, because while in the latter case there is a sharp boundary between the hot H II region and a surrounding cool H I region, high-energy photons from the power-law continuum produce an extended boundary region with declining ionization and residual heating.

3.5   **Abundances: main results**

3.5.1   Solar and local Galactic abundances

- Photospheric absorption lines (Fraunhofer lines) provide element:hydrogen number ratios M/H for commoner elements with suitable features in the optical or IR from neutral atoms, singly charged ions or molecules such as CH, CN, CO, OH, MgH; there have been great improvements in knowledge of oscillator strengths in recent years and abundances thus found for selected elements are given in Table 3.4. He, Ne and Ar are common elements not determinable in this way, and most information on them comes from hot stars or nearby H II regions like Orion. There is but limited information on rare elements such as the actinides (U, Th) and rare earth elements, and on isotope ratios, which are restricted to those that can be determined from molecules, e.g. C and Mg; these all agree with terrestrial ratios. Photospheric analysis shows that the number ratio $(C + N + O)/H$ (the commonest elements after He, not counting Ne) is close to 10^{-3} and that of easily ionized metals (Na, Mg, Al, Si, Ca, Fe) contributing essentially one free electron per atom is 10^{-4}. In terms of mass fractions, $Z/X = 0.024$ and $Z = 0.017$ if one assumes a helium mass fraction $Y = 0.27$ or 0.28. Among fragile light elements, D has been completely destroyed in the photosphere, Li has been reduced by two orders of magnitude and Be and B may have suffered some destruction as well, but only by factors of order 2 which do not exceed uncertainties.

- Emission lines from the solar chromosphere, prominences and corona, especially in the far UV, provide somewhat less accurate indications of abundances (including those of He, Ne and Ar) in the outer solar atmosphere. There is some evidence for element segregation in these regions, including a relative enhancement of elements with a low first ionization potential ('FIP effect'); that effect is also found in high-energy particles from solar flares (sometimes called solar cosmic rays), in the solar wind and in source abundances in Galactic cosmic rays which are determined by correcting for propagation effects in the ISM. The solar wind is a vital source of information on ^{3}He (see Chapter 4).

- The Earth's crust, atmosphere and oceans are strongly influenced by 'differentiation' processes which could have resulted from gravitational separation ('smelting') in an early molten phase of the planet, or from the sequence in which different chemical species condensed from the primitive Solar nebula and were subsequently accreted. Seismology indicates that there is a liquid core (with a solid inner core) with radius 3500 km consisting mainly of iron (with some Ni and FeS) surrounded by a plastic (Fe, Mg silicate) mantle of thickness 2900 km. The solid crust is only 30 to 40 km thick under continents, where it is mainly granite (Al-rich silicates or 'SIAL'), and still thinner (10 km) under the oceans where it is mainly basaltic (Mg-rich silicates or 'SIMA'), and slow convection in the mantle drives plate tectonics. Terrestrial elements are accordingly segregated according to geochemical affinities: 'Atmophile' (*e.g.* N_2, O_2, inert gases, favouring the atmosphere); 'lithophile' (O-rich minerals, Na, K, Ca, rare earths, actinides, favouring silicate rocks); 'chalcophile' (S, Cu, Zn, Sn, Pt, Ag, Cd, Hg, favouring sulphur-rich minerals in ore deposits and the interior); and 'siderophile' (Fe, Ni, Ga, Ge, Ru, Rh, Os, Ir, Pt, favouring metallic iron-nickel phases that have largely sunk down into the core). Moon rocks, returned to Earth by the Apollo astronauts in 1969, are also highly differentiated.

  Terrestrial isotope ratios are mainly unaffected by these processes and therefore provide valid information about standard abundances of individual nuclear species except in special cases where they have been modified by fractionation (e.g. D/H) or radio-active decay, e.g. ^{40}Ar is enhanced relative to ^{36}Ar by ^{40}K decay.

- Meteorites play an essential role in the story. Meteorites come in three main types, stones, stony irons and irons, the latter consisting of metallic iron alloyed with Ni and FeS in various crystalline forms. The stones are divided into 'achondrites', resembling terrestrial rocks, and 'chondrites'. Chondrites contain rounded glassy drop-like inclusions 1 mm or so in diameter called 'chondrules' together with specks of metallic iron and FeS in a surrounding 'matrix' that mainly consists of iron magnesium silicates (olivine and pyroxene). Chondrites differ in such properties as degree of oxidation, metamorphism (by which the chondrules gradually diffuse) and equilibration, i.e. the approach to chemical equilibrium between adjoining mineral phases; but their atomic composition (not counting volatile elements such as H, C, N, O, S ...) is fairly uniform and in agreement with the best determinations from the solar Fraunhofer lines. They probably result from the break-up of small parent bodies (asteroids) that were not subjected to the kind of differentiation that affects major planets like the Earth (and Moon). The least metamorphosed are a subtype known as carbonaceous chondrites e.g. Orgueil (France) 1864, type C1, and Allende (Mexico) 1969, type C3, containing water of crystallization and organic compounds, which implies that they have not been subjected to high temperatures.

CC1 meteorites provide a particularly good match to solar, and are used to fill in the gaps left by unknown or inaccurate photospheric abundances; as the latter have improved over the years, the agreement with CC1 (except in the special situation of Li, Be and B) has become even better, so that it is almost an article of faith that these objects have an atomic composition representing that of the Solar System. The solar iron abundance has played an interesting historical role in this comparison: in the early 1960s, the photospheric estimate of Fe (relative to Si which is the meteoritic standard) was an order of magnitude low compared to meteoritic, which gave rise to much speculation, but this was later found to be a result of bad oscillator strengths. In recent years, the position has been nearly reversed, in that some photospheric determinations based on Fe I are slightly higher than meteoritic, but most people believe that the disagreement (which is in any case under 0.2 dex) is equally spurious, particularly in view of the fact that solar Fe II gives agreement with the meteorites. Table 3.4 gives photospheric and meteoritic (CC1) logarithmic abundances for selected elements. The complete table, together with terrestrial and occasionally meteoritic isotope ratios, forms the basis for the standard local abundance distribution of nuclides illustrated in Fig. 1.4. That distribution is typical (as far as one can tell) of most nearby stars and the local ISM, apart from variations by factors up to 2 or 3 in Fe/H or Z/X, which on average is about 0.1 dex below solar, and minor (but interesting) variations in O/Fe and other elemental ratios that will be discussed in Chapter 8.

3.5.2 Isotopic anomalies in meteorites

Meteorites show a number of isotopic variations or 'anomalies' which are only partially understood and give some tantalising clues as to events in the formation of the

Table 3.4. *Solar abundances relative to* $\log N_H = 12.00$; *meteoritic abundances relative to* $\log N_{Si} = 7.55$

Element		Photospheric	Meteoritic	Element		Photospheric	Meteoritic
1	H	12.00		21	Sc	3.18	3.09
2	He	(10.99)		22	Ti	5.03	4.93
3	Li	1.16	3.30	23	V	4.00	4.01
4	Be	1.15	1.41	24	Cr	5.67	5.68
5	B	2.6:	2.78	25	Mn	5.39	5.52
6	C	8.55		26	Fe	7.50	7.49
7	N	7.97		27	Co	4.92	4.90
8	O	8.87		28	Ni	6.25	6.24
9	F	4.6:	4.47	29	Cu	4.21	4.28
10	Ne	(8.08)		30	Zn	4.60	4.66
11	Na	6.33	6.31	37	Rb	2.6	2.40
12	Mg	7.58	7.57	38	Sr	2.97	2.91
13	Al	6.47	6.48	39	Y	2.24	2.22
14	Si	7.55	7.55	40	Zr	2.60	2.60
15	P	5.45	5.52	56	Ba	2.13	2.21
16	S	7.3	7.19	57	La	1.17	1.21
17	Cl	5.5:	5.27	58	Ce	1.58	1.62
18	Ar	(6.5)		60	Nd	1.50	1.48
19	K	5.1	5.12	63	Eu	0.51	0.54
20	Ca	6.36	6.34	90	Th		0.08

Data are from Anders & Grevesse (1989) updated by Grevesse, Noels & Sauval (1996). Values in brackets are based on solar wind and energetic particles corrected for FIP effect; they are reasonably consistent with results from Galactic nebulae. When two decimal places are given, the error estimate is ± 0.10 or better.

Solar System that may have affected the 'Standard' abundance distribution itself. All meteorites are affected by radio-activity. Long-lived activities enable minerals to be age-dated (see Chapter 10) and short-lived ones lead to fission tracks from ^{244}Pu and to decay products that may significantly enhance the abundance of an intrinsically rare species, e.g. ^{22}Na $\rightarrow ^{22}$Ne, ^{26}Al $\rightarrow ^{26}$Mg, ^{129}I $\rightarrow ^{129}$Xe. Meteoritic isotope ratios are also affected by exposure to cosmic rays.

The type 3 carbonaceous chondrite Allende arrived just at an opportune time when certain laboratories had geared up with ultra-refined mass-spectrometer and related techniques mainly for the purpose of studying lunar samples brought back by the Apollo astronauts. It is highly unequilibrated, containing chondrules and a variety of grains of various minerals that appear to have quite distinct chemical histories. Much attention has been devoted to coarse-grained refractory inclusions, rich in calcium and aluminium, which display a uniform overabundance (relative to Si) of a wide range of refractory elements with quite different chemical properties. This suggests condensation in a high temperature environment without subsequent equilibration with the residual gas, in contrast to the bulk of solar-system material which is supposed to have

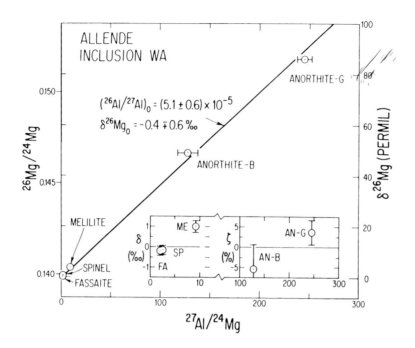

Fig. 3.27. Diagram of ^{26}Al-^{26}Mg evolution for a Ca-Al inclusion from Allende containing minerals with a wide range of Al/Mg. Data taken by an ion microprobe. After T. Lee, D. A. Papanastassiou & G.J. Wasserburg 1977, *Ap. J. Lett.*, **211**, L107. Courtesy G.J. Wasserburg.

undergone some form of condensation sequence under conditions approaching thermal equilibrium. Isotopic anomalies in these inclusions can result either from their having had a separate nucleosynthetic history, e.g. by incorporating dust grains formed in the ejecta from supernovae and other stars in advanced stages of evolution and surviving unchanged in the early Solar System, or merely because their peculiar composition leads to an enhanced sensitivity to events that may have affected the Solar System as a whole.

Substantial abundance anomalies occur among the heavy oxygen isotopes ^{17}O and ^{18}O, which are underabundant by up to about 4 per cent relative to ^{16}O in certain of the Ca-Al-rich inclusions, compared with the bulk composition in which the isotope ratios are close to a terrestrial standard. The intriguing feature of these anomalous ratios is that, in common with some other meteorites, but in contrast to terrestrial samples, the relative deviations of the two heavy isotopes are equal; any normal fractionation process would cause ^{18}O to have twice the anomaly of ^{17}O as indeed is observed in terrestrial samples and more differentiated meteorites, where the anomalies are also usually much smaller. A possible explanation is that there is a substantial admixture of pure ^{16}O from a supernova.

Another isotopic anomaly, discovered in Allende inclusions, concerns magnesium, for which an intrinsically low abundance in these samples makes its isotope ratios

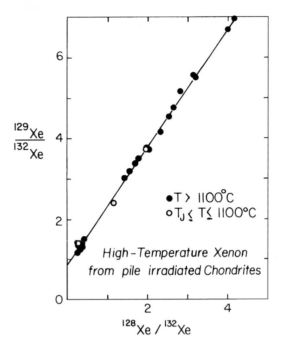

Fig. 3.28. ^{129}Xe/^{132}Xe for several different meteorites irradiated with neutrons from a reactor to produce ^{128}Xe from ^{127}I present in the meteorites. The samples were heated to various temperatures and the Xe isotopic compositions measured. The correlation between ^{129}Xe and iodine demonstrates the presence in the primordial Solar System of ^{129}I (mean life 23 Myr) with an abundance $10^{-4} \times$ that of the stable isotope ^{127}I. Data from Hohenberg, Podosek & Reynolds (1967), adapted from Wasserburg & Papanastassiou (1982).

sensitive to small effects. Certain of the inclusions show a correlation between ^{26}Mg and ^{27}Al, indicating an origin of excess ^{26}Mg from radio-active decay of ^{26}Al (mean life 1 Myr), the existence of which has been postulated as a heat source for meteorite parent bodies (Fig. 3.27).

A few inclusions are exceptional in that they show isotopic anomalies for virtually every element studied; these have been called FUN (fractionation and unknown nuclear) inclusions by Wasserburg and his colleagues and they may contain unmodified stellar ejecta that were incorporated into the solar nebula in a solid form in which they survived.

Meteorites contain traces of inert gases, either from radio-active decay or 'trapped', which can be released by stepped heating, and their isotope ratios can vary by orders of magnitude because of the low abundance of any 'normal' component. The trapped components are classified as 'solar' (resembling the solar wind), 'planetary' (resembling the Earth's atmosphere and related to solar by mass fractionation and radio-active decay) or 'exotic', e.g. Neon E. Neon E is 30 to 100 per cent ^{22}Ne, whereas solar and planetary neon have ^{20}Ne and ^{22}Ne in proportions of about 10:1, and ^{21}Ne occurs in

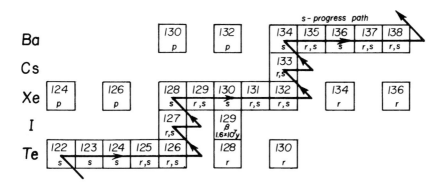

Fig. 3.29. s-process path in the Xe region. Adapted from Anders (1988).

highly variable amounts from spallation. Ne E has been found to have at least two components that are released at different temperatures according to the refractoriness of their mineral carriers: almost pure ^{22}Ne which may result from decay of ^{22}Na, perhaps created in a nova outburst, and a more impure component that could arise from helium-burning on ^{14}N, e.g. in an AGB star (see Chapter 5).

Xenon shows several interesting isotopic variations (see Figs. 3.28, 3.29) including excess abundance of ^{129}Xe, due to radioactive decay of ^{129}I, in some samples (see Chapter 10), and so-called Xe HL with excess abundances by factors of 2 or so of both the heavy (r-process) and light (p-process) isotopes relative to the middle (mainly s-process) isotopes. The r-process nuclei are believed to be largely fission fragments from ^{244}Pu, also formed in the r-process, and a p-process (the exact nature of which is not very clear) could have been associated with this event or could be present in a separate grain component. Another anomalous xenon type, known as Xe S, shows complementary anomalies, i.e. mainly just the s-process isotopes.

While isotopic anomalies among meteoritic inert gases have been known since 1964, their carrier grains were only discovered, by the University of Chicago group, in 1987 (Lewis *et al.* 1987). They are tiny (micron or submicron size) grains embedded in the fine-grained matrix of primitive meteorites and are believed to have acquired their inert gases and other impurities by ion implantation. Several are carbon-bearing: micro-diamonds, silicon carbide SiC, graphite and titanium carbide TiC, formed in a carbon-rich environment (i.e. C/O > 1 as in carbon stars and in certain interior zones of pre-supernovae). There are also non-carbonaceous grains: corundum Al_2O_3 and spinel $MgAl_2O_4$ formed in an oxygen-rich environment (C/O < 1, as is also the case for the coarse refractory inclusions). All are chemically highly resistant and they are extracted using fierce acids and oxidizing agents to remove the carriers of normal-type trapped gases. Their size distribution and other characteristics confirm their extra-solar origin and isotopic anomalies have been found among them for other elements, e.g. D, N and Ba in diamonds, C in graphite and sometimes enormous variations among C, N, Si, Ca, Ti, Sr, Ba, Nd and Sm in silicon carbide. The diamonds may come

from supernovae and the TiC grains probably from AGB carbon stars, while the broad
$^{12}C/^{13}C$ distribution in graphite suggests a variety of sources.

Some implications of long and short-lived radioactivities for the chronology of the
Solar System will be discussed in Chapter 10.

3.5.3   Brief overview of abundances outside the Solar System

3.5.3.1   *Introduction*

Until about 1950, the conventional wisdom was that all stars have the same chemical
composition with the sequence of spectral types caused essentially by differences in
temperature and gravity, with two main exceptions:

- Carbon stars with spectral types R, N (or C) and S. R and N stars are cool giants
  with temperatures roughly corresponding to normal types K and M respectively,
  but which show bands of CH, C_2 and CN instead of TiO, while S stars show
  ZrO bands instead of, or in addition to, TiO. In a classic paper, Russell (1934)
  explained these spectral differences (which also correspond to the nature of
  condensates such as those mentioned in the previous section) on the basis of
  the high degree of stability of the CO molecule, which locks up essentially all
  of either carbon or oxygen according to which has the lower abundance. In M
  stars, which have C/O < 1 like the Sun and other normal stars, some oxygen
  is left over and available to form TiO, VO etc., whereas in carbon stars the
  situation is reversed (as we now believe, owing to nucleosynthesis) and so carbon
  molecules are formed. In S-stars, oxygen is but slightly more abundant than
  carbon, favouring ZrO against the less stable TiO, but as we now know Zr is
  also overabundant as a result of the s-process (cf. Fig. 1.7).
- Wolf–Rayet (WR) stars, which are hot stars of high luminosity like O stars, but
  have peculiar spectra with broad emission bands of He$^+$, N^{++} (WN), C^{++} and
  C^{+++} (WC) or O^{++++} (WO) and with hydrogen weak or absent. These are now
  believed to be massive stars (initially $\geq 40M_\odot$) undergoing extensive mass loss
  in the course of which their surface temperature steadily rises and outer layers
  are successively peeled off revealing more advanced nuclear burning stages.

Progress in understanding stellar evolution and nucleosynthesis, and the discovery
by Merrill (1952) of the unstable element technetium in the S star R Andromedae,
demonstrating the occurrence of stellar nucleosynthesis within a few half lives of Tc
(i.e. < about 1 Myr; cf. Fig. 1.8), has led to acceptance of the idea that abundance
variations among stars are perfectly natural as a consequence of three main effects (see
Fig. 3.30):

(1) Internal nuclear reactions followed by loss of the envelope or mixing to surface
   layers in advanced evolutionary stages, e.g. WR and carbon stars. The Ba
   II and CH stars have anomalies related to those of carbon stars but too low
   luminosities to be expected to have undergone the third dredge-up in thermally

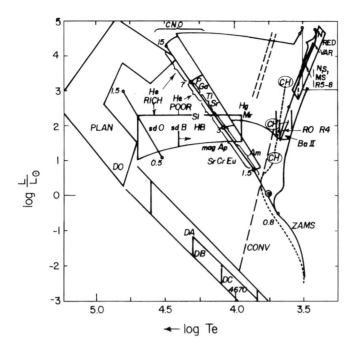

Fig. 3.30. Rough locations of chemically evolved and peculiar stars in the HR diagram. Full lines show the ZAMS with stellar masses indicated and evolutionary tracks for 0.8, 3 and 15 $M_\odot$, the helium main sequence (0.5 to 1.5 $M_\odot$), PN nucleus, horizontal-branch and white-dwarf regions. The dotted line shows a schematic main sequence and evolutionary track for Population II, while various dashed lines show roughly the cepheid instability strip, the transition to surface convection zones and the helium-shell flashing locus for Population I. After Pagel (1977). ©1977 by the IAU. Reproduced with kind permission from Kluwer Academic Publishers.

 pulsing AGB evolution. They are believed to belong to binary systems in which the companion underwent such evolution and shed part of its envelope on to the star that we see. The white dwarfs are classified according to spectral features as DO (He^+), DB (He), DA (H) and DC (continuum sometimes with C_2 bands); their surface composition is affected by gravitational settling and possibly accretion.

(2) Population effects. Stars are born with a composition reflecting that of the ISM at the time and place of formation, which in turn has been enriched in nucleosynthesis products to varying degrees by previous stellar activity. The first stars to have been born in the Galaxy have abundances of heavy elements not produced in the Big Bang that are far below solar, as was first found to be the case in a study by J.W. Chamberlain and L.H. Aller in 1951 of two classical 'subdwarf' stars HD 140283 (actually a subgiant) and HD 19445 which belong to the sparse 'halo' population (see below) and have [Fe/H] values of −2.7 and −2.1 respectively.

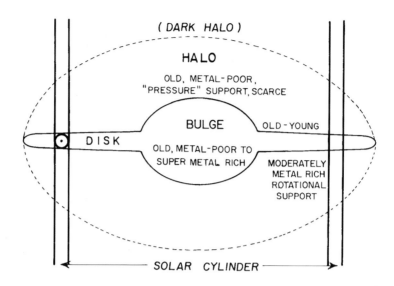

Fig. 3.31.   Schematic cross-section through the Galaxy

(3) Certain stars with $T_{eff} \geq 8000$ K or so, called Chemically Peculiar (CP, also known from their spectral types as Bp, Ap, Am), including the magnetic spectrum variables, display extraordinary over- and under-abundances of different elements, often concentrated in spots that come into and go out of view as the star rotates. Most of these effects are attributed to a combination of gravitational settling and radiative levitation (which favours rare, exotic elements: P, Ga, Hg etc.) modified by the magnetic field, and they are not now considered to have any direct relationship with nucleosynthesis.

3.5.3.2 *Metallicities and stellar populations*

Following on from his success in resolving the nucleus of the Andromeda galaxy M 31 using red-sensitive photographic plates under blackout conditions on Mount Wilson in 1944, Walter Baade introduced in 1948 the influential idea of two stellar populations, Pop I and Pop II, distinguished by age, kinematics and chemical composition. Population I was identified with the gas-rich disk of the Galaxy (see Fig. 3.31) containing young stars and with the light dominated by blue stars on or above the upper main sequence; the associated stellar orbits are nearly circular and close to the Milky Way plane, leading to low velocity dispersion and low velocities relative to the Sun which forms part of the system. Population II was identified with the spheroidal system or halo, sparse in the solar neighbourhood but more concentrated to the centre of the Galaxy; this population is traced by globular clusters and high-velocity stars with more elongated and inclined orbits. Here all the stars are old, the more massive stars have left the main sequence and the light is dominated by red giants. The distinction that he then made between classical cepheids (Pop I) and RR Lyrae variables (Pop II) played an important role is expanding the adopted extragalactic distance scale.

Pop I was considered to be metal-rich, like the Sun and the Hyades open cluster, and Pop II metal-poor by factors of up to 100 or more like certain globular clusters and subdwarfs, resulting in especially luminous red giants that Baade had resolved in M 31. However, as is illustrated in Fig. 3.31, the situation is by no means so simple. Stars in the bulge of our Galaxy and of M 31 are mostly metal-rich, although old, and so there are large-scale spatial gradients in metallicity as well as age effects.

Locally, however, there are significant correlations between kinematics and metallicity: high-velocity stars (relative to us) tend to have [Fe/H] < 0 and there is a loose correlation with age measured from isochrones based on theoretical stellar evolutionary tracks. Eggen, Lynden-Bell & Sandage (1962, ELS), in a classic discussion, presented UV excesses and orbital characteristics (eccentricity, specific angular momentum and velocity at right angles to the plane, from which the maximum height above the plane can be estimated) for stars from two catalogues, one of ordinary nearby stars and one of high-velocity stars selected by proper motion, i.e. angular motion across the sky (see Fig. 3.32). Taking UV excess (equivalent to [Fe/H]) as an indicator of age, they put forward the theory that the Galaxy formed by collapse on a free-fall timescale (then estimated at a few $\times 10^8$ yrs) from a single protogalactic cloud. The first stars were formed in early stages of the collapse, with large orbital energies, i.e. large eccentricities and maximum heights, and have retained these properties (which are adiabatic invariants) forming what we now see as the halo, whereas the remaining gas clouds dissipated their energy by collisions and radiation and settled into a disk with nearly circular orbits from which Pop I stars subsequently formed, inheriting their orbital characteristics from the disk gas.

This picture has given rise to several debates, e.g. how short should the collapse time actually be in relation to age differences among globular clusters, the lack of a radial abundance gradient (to be expected from a dissipative process; see Appendix 5) in the outer halo and the severe difference in angular momentum between the halo (which is virtually non-rotating on average) and the disk (which rotates at about 200 km s^{-1}). Kinematic biases in the catalogues led ELS to overlook a significant population component known alternatively as 'Intermediate Population II' or the 'Thick disk', consisting of stars with modest eccentricities but substantial velocities at right angles to the plane and [Fe/H] ≤ -0.5 or so with a range overlapping halo metallicities, so that it populates the empty top left corner in the left panel of Fig. 3.32. An alternative picture of halo formation has been put forward by Searle & Zinn (1978) who argue that it results from the capture of fragments such as dwarf galaxies over lengthy periods of time — an event that has probably recently been witnessed with the discovery of the Sagittarius dwarf system (Ibata, Gilmore & Irwin 1994). The formation of the thick disk could also have been a consequence of such a merger event.

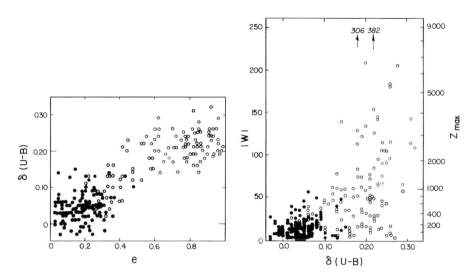

Fig. 3.32. Correlations between UV excess and stellar orbital eccentricity and velocity at right angles to the plane of the Milky Way. Adapted from Eggen, Lynden-Bell and Sandage (1962).

As will be discussed further in Chapter 8, there is probably some truth in both pictures. The inner halo and bulge have properties consistent with a dissipative collapse, while the outer halo is likely to have had a more chaotic origin. The disk may have accumulated over a long period, longer in the outer parts than in the inner parts. The gas fraction (in a column perpendicular to the plane) increases, and abundances in stars and H II regions decrease, with distance from the centre of the Galaxy at a rate of between about 0.05 and 0.1 dex kpc^{-1} (Figs. 3.33, 3.34), although some stellar observations suggest a smaller gradient.

3.5.3.3 *The cold interstellar medium*
The interstellar medium has a complex structure involving hot X-ray gas, warm ionized and neutral gas, molecular clouds, H II regions and diffuse H I clouds; the latter emit 21 cm radio waves from a hyperfine structure transition and produce interstellar absorption lines in the optical and UV. Advances in UV spectroscopy in recent years have provided substantial information on the chemical composition of diffuse clouds in our Galaxy and the Magellanic Clouds and ground-based work has led to corresponding results for intervening absorption-line systems in front of quasars with high red-shifts.

Fig. 3.35 shows that most elements are substantially depleted from the H I gas, due to their being locked on dust grains, with the notable exceptions of sulphur and zinc. The degree of depletion increases fairly smoothly with condensation temperature, consistent with the idea that these elements are ejected from stars in the form of dust in the first place. The degree of depletion is lower in high-velocity clouds (where the line ratio CaII/NaI is greater) and much lower in H II regions and supernova remnants,

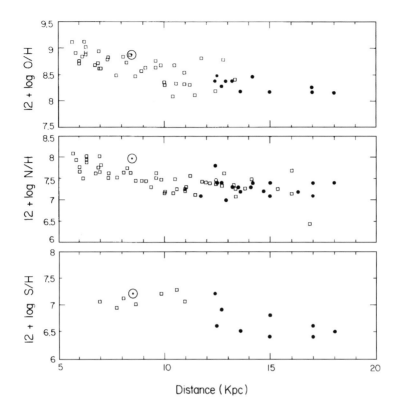

Fig. 3.33. Oxygen abundances in Galactic H II regions, as a function of galactocentric distance, with the Sun shown for comparison. Adapted from Shaver *et al.* (1983), Fitch & Silkey (1991) and Vilchez & Esteban (1996).

owing to the destruction of grains by various processes, although Fe and Si are still substantially depleted in the Orion nebula.

The warm halo gas has been extensively studied with the International Ultraviolet Explorer (IUE) satellite launched in 1978 and shows high ionization states such as C IV and N V which are also seen in high red-shift intervening systems on the line of sight to QSOs, where they may also come from galactic halos. Intervening systems with large H I column densities in the range 10^{20} to 10^{21} cm^{-2} (damped Lyman-α systems) more closely resemble the local diffuse clouds; they are usually metal-deficient as judged from undepleted S and Zn, typically $[M/H] \sim -1$ to -2, and they show milder depletion effects for other elements. Molecular hydrogen, common in local diffuse clouds, is absent in intervening systems except in a few cases that seem to be associated with the QSO itself and also have high metallicities. Abundances at high red-shift are discussed further in Chapter 12.

Dense molecular clouds are too dusty to be studied by optical means, but radio and mm wave observations reveal numerous molecules such as CO, CH$_2$O, HCN, NH$_3$

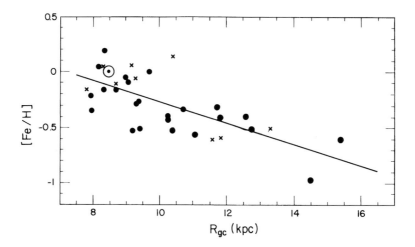

Fig. 3.34. Iron abundances in intermediate-age open Galactic clusters as a function of galacto-centric distance, with the Sun shown for comparison. The line corresponds to a gradient of -0.1 dex kpc^{-1}. Adapted from Friel & Janes (1993).

etc. which give valuable information on isotope ratios across the Galaxy and in other galaxies (see Fig. 3.36). The increase in relative abundance of ^{13}C and ^{18}O towards the Galactic centre indicates a more advanced degree of what is sometimes known as 'secondary processing', where the production rate of these species is enhanced by an increased abundance of their 'primary' progenitors ^{12}C and ^{16}O; but time delays in the evolution of lower-mass stars can also be involved and the nitrogen isotope ratio does not fit this pattern very well anyway. ^{18}O/^{17}O has a constant value of 3.2 across the Galaxy, consistent with both of these isotopes being secondary, but the solar-system ratio is somewhat higher at 5.5.

3.5.3.4 *Abundances in external galaxies*
Apart from spectroscopy of individual supergiant stars in the Magellanic Clouds, most information on abundances in galaxies comes from emission lines from H II regions and supernova remnants in gas-rich systems and from features like Mg_2 in the integrated spectra of old stellar populations. Another source of information in nearby dwarf spheroidals is the location of the red giant branch of old stars in the HR diagram, which becomes brighter and bluer with decreasing metallicity.

Broadly speaking, the abundances in nearby galaxies show two main features:

- There is a general tendency for average abundances to increase with increasing luminosity of the parent galaxy (see Fig. 3.37), ranging from slightly above solar ('super metal-rich') in the dominant component of the nuclear regions of the biggest spirals and ellipticals to a factor of about 100 below solar in dwarf spheroidals and blue compact gas-rich dwarfs, roughly corresponding to

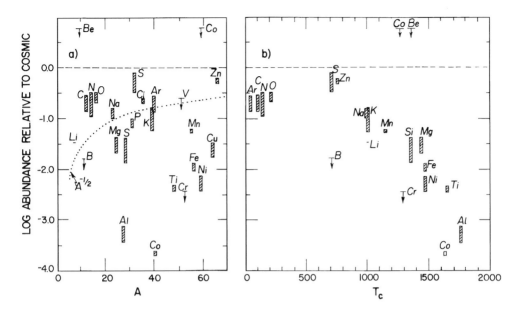

Fig. 3.35. Depletion below solar abundances of elements in the H I gas towards ζ Ophiuchi plotted against atomic mass number in (a) and condensation temperature in (b), based in part on the curve of growth shown in Fig. 3.11. Vertical boxes indicate error bars. The dotted curve in the left panel represents an $A^{-1/2}$ dependence expected for non-equilibrium accretion of gas on to grains in the ISM. The condensation temperature gives a somewhat better, though not perfect, fit, suggesting condensation under near-equilibrium conditions at a variety of temperatures either in stellar ejecta or in some nebular environment. Note the extreme depletion of Ca ('Calcium in the plane stays mainly in the grain'). After L. Spitzer & E.B. Jenkins (1975), with permission, from the *Annual Review of Astronomy and Astrophysics*, Vol. **13**, p. 133, ©1975 by Annual Reviews, Inc.

the relation

$$<Z> \propto L_B^{0.3} \propto M_S^{0.25}, \tag{3.71}$$

where $<Z>$ is the luminosity-weighted mean 'metallicity' and M_S the total mass of stars (living or dead) and gas in the galaxy. This may reflect the ability of the gravitational potential well to retain processed material expelled from supernovae, and/or bursts of star formation and concomitant 'metal' enrichment accompanying mergers of small systems to form larger ones (see Chapter 11).

• Most spirals and many ellipticals display a large-scale radial abundance gradient (see Figs. 3.38, 3.39) analogous to what seems to be the case in the Milky Way (cf. Figs. 3.33, 3.34). Such gradients may result from a combination of many factors including differing time scales of evolution, differing exposure to inflowing material and slow inward radial flows of gas due to energy dissipation,

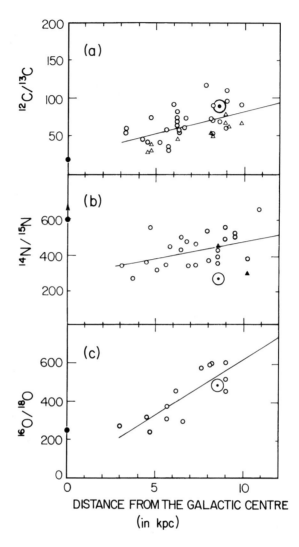

Fig. 3.36. Variation of CNO isotopic ratios with Galactocentric distance, deduced from mm wave measurements of molecules in molecular clouds. (a) $^{12}C/^{13}C$ from CO (triangles) and formaldehyde CH_2O (circles). (b)$^{14}N/^{15}N$ from HCN (circles) and NH_3 (triangles). (c)$^{16}O/^{18}O$ from formaldehyde. Solar-System values are indicated by the $\odot$ sign and Galactic centre values (or a lower limit in the case of $^{14}N/^{15}N$) by a filled circle. Adapted from Wilson & Rood (1994).

viscosity or other effects. The gradients tend to be lower in barred spirals where the non-axisymmetric potential can give rise to rapid radial flows and mixing.

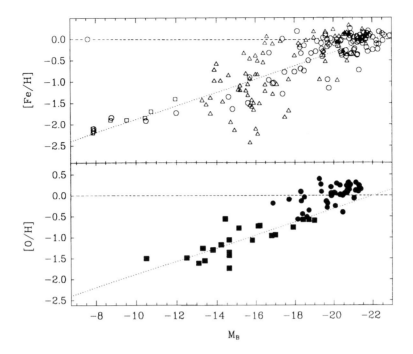

Fig. 3.37. Metallicities in gas-poor galaxies (open symbols) and oxygen abundances at a representative radius in gas-rich disk galaxies (filled symbols), as a function of galaxy luminosity in blue light. The dotted lines in each panel represent identical trends for '[Fe/H]' and [O/H] and the ordinate 0.0 represents solar composition. Adapted from Zaritsky, Kennicutt & Huchra (1994).

Notes to Chapter 3

More information about the topics discussed in this chapter can be found in the following textbooks:-

H.G. Kuhn, *Atomic Spectra*, Longman 1971.

D. Mihalas, *Stellar Atmospheres*, W.H. Freeman & Co., San Francisco, 1970, 1978. The first edition of this classic text deals with radiative and convective equilibrium and line formation in normal stellar atmospheres, while the second treats non-LTE effects in more detail.

D.F. Gray, *The Observation and Analysis of Stellar Photospheres*, Wiley, New York 1976, gives many useful practical details in this field, including a description of Fourier analysis of line profiles.

D.E. Osterbrock, *Astrophysics of Gaseous Nebulae and Active Galactic Nuclei*, University Science Books, Mill Valley, Cal., 1989, is another classic text, indispensable for studies of emission nebulae.

C.W. Allen, *Astrophysical Quantities (AQ)*, Athlone Press, London 1981, is an invaluable and instructive source of atomic and astronomical data and formulae.

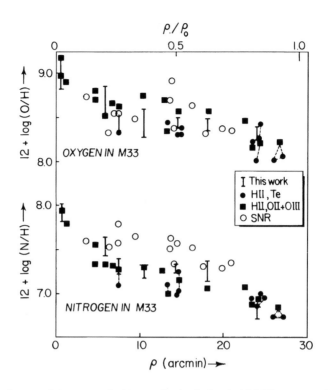

Fig. 3.38. Abundances of oxygen and nitrogen in the Scd spiral M 33, measured in H II regions and supernova remnants, as a function of galactocentric distance; 1 arcmin $\equiv$ 200 pc and ρ_0 is the de Vaucouleurs radius corresponding to a visual surface brightness of 25^m arcsec^{-2}. After J.M. Vílchez, B.E.J. Pagel, A.I. Díaz, E. Terlevich & M.G. Edmunds (1988), *MNRAS*, **235**, 633.

C.R. Cowley, *An Introduction to Cosmochemistry*, Cambridge University Press 1995, covers similar ground to some of this book, but with different emphasis, giving more details about solar-system chemistry, atomic and molecular spectra and chemically peculiar stars.

J. Binney & S. Tremaine, *Galactic Dynamics*, Princeton University Press 1987, is a third classic text, recommended for background reading on Galactic aspects of abundances.

Some important references for atomic and molecular spectra are:

R.D. Cowan, *The Theory of Atomic Structure and Spectra*, University of California Press, Berkeley 1981.

G.R. Harrison, *MIT Wavelength Tables*, Wiley, New York, 1939.

C.E. Moore, *A Multiplet Table of Astrophysical Interest*, Revised Edition *(RMT)*, *NSRDS-NBS* **40**, Nat. Bur. Stand. Washington 1972.

C.E. Moore, *An Ultraviolet Multiplet Table*, NBS Circular no. 488, Nat. Bur. Stand. Washington 1962.

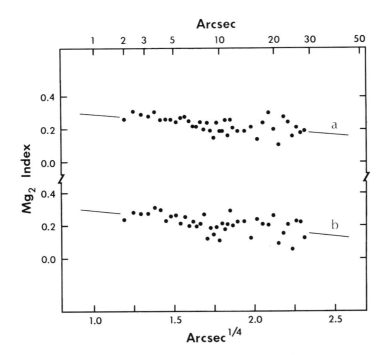

Fig. 3.39. Radial trends of surface brightness and the Mg$_2$ index in the elliptical galaxy NGC 4881, after B. Thomsen & W.A. Baum (1987), *Ap. J.*, **315**, 460. 1 arcsec $\equiv$ 500 pc and 'a' and 'b' refer to observations taken on separate dates. Courtesy Bjarne Thomsen.

W.F. Meggers, C.H. Corliss & B.F. Scribner, *Tables of Spectral Line Intensities*, NBS Monograph **145**, Govt. Printing Office, Washington 1975.

G. Herzberg, *Molecular Spectra and Molecular Structure I. Spectra of Diatomic Molecules*, Van Nostrand, Princeton 1950.

G. Herzberg, *Molecular Spectra and Molecular Structure II. Infra-red and Raman Spectra of Polyatomic Molecules*, Van Nostrand, Princeton 1945.

G. Herzberg, *Molecular Spectra and Molecular Structure III. Electronic Spectra and Electronic Structure of Polyatomic Molecules*, Van Nostrand, New York 1966.

G. Herzberg, *The Spectra and Structures of Simple Free Radicals*, Cornell University Press, Ithaca, 1971.

W. Gordy & R.L. Cook, *Microwave Molecular Spectra*, Wiley, New York 1984.

M.S. Cord, J.D. Peterson, M.S. Lojko & R.H. Haas, *Microwave Spectral Tables V. Spectral Line Listing*, Govt. Printing Office, Washington 1968.

Reports of IAU Commissions 14 and 29 in *Transactions of the International Astronomical Union*, published triennially by Kluwer, Dordrecht.

The literature on transition probabilities (or oscillator strengths) is vast, rapidly growing and difficult to summarise. A small selection for atoms and molecules is given in Allen, *AQ* and larger selections in

W.L. Wiese & G.A. Martin, *Wavelengths and Transition Probabilities for Atoms and Atomic Ions*, NSRDS-NBS68, 1980, and

W.L. Wiese, J.R. Fuhr & T.M. Dieters, *Atomic Transition Probabilities for Carbon, Nitrogen and Oxygen: A Critical Data Compilation, J. Phys. Chem. Ref. Data*, Monograph 7, 1996.

Wave-lengths and oscillator strengths for ultra-violet lines relevant to interstellar work have been given by

D.C. Morton, *Astrophys. J. Suppl.*, **77**, 119, 1991.

A vast and very complete data set for atoms and ions, with very variable accuracy (but sufficient for statistical calculations of line blanketing for model atmospheres in which molecules are not too important), is given by

R.L. Kurucz & E. Peytremann, 'A Table of Semiempirical *gf* Values', *Smithsonian Astrophys. Obs. Special Rep.*, **362**, 1975.

A guide to the literature is provided in

W.C. Martin, 'Sources of Atomic Spectroscopy Data for Astrophysics', in P.L. Smith & W.L. Wiese (eds.), *Atomic and Molecular Data for Space Astronomy: Needs, Analysis and Availability*, Springer, Berlin 1992.

A list of 'solar oscillator strengths' for selected lines (based on a solar model with assumed abundances) is given in

T. Meylan, I. Furenlid, M.S. Wiggs & R.L. Kurucz, *Astrophys. J. Suppl.*, **85**, 163, 1993.

This does not really replace the classic and indispensable revision of what was originally Rowland's table of spectral lines from the centre of the solar disk:

C.E. Moore, M. Minnaert & J. Houtgast, 'The Solar Spectrum 2935 Å to 8770 Å', *Nat. Bur. Stand. Monograph* **61**, Washington 1966.

The best theoretical data for oscillator strengths and other properties of lighter atoms are becoming available from the Stellar Opacity Project (Seaton 1987). Results are given in

M.J. Seaton, *The Opacity Project, Vol. I: Selected Research Papers, Atomic Data Tables for He to Si;* Vol II: *Atomic Data Tables for S to Fe, Photo-ionization Cross-sections Graphs*, Inst. of Phys. Publ., Bristol, 1996.

The art of computing model stellar atmospheres has progressed rapidly with increases in computer power. A guide to atmospheres of cool stars is given by

B. Gustafsson, 'Chemical Analyses of Cool Stars', *Ann. Rev. Astr. Astrophys.*, **27**, 701, 1989.

LTE line-blanketed models and spectrum synthesis programs for stars hotter than $T_{\text{eff}} = 5500$ K are under continuing development by R.L. Kurucz at the Smithsonian Astrophysical Observatory, Cambridge, Mass, and available from their author in CD rom. A sample publication is

R.L. Kurucz, 'Model Atmospheres for G, F, A, B and O Stars', *Astrophys. J. Suppl.*, **40**, 1, 1979.

Models for the hottest stars, taking into account non-LTE and dynamical effects, are described in

R.F. Kudritsky & D.G. Hummer, 'Quantitative Spectroscopy of Hot Stars', *Ann. Rev. Astr. Astrophys.*, **28**, 303, 1990.

Abundances in our own and other galaxies are discussed in:

B.E.J. Pagel & M.G. Edmunds, 'Abundances in Stellar Populations and the Interstellar Medium in Galaxies', *Ann. Rev. Astr. Astrophys.*, **19**, 77, 1981.

G. Gilmore, R.F.G. Wyse & K. Kuijken, 'Kinematics, Chemistry and Structure of the Galaxy', *Ann. Rev. Astr. Astrophys.*, **27**, 555, 1989.

G.A. Shields, 'Extragalactic H II Regions', *Ann. Rev. Astr. Astrophys.*, **28**, 525, 1990.

Problems

1. Check the integrations leading to Eqs. (3.41) and (3.54) for the damping branches of simple curves of growth. Obtain an expression for the ratio of one to the other, taking $\eta_0 \equiv N\alpha_0$.

2. The equivalent width of Lyman-β, λ 1025, in Fig. 3.8 is 1.14 Å. Use the damping formula Eq. (3.42) to estimate the H I column density, given the parameters $f = 0.079$, $\gamma = A_{3p,1s} + A_{3p,2s} = 1.87 \times 10^8$ s^{-1}.

3. The Na I D-lines have wavelengths and oscillator strengths $\lambda_1 = 5896$ Å, $f_1 = 1/3$, and $\lambda_2 = 5889$ Å, $f_2 = 2/3$. In a certain interstellar cloud, their equivalent widths are measured to be 230 mÅ and 370 mÅ respectively, with a maximum error of ± 30 mÅ in each case. Assuming a single cloud with a gaussian velocity dispersion, use the exponential curve of growth to find preferred values of Na I column density and b, and approximate error limits for each of these two parameters. (Doublet ratio method.)

4. Use the solar Fe I curve of growth in Fig. 3.12 to deduce the solar iron abundance, using $\theta_{ion} = 0.9$ and given that Fe I has an ionization potential of 7.87 V and that the partition function of Fe II is 42. Compare the result with the one in Table 3.4.

5. Use the differential curve of growth in Fig. 3.13 to estimate [Fe/H] for μ Cas.

6. From the analogue of Saha's equation for the dissociation of a diatomic molecule AB, one can define a dissociation 'constant' depending only on temperature

$$K_{AB} = \frac{p_A p_B}{p_{AB}}, \tag{3.72}$$

where the ps represent partial pressures. Show that, in a situation where the atomic abundances of carbon and oxygen are significantly depleted by CO formation, but other molecules are negligible and the hydrogen pressure is P_H, the partial pressures of oxygen and carbon are given by

$$2\frac{p_O}{P_H} = \left(\frac{O}{H} - \frac{C}{H}\right) - \frac{K_{CO}}{P_H} + \left[\left(\frac{K_{CO}}{P_H} + \frac{O}{H} + \frac{C}{H}\right)^2 - 4\frac{O}{H}\frac{C}{H}\right]^{1/2} \tag{3.73}$$

and

$$2\frac{p_C}{P_H} = \left(\frac{C}{H} - \frac{O}{H}\right) - \frac{K_{CO}}{P_H} + \left[\left(\frac{K_{CO}}{P_H} + \frac{O}{H} + \frac{C}{H}\right)^2 - 4\frac{O}{H}\frac{C}{H}\right]^{1/2}. \tag{3.74}$$

Deduce that the depletion of the free atoms becomes appreciable when K_{CO} gets down to values comparable to the (undepleted) partial pressure of the more abundant element.

7. Given that the transition probability for the nebular [O III] line λ 4959 is 7×10^{-3} s^{-1}, estimate the critical density if the electron temperature is 10^4 K and $\Omega_{12} = 2$. The upper and lower levels of the line have $J = 2$ and $J = 1$ respectively.

4 Cosmological nucleosynthesis and abundances of light elements

4.1 Introduction

The 'Hot Big Bang' theory of the universe was pioneered by George Gamow, R.A. Alpher and R.C. Herman in the late 1940s and early 50s. They supposed that during the first few minutes of the (then radiation-dominated) universe, matter was originally present in the form of neutrons and that, after some free decay, protons captured neutrons and successive captures built up all the elements.

C. Hayashi first put the theory on a sound physical basis by pointing out that, at the high densities and temperatures involved, there would be thermal equilibrium between protons and neutrons at first, followed by a freeze-out, and this intensified the difficulty already inherent in that theory that the absence of stable nuclei at mass numbers 5 and 8 would prevent significant nucleosynthesis beyond helium. In the meantime, progress in the theory of stellar evolution and nucleosynthesis (see Chapter 5) led to comparative neglect of Big Bang nucleosynthesis theory (BBNS)[1] until the discovery by A.A. Penzias and R. Wilson in 1964 of the microwave background radiation, existence of which Gamow and his colleagues had predicted. The 'standard Big-Bang nucleosynthesis theory' (SBBN), developed subsequently by P.J.E. Peebles, R.V. Wagoner, W.A. Fowler, F. Hoyle, D.N. Schramm, G. Steigman and others, which assumes standard cosmology and particle physics and a uniform baryon density, has been very successful in several respects, e.g.:

- It accounts for what can be deduced from astrophysical and cosmochemical measurements about the primordial abundances of ^2D, ^3He, ^4He and ^7Li, over 9 orders of magnitude in abundance.
- It led to a prediction that the number of different sorts of neutrino (equivalent in standard particle physics to the number of families of quarks and leptons) is less than 4 and probably no more than 3. This prediction was subsequently confirmed (subject to slight reservations about differences between effective numbers of neutrino species in the laboratory and in the early universe) by measurements of the width or lifetime of the Z^0 boson at CERN in 1990.

[1] Apart from a prescient paper by Hoyle & Tayler (1964).

- Until 1989, measurements of the neutron half-life had given somewhat discrepant results. SBBN predicted that the half-life should not exceed 10.4 minutes and the currently accepted value is 10.28 minutes, considerably less than had been assumed some years earlier.

At the same time, questions have been raised about SBBN. One feature found unattractive by some scientists has been that it leads to a low density of baryonic matter, $\Omega_{b0} < 0.1$, where Ω_{b0} is the smoothed-out cosmological density of baryons at the present time in units of the closure density corresponding to the Einstein–de Sitter world model favoured by many cosmologists on the basis of the inflationary scenario or for other reasons. This gap is usually bridged by assuming the existence of substantial non-baryonic dark matter in the form of massive neutrinos and/or exotic particles ('cold dark matter') predicted by super-symmetry or other theories of particle physics, but there have also been many attempts to modify BBNS theory to allow higher baryonic densities, e.g. by invoking baryonic density fluctuations due to earlier cosmological phase transitions or subsequent modification of BBNS products by high-energy processes. Suggestions have been made from time to time that there could be detectable primordial abundances of elements such as ^{9}Be, not expected from SBBN, that might show up in the spectra of stars with extremely low metallicity. However, detailed investigations have not enhanced the attractions of these alternatives to SBBN and that theory can be considered to be in a pretty healthy state; at the same time, the questions raised have been a valuable spur to related astrophysical investigations of nucleosynthesis, stellar abundances, cosmic rays, γ-ray background etc.

4.2   Background cosmology

The basic assertion of the 'Hot Big Bang' theory is that, as one goes back to early times, one encounters almost indefinitely high temperatures and densities. The basic evidence for this is threefold:-

- The Hubble expansion law

$$cz \simeq V = H_0 D; \ H_0 = 100h \, \text{km s}^{-1} \text{Mpc}^{-1} = (10^{10} \text{yr})^{-1} h \qquad (4.1)$$

where the dimensionless number h, somewhere between 0.5 and 0.8, expresses our uncertainty in the value of the Hubble constant H_0. Expansion gives rise to a red-shift z such that

$$1 + z \equiv \frac{\lambda}{\lambda_{\text{lab}}} = \frac{R_0}{R_z} \qquad (4.2)$$

where R_0 is the scale factor of the universe now and R_z the scale factor at the time when the radiation was emitted.
- The microwave background radiation has very precisely the spectrum of a black body with a temperature

$$T_{\gamma 0} = 2.73 \pm 0.01 \, \text{K} \qquad (4.3)$$

leading with adiabatic expansion to temperatures at earlier times

$$T_\gamma(z) = T_{\gamma 0} R_0 / R = T_{\gamma 0} (1 + z). \tag{4.4}$$

The combination of high temperatures and densities at early times leads to the existence of a phase close to thermal equilibrium, when the significant particle reaction rates dominated over the expansion rate. This enables precise calculations to be made.

- Abundances of light elements. In particular, the very existence of substantial deuterium is an argument for BBNS because nuclear reactions in stars destroy it and no plausible high-energy effects are known that could create D without over-producing other light elements (Epstein, Lattimer & Schramm 1976). This argument was already used by Gamow at a time when only the terrestrial abundance was known. Another consideration is that the mass fraction Y of ^{4}He is more than 0.2 in virtually all objects investigated, even when abundances of all elements from carbon upwards are very low. Finally, the existence of primordial ^{7}Li was a prediction of the theory that was subsequently confirmed.

SBBN assumes

- Normal physical laws (including standard particle physics).
- A small degree of matter–antimatter asymmetry, with a baryon number B (ratio of net number of baryons $N_B - N_{\bar{B}}$ in a co-moving volume to the entropy S) in the range 10^{-11} to 10^{-8}.
- Absence of electron or neutrino degeneracy, corresponding to lepton numbers L_e, L_μ, L_τ that are zero or at least not large compared to B, where

$$L_e \equiv (N_{e^-} - N_{e^+} + N_{\nu_e} - N_{\bar{\nu}_e})/S, \tag{4.5}$$

$$L_{\mu,\tau} \equiv (N_{\nu_{\mu,\tau}} - N_{\bar{\nu}_{\mu,\tau}})/S \tag{4.6}$$

(after μ, τ decay), together with zero charge number. S is related to the number of photons by a factor related to g^* in Table 4.1, which is nearly constant over substantial periods, so that in practice one can use number-density ratios n/n_γ as an approximation to the baryon and lepton numbers.

- Gravitation described by General Relativity.
- A large measure of thermal equilibrium in the early universe, which implies, roughly speaking, that particles and anti-particles with $mc^2 < kT_\gamma$ are present in comparable numbers to photons, whereas when kT_γ falls below mc^2, they annihilate and exist only in trace quantities.
- The 'cosmological principle' which states that the universe is always homogeneous and isotropic and leads to the Robertson–Walker metric

$$ds^2 = c^2 dt^2 - R^2(t) \left[\frac{dr^2}{1 - kr^2} + r^2(d\theta^2 + \sin^2\theta \, d\phi^2) \right]. \tag{4.7}$$

Here r, θ, ϕ are dimensionless comoving coordinates attached to fundamental observers and $R(t)$ a scale factor with a dimension of length depending only on cosmic time t. k is the curvature constant, which with suitable choice of units takes one of the three values $+1$ (closed world-model with positive curvature), 0 (flat, open model) or -1 (open model with negative curvature). Some consequences of Eq. (4.7) are the relation between red-shift and scale factor Eq. (4.2) and the variation of temperature

$$RT_i = \text{const.} \tag{4.8}$$

for adiabatic expansion of a relativistic gas i with $\gamma = 4/3$.

The expansion of the universe is governed by the field equations of General Relativity for isotropic expansion or contraction:

$$8\pi G \rho R^2 = 3kc^2 + 3\dot{R}^2 - \Lambda R^2, \tag{4.9}$$

where ρc^2 is the total energy density. Since the 'cosmological constant' Λ is not very greatly dominant now (although possibly non-zero), it is negligible at early times when the total density ρ in Eq. (4.9) was much greater; so we can take $\Lambda = 0$ in that equation and obtain the Friedman equation

$$H^2 \equiv \left(\frac{\dot{R}}{R}\right)^2 = \frac{8\pi G \rho}{3} - \frac{kc^2}{R^2}. \tag{4.10}$$

Thus $k = 0$ if

$$\rho = \rho_{\text{crit}} \equiv 3H^2/8\pi G, \tag{4.11}$$

the critical density, which has the present-day value

$$\rho_{\text{crit},0} = 1.88h^2 \times 10^{-29} \text{ gm cm}^{-3}, \tag{4.12}$$

and one can define a dimensionless density parameter Ω by

$$\Omega \equiv \rho/\rho_{\text{crit}}. \tag{4.13}$$

From the Friedman equation, because ρ varies as R^{-4} (radiation-dominated) or at least as R^{-3} (matter-dominated), $\Omega = 1$ at early times, to a very good approximation. (If it is minutely below 1, it becomes very much less than 1 at later times, and conversely; only if it is very exactly 1 does it remain so.) With $\Omega = 1$, Eq. (4.10) becomes

$$V^{-1}\frac{dV}{dt} \equiv 3R^{-1}\dot{R} = (24\pi G \rho)^{1/2} \tag{4.14}$$

for the physical volume occupied by a given set of particles (fixed comoving volume).

Another equation from General Relativity (resembling energy conservation) is

$$\frac{d}{dt}(\rho V) + \frac{P}{c^2}\frac{dV}{dt} = 0. \tag{4.15}$$

At early times, dominated by radiation and relativistic particles, $3P = \rho c^2$, so that

$$\rho \propto R^{-4}, \text{ or } R \propto \rho^{-1/4}. \tag{4.16}$$

Substituting (4.16) in (4.14) with $\rho(0) \to \infty$ gives the density as a function of time, i.e. the expansion history:

$$\rho = \frac{3}{32\pi G} t^{-2}. \tag{4.17}$$

4.3 Thermal history of the universe

Given the total density from Eq. (4.17), the temperature follows from the equation of state which depends in turn on what particles are present. For any one species i, with temperature T_i, we have from the Fermi–Dirac or Bose–Einstein distribution, Eq. (2.40),

$$c^2 \rho_i = 4\pi g_i k T_i \left(\frac{kT_i}{hc}\right)^3 \int_{x_i}^{\infty} \frac{x^2(x^2 - x_i^2)^{1/2} \, dx}{e^{x-y_i} \pm 1}, \tag{4.18}$$

where $x \equiv \tilde{E}/kT_i$, $x_i \equiv m_i c^2/kT_i$, $y_i \equiv \mu_i/kT_i$ and the $+$ and $-$ signs refer to fermions and bosons respectively. In the fully relativistic ($x_i \to 0$) and non-degenerate ($y_i \to 0$) limits, the integration gives

$$c^2 \rho_i = \frac{\pi^2}{30} g_i k T_i \left(\frac{kT_i}{\hbar c}\right)^3 \quad \text{(bosons)} \tag{4.19}$$

$$= \frac{7}{8} \left[\frac{\pi^2}{30} g_i k T_i \left(\frac{kT_i}{\hbar c}\right)^3\right] \quad \text{(fermions)}. \tag{4.20}$$

At each stage, particles coupling to photons ($T_i = T_\gamma$) with $m_i c^2 < kT_\gamma$ are relativistic and present in comparable numbers to photons. When kT_γ drops to $m_i c^2$, they annihilate with their anti-particles and/or decay, or if the coupling is so weak that $T_i \ll T_\gamma$, they contribute little mass-energy and are 'suppressed'. The temperature is then fixed as a function of density, and hence time, by the relation

$$\frac{\pi^2}{30} g_* \frac{(kT_\gamma)^4}{(\hbar c)^3} \equiv \frac{1}{2} g_* a T_\gamma^4 = \rho c^2, \tag{4.21}$$

where g_* is the effective number of degrees of freedom given by

$$g_* = \sum_{\text{bos}, x_i \ll 1} g_i \left(\frac{T_i}{T_\gamma}\right)^4 + \frac{7}{8} \sum_{\text{ferm}, x_i \ll 1} g_i \left(\frac{T_i}{T_\gamma}\right)^4 \tag{4.22}$$

and a is the normal Stefan radiation-density constant. The resulting thermal history of the universe is sketched in Fig. 4.1 and Table 4.1.

During most of the first 0.1 second after the Big Bang (ABB), the relativistic particles are photons, electrons, positrons and N_ν species of neutrinos and anti-neutrinos; N_ν is expected to be 3, from ν_e, ν_μ and ν_τ. There is a sprinkling of non-relativistic protons

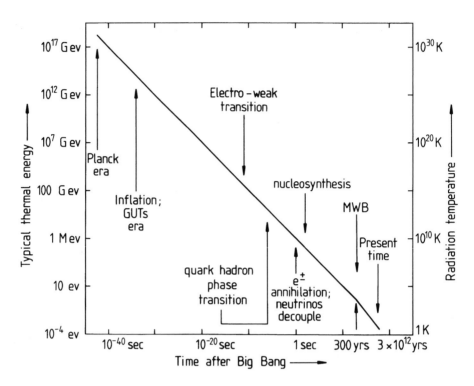

Fig. 4.1. Schematic thermal history of the universe showing some of the major episodes envisaged in the standard model. GUTs is short for grand unification theories and MWB is short for (the last scattering of) the microwave background radiation. The universe is dominated by radiation and relativistic particles up to a time a little before that of MWB and by matter (possibly including non-baryonic matter) thereafter.

Table 4.1. *Brief thermal history of the universe*

	Time	kT_γ	T_γ	g_*	
Un- cer- tain physics	10^{-42} s 10^{-35} s 10^{-9} s 10^{-4} s	10^{19} GeV 10^{15} GeV 100 GeV 150 MeV		~ 100 20	Planck era; quantum grav. GUTs, Infl., Primord. fluct. El. weak trans. Q-H trans.; meson decay
Physics fairly well known	1s 100 s	1 MeV 0.1 MeV	10^{10} K 10^9 K	43/4 3.36	Weak interaction decoupl.; $e^\pm$ annihilation; $T_\nu < T_\gamma$ BBNS (D, ^{3}He, ^{4}He, ^{7}Li)
Un- certain details	10^5 yr 10^{10} yr	0.2 eV 2×10^{-4} eV	3000 K 3 K		Matter domination (Re)comb.; MWB last scat. Structure formation Present

and neutrons which make a completely negligible contribution to the energy density. The temperature is then given by

$$\rho c^2 = c^2 \sum \rho_i = \frac{1}{2} g_* a T_\gamma^4 \qquad (4.23)$$

$$= a T_\gamma^4 \left[1 + \frac{7}{4} + \frac{7}{8} N_\nu \left(\frac{T_\nu}{T_\gamma} \right)^4 \right], \qquad (4.24)$$

where the 1 in Eq. (4.24) comes from the photons, the 7/4 from electrons and positrons and the $(7/8)N_\nu$ from the neutrinos ($g_i = 1$ for each neutrino species because of their helicity). Before weak-interaction decoupling, $T_\nu = T_\gamma$, and so from Eqs. (4.17) and (4.24),

$$T_\gamma = T = 0.96 \times 10^{10} t^{-1/2} \text{ K}, \qquad (4.25)$$

or

$$kT = 0.8 t^{-1/2} \text{ MeV}. \qquad (4.26)$$

After 1s or so, the neutrinos decouple from the cosmic plasma and expand isentropically for ever afterwards remaining as 'microwave neutrinos' with

$$RT_\nu = \text{const.} \qquad (4.27)$$

(assuming $m_\nu = 0$). Photons, electrons and positrons do the same for the time being with $T_\gamma = T_\nu$ until $e^\pm$ annihilation several seconds ABB, which transfers entropy to the photon gas. Since the photon entropy density $s_\gamma \propto T_\gamma^3$ and s_γ is multiplied by 11/4 at $e^\pm$ annihilation according to Eq. (4.24), one has subsequently

$$T_\gamma^3 = \frac{11}{4} T_\nu^3 \qquad (4.28)$$

(which implies $T_{\nu 0} = 0.7 T_{\gamma 0}$ for massless neutrinos). Thereafter (and in particular during nucleosynthesis), photons and nucleons are both conserved, so the ratio of their number densities

$$\eta \equiv n_b/n_\gamma = \text{const.} \sim 10^{-10} \text{ to } 10^{-9}. \qquad (4.29)$$

The number η, together with the known background temperature $T_{\gamma 0}$, measures the cosmological baryon density today:

$$\Omega_{b0} h^2 = 3.65 \times 10^{-3} (T_{\gamma 0}/2.73 \text{ K})^3 \eta_{10} \qquad (4.30)$$

where η_{10} is η in units of 10^{-10}.

Since density and temperature are fixed functions of time during BBNS, *the outcome of the relevant nuclear reactions is a function of the single cosmological parameter η.*

4.4   Neutron:proton ratio

Before ν_e decoupling at around 1 MeV, neutrons and protons are kept in mutual thermal equilibrium through charged-current weak interactions:

$$n + e^+ \longleftrightarrow p + \bar{\nu}_e \tag{4.31}$$

$$p + e^- \longleftrightarrow n + \nu_e \tag{4.32}$$

$$n \longleftrightarrow p + e^- + \bar{\nu}_e. \tag{4.33}$$

Equilibrium in the two-body reactions implies a relation between the chemical potentials:

$$y_p - y_n = y_{e^+} - y_{\bar{\nu}_e} = y_{\nu_e} - y_{e^-}. \tag{4.34}$$

Since the chemical potentials of particles and anti-particles that are in equilibrium with photons are equal and opposite, it can be shown from consideration of the integral in Eq. (4.18) in the relativistic limit ($x_i \to 0$) that for any such pair, to order of magnitude,

$$y_i \sim (n_i - n_{\bar{i}})/n_\gamma \quad \text{if} \quad |y_i| \leq 1. \tag{4.35}$$

Because of charge neutrality, $y_{e^-} \sim \eta \sim B$ and is consequently negligible. That the same thing holds for y_{ν_e} (and for $y_{\nu_{\mu,\tau}}$) is a postulate, but a very plausible one because otherwise we would have neutrino or anti-neutrino degeneracy with $|L_\nu| \gg B$. The upshot is that the chemical potentials of neutrons and protons are equal and so from Eq. (2.43) one has a simple Boltzmann-type equilibrium ratio

$$(n/p)_{\text{eq}} = e^{-(m_n - m_p)c^2/kT} = e^{-1.29\,\text{MeV}/kT}. \tag{4.36}$$

Now the reaction rate per neutron is $n_{\nu_e} c <\sigma>$ where the weak-interaction cross-section $<\sigma>$ is proportional to the square of the energy (cf. Eq. 2.103), i.e. to T^2, and to the reciprocal neutron half-life for free decay, $\tau_{1/2}^{-1}$, which measures the intrinsic strength of the interaction. Since $n_{\nu_e} \propto T^3$, the reaction rate varies as $T^5/\tau_{1/2}$ and declines steeply with decreasing temperature. The expansion rate, on the other hand, varies as

$$\rho^{1/2} \propto \left(\frac{11}{4} + \frac{7}{8}N_\nu\right)^{1/2} T^2 \tag{4.37}$$

and thus declines much more slowly. Thus, as T goes down, there comes a point at $kT_d \simeq 0.8$ MeV where the weak-interaction rate falls rather suddenly below the expansion rate and the ratio n/p is frozen (apart from free decay and some residual weak interactions) at a value

$$e^{-1.29\,\text{MeV}/kT_d} \simeq 0.2, \tag{4.38}$$

which later fixes the primordial helium mass fraction Y_P because virtually all surviving neutrons are soaked up to make ^{4}He. Hence (after free decay for about 300 s)

$$Y_P = \frac{2n}{n+p} \simeq 0.24. \tag{4.39}$$

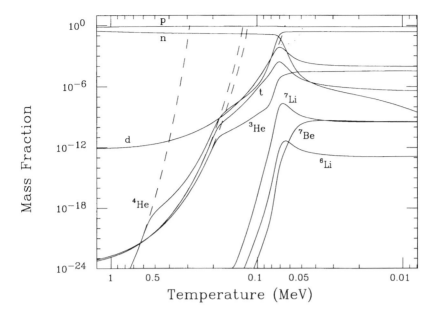

Fig. 4.2. Evolution of light-element abundances with temperature, for $\eta_{10} = 3.16$. The dashed curves give the nuclear statistical equilibrium abundances for ^{4}He, ^{3}He, ^{3}H and ^{2}D respectively; the dotted curve for ^{2}D allows for the diminishing number of free neutrons. After M.S. Smith, L.H. Kawano & R.A. Malaney 1993, *Ap. J. Suppl.*, **85**, 219. Courtesy Michael Smith.

The precise value of the decoupling temperature T_d depends on the two physical constants N_v and $\tau_{1/2}$:

$$T_d{}^3 \propto \tau_{1/2} \left(\frac{11}{4} + \frac{7}{8} N_v\right)^{1/2}, \tag{4.40}$$

implying that larger values of either of these constants lead to higher T_d, higher n/p and hence more primordial helium, for given η.

4.5 Nuclear reactions

During the first few seconds ABB, nuclear abundances are in statistical equilibrium (i.e. thermal equilibrium subject to the appropriate n/p ratio), but the corresponding abundance of any nucleus above neutrons and protons is very low because of the low value of η or the high entropy per baryon. When kT falls to 0.3 MeV, the equilibrium mass fraction of ^{4}He reaches about 0.15 (for $n/p \simeq 0.2$), but by this time equilibrium no longer applies: nuclear reactions have become too slow, partly from Coulomb barriers and partly because of low (still near-equilibrium) abundances of the lighter nuclei D, ^{3}H and ^{3}He. Only when the d/p ratio (depending on the balance between $p-n$ captures and photodisintegration) has built up to a value of order 10^{-5}, at $kT \simeq 0.1$ MeV, do nuclear reactions effectively build up to ^{4}He, which then soaks up virtually all the neutrons remaining from freeze-out and subsequent decay (see Fig. 4.2). After the formation of ^{4}He, traces of the lighter elements survive because nuclear reactions are frozen out by low density and temperature before their destruction is complete, and still smaller traces of ^{7}Li and ^{7}Be (which later decays by K-capture to ^{7}Li) are formed and survive. The outcome of primordial nucleosynthesis is calculated by numerically following a series of nuclear reactions including in particular the following (and their reverse):

$$p + n \longrightarrow d + \gamma \tag{4.41}$$

$$d + d \longrightarrow {}^3\text{He} + n, \ {}^3\text{H} + p \tag{4.42}$$

$${}^3\text{He} + d \longrightarrow {}^4\text{He} + p \tag{4.43}$$

$${}^3\text{H} + d \longrightarrow {}^4\text{He} + n \tag{4.44}$$

$${}^3\text{H} + \alpha \longrightarrow {}^7\text{Li} + \gamma \tag{4.45}$$

$${}^3\text{He} + \alpha \longrightarrow {}^7\text{Be} + \gamma \tag{4.46}$$

$${}^7\text{Be} + n \longrightarrow {}^7\text{Li} + p \tag{4.47}$$

$${}^7\text{Li} + p \longrightarrow 2\,{}^4\text{He} \tag{4.48}$$

which are followed by ^{3}H decay to ^{3}He ($\tau_{1/2} = 12.2$ yr) and ^{7}Be decay to ^{7}Li (by K-capture, after recombination).

The final outcome of these reactions, as a function of η or equivalently $\Omega_{b0}h^2$, is shown in Fig. 4.3. The primordial helium mass fraction Y_P, shown on a large scale, is not very sensitive to η, since this parameter only affects the time for neutron decay before nucleosynthesis sets in, and it can be fitted by the relation

$$Y_P = 0.226 + 0.025 \log \eta_{10} + 0.0075(g_* - 10.75) + 0.014(\tau_{1/2}(n) - 10.3 \, \text{min}). \tag{4.49}$$

The deuterium abundance, on the other hand, is a steeply decreasing function because it is destroyed by two-body reactions with p, n, D and ^{3}He. ^{3}He (not shown separately in the figure) declines more gently because this nucleus is more robust. ^{7}Li has a bimodal behaviour because at low baryon densities it is synthesised from ^{3}H by reaction (4.45) and both nuclei are destroyed by two-body reactions, whereas at higher densities it results from reaction (4.46) on ^{3}He to make ^{7}Be and the comparative robustness of both these nuclei causes an increase with η. The comparison with observations, and deductions therefrom, are discussed in the following sections.

4.6   Deuterium and ^{3}He

Essentially all deuterium in the universe is believed to come from BBNS, because thermonuclear reactions in stars only cause net destruction of D and it is vastly more abundant than other light nuclei like ^{6}Li or ^{9}Be that are believed to result from spallation processes. However, deuterium in both terrestrial and meteoritic water is enhanced by a factor of about 6 owing to fractionation. The first evidence for

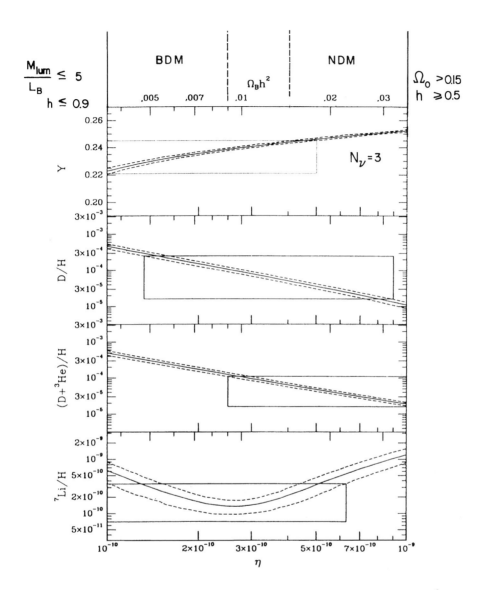

Fig. 4.3. Primordial abundances from SBBN as functions of η or, equivalently, $\Omega_{b0}h^2$. Curves show theoretical predictions, with their uncertainty limits represented by dashed lines, while the horizontal error boxes indicate various limits derived from observation. Vertical broken lines at the top of the figure indicate the optimal region of concordance with data on the various light elements discussed in the text, and 'BDM' and 'NDM' indicate the gaps between the likely range of Ω_{b0} thus derived and other evidence on Ω_0 (from luminous matter in galaxies) and Ω_0 (total) that are believed to be filled by baryonic and non-baryonic dark matter respectively. Adapted from Copi, Schramm & Turner (1995).

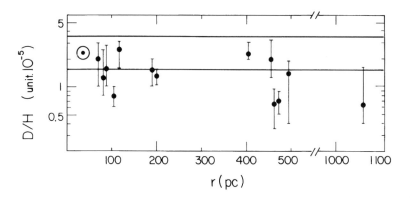

Fig. 4.4. D/H ratios measured along the lines of sight to hot stars in the Galaxy. Filled circles with error bars are data points from Laurent (1983), adapted from Boesgaard & Steigman (1985). The horizontal band shows error limits to the proto-solar D/H ratio after Geiss (1993).

'cosmic' deuterium arrived in the early 1970s from studies of the solar wind, planetary atmospheres and the interstellar medium.

The composition of the solar wind is studied by ion counters on space probes, and by analysis of samples collected on metal foils placed on the Moon and from the outer layers of gas-rich meteorites. These give an average ^{3}He/^{4}He ratio of 4.1×10^{-4}, equivalent to ^{3}He/H $\simeq 4.1 \times 10^{-5}$, since for the Sun He/H $\simeq 0.1$. D.C. Black in 1971 and J. Geiss and H. Reeves in 1972 noted that helium gas released by heating carbonaceous chondrites had a lower ^{3}He/^{4}He ratio of about 1.5×10^{-4}, which they identified as the proto-solar ^{3}He abundance present in the wind from the primitive Sun, attributing the excess in the present-day solar wind to proto-solar deuterium that had been burned to ^{3}He during the Sun's evolution. The proto-solar deuterium abundance is thus about 2.6×10^{-5}, which is in fair agreement with observations of deuterated molecules (HD and CH_3D) in the atmospheres of the major planets.

Searches for the D I hyperfine structure line transition at 91.6 cm wavelength from the Galaxy have been unsuccessful, but in the early 70s interstellar deuterium was detected by Penzias, Wilson and others in the form of DCN and DCH$^+$ from microwave transitions in molecular clouds, and by L. Spitzer and others who found UV transitions of HD and D I (Lyman series) caused by diffuse clouds on the line of sight to hot stars observed with the *Copernicus* satellite (cf. Fig. 3.8). Whereas quantitative interpretation of the molecular lines is greatly complicated by various fractionation effects (which can either increase or reduce the D/H ratio according to circumstances), the atomic lines give reasonably accurate column densities from which an average interstellar D/H ratio of about 1.5×10^{-5} is inferred, marginally below (but not really inconsistent with) the proto-solar value (see Fig. 4.4). The nearby star Capella has Lyman-α in emission with interstellar absorption components superposed that have been observed and analysed using the Hubble Space Telescope and lead to a

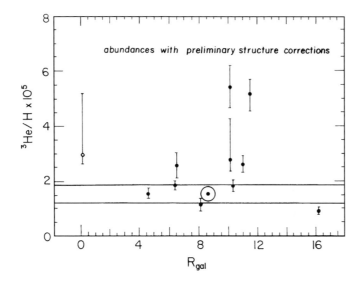

Fig. 4.5. ^{3}He abundances in Galactic H II regions. The horizontal band shows error limits for the protosolar ^{3}He/H after Geiss (1993). Adapted from Balser *et al.* (1994) and Rood *et al.* (1995).

D/H ratio of $(1.65 \pm 0.15) \times 10^{-5}$ (Linsky *et al.* 1993), but there could be some minor variations in D/H along different sight-lines.

Because of the destruction of D when interstellar gas is recycled through stars, this number is a firm lower limit to the primordial abundance and thus places an upper limit on the baryonic density parameter,

$$\eta_{10} < 8.5, \quad \text{or} \quad \Omega_{b0}h^2 < 0.032 \quad \text{and} \quad \Omega_{b0} < 0.13 \tag{4.50}$$

assuming $h \geq 0.5$. Qualitatively, the destructive effect of recycling through stars, or 'astration', is supported by the fact that emission lines from deuterated molecules are relatively weaker in the Sagittarius B2 clouds near the Galactic centre, where stellar evolution and nucleosynthesis are believed to have gone further than in the solar neighbourhood. Thus there could be a positive gradient in D/H with galactocentric distance corresponding to the negative gradient in metallicity and $^{13}C/^{12}C$, but the precise factor by which D/H has been reduced in this way is model-dependent and correspondingly uncertain. A possible way to determine that factor is provided by observations of intervening systems at high red-shift, with low metallicities. In 1995-96 quite contradictory estimates of the relevant destruction factor appeared, ranging from factors of little above 1 to 10 or more, and probably not very precise because of the difficulty in sorting out the cloud structure and kinematics in these remote systems. However, some of the larger estimates were later withdrawn. Some plausible (if not completely compelling) arguments on this issue can also be derived from meteoritic ^{3}He, discussed below.

Table 4.2. *Interstellar and proto-solar deuterium and ^3He*

	ISM	Pre-solar
10^5 (D+^{3}He)/H	3.7 ± 0.9	$y_{23} = 4.1 \pm 1.0$
10^5 D/H	1.6 ± 0.2	$y_2 = 2.6 \pm 1.0$
10^5 ^{3}He/H	2.1 ± 0.9	$y_3 = 1.5 \pm 0.3$

Solar System data from Geiss (1993); ^{3}He in ISM from Gloeckler & Geiss (1996).

^{3}He has been detected in Galactic H II regions and some planetary nebulae using the hyperfine transition of ^{3}He$^+$ at 3.46 cm. In PN, the ^{3}He/H ratio can be quite high, of the order of 10^{-3}, whereas in H II regions it is more than an order of magnitude lower, apparently with large scatter (see Fig. 4.5) and no clear dependence on galactocentric distance that might have been a clue to the net effect of stellar evolution and astration following primordial nucleosynthesis. The two largest abundances are found in relatively small H II regions ($M_{\mathrm{HII}} < 100 \, M_\odot$), which has given rise to a suggestion that these regions are 'self-polluted' by winds from evolving stars within them, but this is far from certain. The PN data support theoretical predictions that there is net production of ^{3}He in low-mass stars; in high-mass stars, which are more likely to have undergone some evolution in an H II region, one expects net destruction overall, but it is possible that some high-mass stars emit winds enriched in ^{3}He at early evolutionary stages. It is not yet possible to make any very useful deductions about primordial ^{3}He from these data. However, the *Ulysses* spacecraft has been used to measure the ^{3}He abundance in the local interstellar medium by sampling 'pick-up ions' which enter the Solar System as neutral atoms and are then ionized and swept outwards by the solar wind (Gloeckler & Geiss 1996).

It was pointed out by Yang *et al.* (1984) that an upper limit can be derived for the sum of primordial D and ^{3}He abundances from the proto-solar abundances by assuming that a reasonable fraction of ^{3}He survives stellar processing. The symbols and data are summarised in Table 4.2.

The argument assumes that a fraction g of ^{3}He resulting from BBNS or from subsequent D-burning survives astration through a typical generation of stars, which may also lead to some fresh production. (Some gas will have been recycled more than once and some not at all; the assumption that the average can be represented by one cycle corresponds to a probability of e^{-1} for gas to remain un-cycled, which is a fair approximation to what is expected from more detailed models.) We then have, for the proto-Sun,

$$y_2 = y_{2P}(1 - a), \tag{4.51}$$

say, and

$$y_3 = g(y_{3P} + ay_{2P}) + y_{3\,\mathrm{fresh}}, \tag{4.52}$$

where the suffix p denotes primordial values. Combining these equations,

$$y_{23P} = y_2 + g^{-1}(y_3 - y_{3\,\text{fresh}}) < y_2 + g^{-1}y_3. \tag{4.53}$$

Taking $g > 0.25$ (a 'safe bet' according to Yang *et al.*), one derives

$$y_{23P} < 12, \tag{4.54}$$

whence it follows from Fig. 4.3 that

$$\eta_{10} > 2.5; \quad \Omega_{b0}h^2 > 0.009; \quad \Omega_{b0} > 0.011 \tag{4.55}$$

assuming $h < 0.9$. This limit, shown by the left dashed vertical line at the top of Fig. 4.3, implies that $y_{2P} < 10$ and the corresponding astration factor less than about 6, which is also a generous upper limit from the point of view of conventional models of Galactic chemical evolution (cf. Chapter 9). The argument receives some support from the near-equality of $(D + {}^3\text{He})/H$ in the present-day ISM and the proto-solar system, but at the same time it is clear from the PN and H II region data that there are aspects of the chemical evolution of ${}^3\text{He}$ that are not well understood, and there could have been more complex mechanisms of both creation and destruction in the early Galaxy than are envisaged in such a simple picture.

4.7   Helium

Helium is the second most abundant element in the visible universe and accordingly there is a mass of data from optical and radio emission lines in nebulae, optical emission lines from the solar chromosphere and prominences and absorption lines in spectra of hot stars. Further estimates are derived more indirectly by applying theories of stellar structure, evolution and pulsation. However, because of the relative insensitivity of Y_P to cosmological parameters, combined with the need to allow for additional helium from stellar nucleosynthesis in most objects, the requirements for accuracy are very severe: better than 5 per cent to place cosmological limits on N_ν and better still to place interesting constraints on η or Ω_{b0}. One can, however, assert with confidence that there is a universal 'floor' to the helium abundance in observed objects corresponding to $0.22 \leq Y_P \leq 0.25$.

The historical development has been quite interesting. In the early 1950s, data were available only for a few representative B-type stars and planetary nebulae, consistent with $y \equiv \text{He}/\text{H} \simeq 0.1$ or $Y \simeq 0.3$, and in fair agreement with early BBNS predictions by Hayashi (1950) and by Alpher, Follin and Herman (1953); at that time it was widely taken for granted that this abundance was universal, in common with others. With the arrival of evidence for nucleosynthesis in stars, and the development of the related theory, it became fashionable to suppose that the oldest, lowest-metallicity objects would also have a low helium abundance, although B^2FH in 1957 wisely left open the possibility that there might be 'some helium in the initial matter of the Galaxy' and several authors noted at various times that a purely stellar origin for the large present-day helium abundance implied a very high luminosity for galaxies in the past. In 1963, C.R. O'Dell, M. Peimbert and T.D. Kinman at the Lick Observatory found a normal helium abundance in a planetary nebula belonging to the globular cluster M15 (in which the stars have [Fe/H] $\simeq -2$). This discovery stimulated Hoyle and R.J. Tayler to point out in a very prescient paper in 1964 (just before discovery of the microwave background) the crucial importance to cosmology of the presence or absence of a universal helium 'floor', and to give a fresh calculation of Y_P from BBNS in which they first pointed out the influence of N_ν; the ν_μ had then been recently discovered. After the microwave background observation, opinion gradually came round in favour of a helium floor, based on several arguments: R.F. Christy's studies of the pulsation of RR Lyrae stars, energised by He^+ in the envelope; J. Faulkner and I. Iben's modelling of stellar evolution in globular clusters (see Fig. 4.6) and the relative lifetimes on the red giant branch (hydrogen shell burning) and on the horizontal branch (helium core burning); and the masses, luminosities and effective temperatures of low-metallicity subdwarfs, which lie somewhat below the Population I lower main sequence in the HR diagram, but have larger luminosities for the same mass (see Chapter 5). Some contrary indications also came up, in particular the so-called subdwarf B-stars which lie at the extreme blue end of the horizontal branch: their atmospheres are helium-deficient, but it was shown fairly soon afterwards that they also have other anomalies similar to those of CP stars (see Chapter 3), leading to the conclusion that the helium deficiency is just a superficial effect due to gravitational settling. (The same phenomenon seems to have occurred, to a minor extent, even in the Sun.) The argument was finally settled, to every reasonable person's satisfaction, in 1972, when L. Searle and W.L.W. Sargent discovered that two blue compact galaxies previously discovered by Fritz Zwicky, II Zw 40 and I Zw 18, were dominated by giant H II regions with low abundances of oxygen and other heavy elements (I Zw 18 still holds the record for low abundances in extragalactic H II regions with [O/H] $\simeq -1.7$), but nearly normal helium. This discovery set the stage for the modern era, marked by systematic attempts to estimate Y_P by measuring or estimating helium abundances in a variety of objects with differing metallicities and extrapolating to zero metallicity.

Table 4.3 shows some estimates of helium abundance in various objects. The value for the interior of the proto-Sun is based on theoretical models of its evolution (see Chapter 5) constrained by its known mass, age, chemical composition (apart from Y) and present-day luminosity and radius. Helium, through its effect on the mean molecular weight, influences the structure of a star and speeds up its evolution. Uncertainties include the opacity κ, the precise equation of state, the treatment of convection and the deficit of solar neutrinos relative to prediction, which may in fact be due to neutrino physics rather than weaknesses in the solar models. Solar seismology (based on observations of minute oscillations with periods of the order of 5 minutes) has led to greatly improved understanding of the Sun's internal structure in recent years, but it also indicates an appreciably lower helium abundance ($Y = 0.25$) in the convective envelope, which is attributed to gravitational

Table 4.3. *Estimates of primordial helium mass fraction*

Objects	Y_P	Method	Ref.	Problems
Sun	$< .28 \pm .02$	Interior	1	κ; Eq. of st; ν prob.
" "	$< .28 \pm .05$	Prom. HeI	2	Level pops.
B stars	$< .30 \pm .04$	Abs. lines	3	Precision
Field sd	$.19 \pm .05$	Main seq.	4	Plx; T_{eff}; conv.
Glob-	$.23:$	RR, Δm	5	Physical
ular	$.23 \pm .02$	N(HB)/N(RG)	6	basis of
clus-	$.23 \pm .02:$	M15 HB	7	stellar
ters	$\leq .24 \pm .02$	47 Tuc HB	8	evolution
Gal.	$.22 \pm .02$	Plan. neb.	9	Self + gal enr.
neb.	$.22:$	H II reg.	10	He^0; gal. enr.
Ex-gal	$.233 \pm .005$	Irr.+BCG	11	He^0; data
HII reg.	$< .243 \pm .010$	BCG	12	II Zw 40

References

1. Turck-Chièze & Lopez (1993). 2. Heasley & Milkey (1978). 3. Kilian (1992). 4. Carney (1983). 5. Caputo, Martínez Roger & Páez (1987). 6. Buzzoni *et al.* (1983). 7. Dorman, Lee & VandenBerg (1991). 8. Dorman, VandenBerg & Laskarides (1989). 9. Peimbert (1983). 10. Mezger & Wink (1983). 11. Lequeux *et al.* (1979). 12. Kunth & Sargent (1983).

settling. Estimates of helium abundance in the chromosphere have been made on the basis of hydrogen and helium emission lines from prominences, but these are quite imprecise, owing to the difficulty in modelling radiative transfer effects and the populations of atomic levels. Finally, because there has been net production of helium during the evolution of the Galaxy, and the Sun is a relatively young object, its helium abundance is expected to exceed the primordial value by several hundredths.

The B stars, with $10^4 \, \mathrm{K} \leq T_{\mathrm{eff}} \leq 3 \times 10^4 \, \mathrm{K}$, show He I lines in absorption, but again these are young objects and the inferred helium abundances are not yet very precise.

The most metal-deficient stars comprise field stars in the solar neighbourhood (where in some cases distances and luminosities can be found from parallaxes) and stars in globular clusters where the morphology of the HR diagram can be studied (Fig. 4.6). Such stars are of particular interest because their content of heavy elements (synthesised in still earlier generations of stars) is so low that they can be considered to have been born with essentially the primordial helium abundance. That abundance influences certain of their structural properties: the location of the lower main sequence in absolute magnitude, the blue edge of the instability strip occupied by RR Lyrae variables, the luminosity along the 'zero-age' horizontal branch ZAHB (which can be

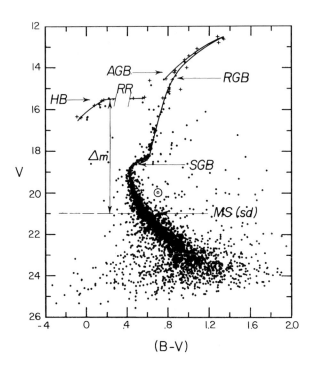

Fig. 4.6. Colour-magnitude (HR) diagram of the globular cluster Messier 68 with [Fe/H] $\simeq -2$. In order of successive evolutionary stages, MS (sd) indicates the main sequence occupied by cool subdwarfs, with the position of the Sun shown for comparison, SGB indicates the subgiant branch, RGB the red giant branch, HB the horizontal branch including a gap in the region occupied by RR Lyrae pulsating variables and AGB the asymptotic giant branch. Adapted from McClure *et al.* (1987).

estimated from Δm, the magnitude difference from a fiducial point along the main sequence), the relative numbers (i.e. lifetimes) along the HB and the RGB and (in some cases) the detailed distribution in the diagram of stars evolving away from the ZAHB. Some estimates of Y_P derived from these various properties are shown in Table 4.3, but the errors are hard to quantify because many features of stellar evolution are sensitive to uncertainties in the input physics and the helium abundance is generally only one of many factors influencing stellar structure at different stages of evolution (cf. Chapter 5).

The most accurate helium abundances come from measurements of recombination lines of hydrogen and helium in the emission spectra of planetary nebulae and H II regions (cf. Fig. 3.20), using theoretical effective recombination coefficients calculated from quantum mechanics. The He I term scheme and several optical transitions of interest are shown in Fig. 4.7; in Galactic H II regions it is also possible to measure radio recombination lines, which is especially useful when the optical spectrum is obscured by dust. Silvia and Manuel Peimbert in 1974 outlined a programme whereby measurements

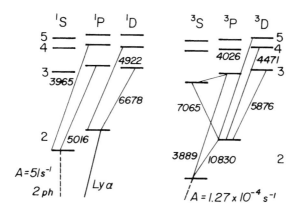

Fig. 4.7. Term scheme for He I, showing some of the transitions of interest.

in Galactic and extragalactic H II regions having different Z's (represented mainly by oxygen) could be used to plot a regression represented by

$$Y = Y_\mathrm{P} + Z\frac{\Delta Y}{\Delta Z} = Y_\mathrm{P} + (\mathrm{O/H})\frac{\Delta Y}{\Delta(\mathrm{O/H})}. \tag{4.56}$$

Extrapolation to $Z = (\mathrm{O/H}) = 0$, would give the primordial helium abundance Y_P, while the slope of the regression, $\Delta Y/\Delta Z$ (not necessarily a constant), would place a significant constraint on stellar nucleosynthesis by giving the rate at which helium is freshly synthesised and ejected into the ISM compared to the corresponding rate for carbon and heavier elements. The application of this idea to objects in our own Galaxy suffers from the fact that the abundances represented by Z are so high that rather a large extrapolation is needed; in addition, planetary nebulae need a small correction for helium dredged up to the surface during prior stellar evolution, while Galactic H II regions mostly have ionizing stars that are not hot enough to guarantee the absence of undetectable neutral helium, He^0, especially with the limited angular resolution of radio telescopes.

Consequently the best objects in which to apply this idea are extragalactic H II regions in dwarf irregular galaxies like the Magellanic Clouds and in blue compact (or H II) galaxies where the abundances of oxygen and other heavy elements are low and the ionizing stars often very hot, so that the regions of ionized hydrogen and helium coincide within 1 per cent or so. The last two lines of Table 4.3 summarise the outcome of two of the more accurate early investigations in this field, which led to slightly contradictory results: Lequeux *et al.* (1979) found a linear regression with a slope $\Delta Y/\Delta Z = 3$, whereas Kunth & Sargent (1983) found no evidence for a slope, but their results were probably somewhat biased by the high weight given to II Zw 40, for which the yellow helium line λ 5876 is red-shifted to coincide with, and be somewhat absorbed by, interstellar sodium λ 5889 in the Milky Way.

In recent years, intensive efforts have been made to improve the determination of the

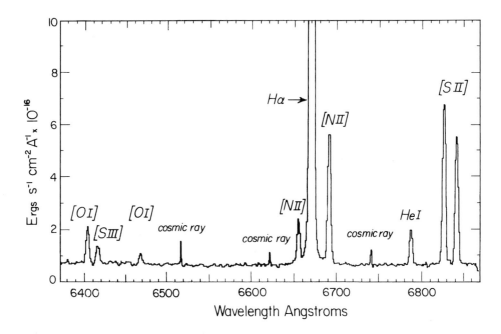

Fig. 4.8. Portion of the red spectrum of the H II galaxy Tololo 0633-415 with a red-shift of 0.016, showing diagnostic features for helium (Hα and λ 6678), electron density ([S II]) and ionization ([S III]). The features marked 'cosmic ray' are due to impacts of charged particles on the CCD detector. After B.E.J. Pagel *et al.* (1992), *MNRAS*, **255**, 325.

pre-galactic helium abundance (generally assumed to be the same as Y_P) and $\Delta Y/\Delta Z$ from observations of extragalactic H II regions. Advantages of this method are

- The theory of recombination for hydrogen and helium seems to be well under-stood, at least to 1 or 2 per cent, and the line intensities are not very sensitive to temperature or density; in particular, there is no exponential factor.
- Low abundances (extending down to [O/H] $\simeq$ −1.7 in the case of I Zw 18) ensure that little extrapolation to $Z = 0$ is needed.
- One can assure oneself that neutral helium is negligible by assessing that the effective temperatures of the ionizing stars are high enough (say, above 40 000 K). One way to do this is to compare the degrees of ionization of oxygen (O II/O III) and sulphur (S II/S III), which have widely differing ionization potentials, with predictions from numerical photo-ionization models.

There are also difficulties:

- Experimental problems: signal:noise, detector linearity (Hα/λ 6678 is typically about 100), calibration, reddening.
- Underlying absorption lines in the stellar continuum.
- Line absorption in the intervening interstellar gas, or Earth's atmosphere.

- Fluorescence and collisional excitation, arising primarily from the metastability of the 2^3S level (see Fig. 4.7), in which consequently a high population accumulates which can cause additional emission from lines such as λ 4471, λ 5876 by either collisional excitation or radiative transfer effects following absorption of higher lines in the 2^3S $- n^3$P series. The singlet line λ 6678 can also be enhanced by collisional excitation from 2^3S. The collisional effects can be calculated from the known electron temperature and density, and are quite small at typical electron densities of the order of 100 cm^{-3} except in the case of λ 7065. The latter line, once collisional effects are allowed for, can also be used to gauge the influence of radiative transfer effects, which are normally found to be small for extragalactic H II regions.

- While the triplet levels exist only from $n = 2$ upwards, the singlets have an analogue to the hydrogen Lyman-α line and consequently, like hydrogen, have Cases A and B (or intermediate). The assumption of Case B, which is usually made, could be slightly inaccurate if Ly-α were to either partially escape or be absorbed by small amounts of dust in the nebula. However, the above-mentioned lines are not very sensitive to Case A or B, while λ 5016, which is very sensitive because of the strong transition from its upper state 3^1P to the ground state 1^1S, is found to be in agreement with Case B whenever it is measured.

- $\Delta Y / \Delta Z$ or $\Delta Y / \Delta(\text{O/H})$ may vary, either systematically as a function of Z or randomly, e.g. if some H II regions are self-polluted with helium and other elements (e.g. N) ejected in winds from massive embedded stars. There is some evidence that this actually happens in a few cases such as the nucleus of NGC 5253, where the continuum shows a strong, broad feature at the He$^+$ wavelength λ 4686, due to Wolf–Rayet stars.

Fig. 4.8 shows part of a high-resolution high signal:noise spectrum typical of more recent attempts to determine helium abundance in low-Z H II regions, and Fig. 4.9 shows resulting regressions against oxygen and nitrogen. Regressions resulting from data such as these (Olive & Steigman 1995) give the result

$$Y_\text{P} = 0.232 \pm 0.003\,(1\sigma) \pm 0.005\,(\text{syst.}) \leq 0.243 \tag{4.57}$$

with 95 per cent confidence, which limits η_{10} to < 3.9 and $\Omega_{b0}h^2$ to < 0.015, shown by the right-hand vertical broken line at the top of Fig. 4.3. Assuming that the errors have not been underestimated, this result gives the most restrictive upper limit to the cosmic density of baryonic matter ($\Omega_{b0} < 0.06$). Since dynamical evidence indicates larger values for Ω_0 from galaxy motions, even if one discounts the arguments that $\Omega = 1$, there is a gap of at least about a factor of 3 that needs to be filled by non-baryonic dark matter, whereas the conservative limit on Ω_{b0} derived from deuterium assuming negligible destruction (eq 4.50) is just barely compatible with existing dynamical data on Ω_0.

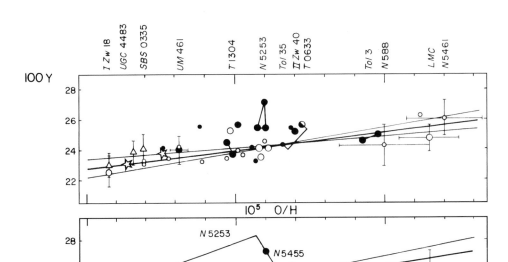

Fig. 4.9. Regressions of helium against oxygen and nitrogen in H II galaxies and some H II regions in irregular galaxies and spirals. Filled and open circles (representing respectively objects with and without definite WR features and with different sizes according to weight) are from Pagel *et al.* (1992); full and broken lines show their maximum likelihood regressions (rejecting the filled points) and equivalent $\pm 1\sigma$ limits. Some later data from other authors are shown by stars and triangles. After B.E.J. Pagel, 'Helium in H II Regions and Stars', in P. Crane (ed.), *The Light Element Abundances*, p. 160, Fig. 1, ©Springer-Verlag Berlin Heidelberg 1995.

4.8   Lithium 7

The lithium resonance doublet line λ 6707 is fairly easy to observe in cool stars of spectral types F and later, and it has also been detected in diffuse interstellar clouds. There is thus an abundance of data, although in the ISM the estimation of an abundance is complicated by ionization and depletion on to dust grains. The youngest stars (e.g. T Tauri stars that are still in the gravitational contraction phase before reaching the main sequence) have a Li/H ratio that is about the same as the solar-system ratio derived from meteorites, $Li/H = 2 \times 10^{-9}$, which is thus taken as the Population I standard.

As stars become older, lithium at their surface becomes gradually depleted by mixing with deeper layers at temperatures above 2.5×10^6 K where it is destroyed by the (p, α) reaction, Eq. (4.48). This destruction takes place more rapidly in cooler stars with deeper outer convection zones, so that there is a trend for lithium abundance to decrease with both stellar age and diminishing surface temperature; in cooler stars some depletion takes place already in the pre-main-sequence stage. Thus, in the young Pleiades cluster ($\sim 10^8$ yr), lithium has its standard abundance down to $T_{\mathrm{eff}} = 5500$ K (type G5), whereas in the older Hyades cluster ($\sim 6 \times 10^8$ yr), it is noticeably depleted below $T_{\mathrm{eff}} = 6300$ K (F7) and also in a narrow range of effective temperatures around 6600 K. ^{6}Li is still more fragile than ^{7}Li and is accordingly virtually absent in most stars, although it is detected in interstellar clouds when the spectral resolution and sensitivity are high enough. When stars evolve up the giant branch, their surface ^{7}Li is diluted by the dredge-up of Li-free material from below, but fresh lithium is created in some AGB stars, including 'super lithium-rich' carbon stars such as WZ Cassiopeiae and a number of giants in the Magellanic Clouds, so that astration of the ISM leads to both destruction and creation of ^{7}Li. However, the destruction mechanisms are not completely understood, notably in the case of the Sun where the depletion (by a factor of 150) has gone much further than standard models predict, and there is presumably some special mechanism not taken into account in those models, e.g. additional mixing due to processes associated with the loss of rotational angular momentum (spin-down).

Against this background, it occasioned some surprise when Spite & Spite (1982) discovered lithium in Population II stars, with a quite uniform abundance of the order of 0.1 of the standard Population I value, although their metallicities are down by factors of 100 or more. Fig. 4.10 shows an ensemble of data which exhibits a lithium 'plateau' (for stars with $T_{\mathrm{eff}} > 5500$ K or so) extending down to [Fe/H] ≤ -4. Apart from a few exceptions not shown in the diagram, the Population II stars have a much smaller range in lithium abundance than do Population I stars, suggesting that depletion effects are mild or absent (in agreement with standard stellar models) and that here one is indeed seeing the primordial abundance or a close approximation thereto (represented by a rectangular error box in Fig. 4.3). The upper envelope for Population I stars presumably represents a net increase due to AGB stars and/or other Galactic sources. The range of abundances on the plateau (where at least a part of the spread is due to errors in the determinations) is in excellent accord with the range of baryonic densities deduced from helium and deuterium; it gives an upper limit not much greater than the one deduced from ^{4}He and a lower limit corresponding to the high D/H values that have sometimes been claimed on the basis of some observations of high red-shift absorption-line systems. The true primordial lithium abundance is most unlikely to be lower than the lower bound of the plateau at $12 + \log{(\mathrm{Li/H})} = 2.0$, because these stars are evidently formed from material that had undergone only minute amounts of processing from stellar nucleosynthesis. It could be higher if depletion is a significant factor, which it might possibly be as a result of non-standard processes similar to those that have evidently affected the Sun.

Some direct evidence that surface depletion of ^{7}Li is not an important factor along the plateau is provided by the tentative detection of a small amount (about 5 per cent) of ^{6}Li in two stars (Fig. 4.11). ^{6}Li is not produced by BBNS, and its main source in the Galaxy (e.g. in meteorites where ^{7}Li/^{6}Li $= 12$) is believed to be spallation of

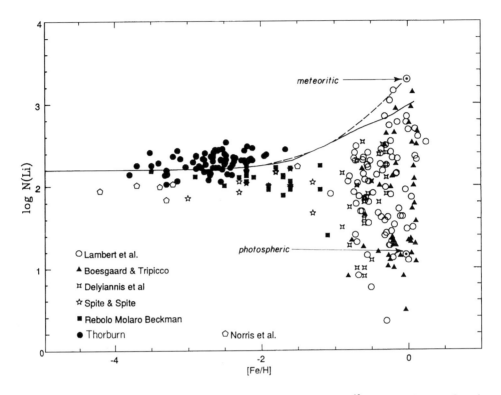

Fig. 4.10. Stellar lithium abundances (log of the number per 10^{12} H atoms) as a function of metallicity. The full-drawn curve shows the prediction of a numerical Galactic chemical evolution model, while the broken-line curve gives the sum of a primordial component and an additional component proportional to iron and normalised to meteoritic abundance. Adapted from Matteucci, D'Antona & Timmes (1995).

CNO nuclei and $\alpha - \alpha$ fusion reactions at high energies either from cosmic rays in the interstellar medium or from effects of particle acceleration in supernova remnants, stellar flares or elsewhere. Such processes might be expected to produce ^{6}Li along with beryllium (which is known to be present) in the young Galaxy, resulting in an abundance increasing roughly in proportion to metallicity. One may therefore expect a ^{6}Li/^{7}Li ratio an order of magnitude below that of meteorites, since the ^{7}Li has been enhanced by about an order of magnitude through BBNS. However, some diffuse clouds show ^{6}Li/^{7}Li ratios considerably higher than what is found in meteorites, so that an initial ratio of a few per cent in halo stars is not inconceivable. ^{6}Li is a more fragile nucleus than ^{7}Li, so that in standard models it is predicted (and it is also found) to be completely absent in the cooler subdwarfs on the ^{7}Li plateau, but it could survive at effective temperatures above 6000 K (close to the turnoff from the main sequence in Population II), and slightly lower temperatures along the subgiant branch. The finding of evidence (see Fig. 4.11) for about 5 per cent ^{6}Li in one, or possibly two, such stars (an observational *tour de force*) is evidence that the standard models are

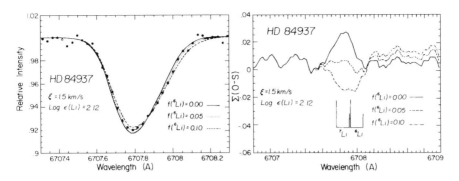

Fig. 4.11. Evidence for the presence of 5 ± 2 per cent ^{6}Li in the warm halo subdwarf HD 84937 with [Fe/H] $= -2.4$, $T_{\text{eff}} = 6090$ K. The left panel shows synthesised profiles for 0, 5 and 10 per cent ^{6}Li with the observational points on an absolute wavelength scale in the rest frame of the stellar photosphere. The right panel shows the central wave-lengths of the doublets of each isotope and a deviation plot for the three hypotheses on ^{6}Li/^{7}Li. Adapted from Smith, Lambert & Nissen (1993).

indeed approximately correct; here evidently ^{6}Li has been depleted by only a modest factor (estimated at about 0.7) and it then follows that ^{7}Li is virtually undepleted. The observation is a delicate one, however, and still needs to be confirmed.

4.9 Non-'standard' BBNS models

The assumptions underlying SBBN theory are very simple and economical, but things could have been more complicated in the early universe and the complications could have left some signature on the results of BBNS, which in turn could modify the conclusions drawn on Ω_b etc. More exotic hypotheses include:

- Anisotropic expansion.
- Neutrino or anti-neutrino degeneracy.
- Existence of massive, non-classical unstable particles, e.g. photinos, massive neutrinos and anti-matter. These could modify the equation of state at the relevant times, but much of parameter space has now been ruled out by experiments with the Large Electron-Positron Collider (LEP) at CERN.
- Massive (> 2 Gev) particles decaying after about 10^5 s into electromagnetic and hadron showers could 'wipe the slate clean', removing the SBBN restrictions on N_v and Ω_b (Dimopoulos *et al.* 1988). This specific theory is not in good accordance with observed ^{6}Li/^{7}Li ratios (Audouze & Silk 1989), but one could take the view that a more fundamental objection to this theory is that it is ugly.

One of the more interesting variants on SBBN considers possible effects of the quark–hadron phase transition which takes place at $kT \simeq 150$ MeV about 5×10^{-5} s ABB, involving confinement of the quark–gluon plasma into mesons and baryons

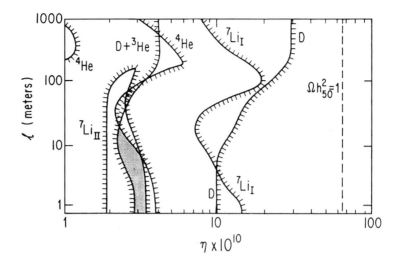

Fig. 4.12. Regions of parameter space allowed by primordial abundances in the framework of density fluctuations presumed to arise from the quark–hadron phase transition. The ordinate is the length scale, small values of which correspond to the 'standard' homogeneous model and the 'prickles' indicate excluded regions. The region allowed by all the light element abundances is shaded. Adapted from D.N. Schramm (1995), 'Primordial Nucleosynthesis and Light Element Abundances', in P. Crane (ed.), *The Light Element Abundances*, p. 51, Fig. 3, ©Springer-Verlag Berlin Heidelberg 1995, updating work by Kurki-Suonio *et al.* (1990). Courtesy D.N. Schramm and C. Copi.

and chiral symmetry breaking. If the transition was first order (which now seems somewhat doubtful), then bubble nucleation might occur leading possibly to quark nuggets (Witten 1984) and to inhomogeneities in η: supercooled quark–gluon droplets would still be relativistic, giving rise later to regions of high baryon density, whereas hadron droplets would be non-relativistic giving rise to regions of low baryon density. Such inhomogeneities by themselves would lead to a mixture of SBBN products corresponding to a range of values of η, which immediately leads to difficulties with ^{7}Li because it is already near the minimum. However, Applegate & Hogan (1985) pointed out that things are more complicated because neutrons diffuse more easily than protons. Thus neutrons would diffuse out of high-density regions leaving them proton-rich (leading to low ^{2}D and ^{4}He), the low density regions would be neutron-rich (leading to the reverse) and there could be many possible consequences: the data might be fitted with $\Omega_b = 1$; additional elements like beryllium, boron and even much heavier nuclei might be created in the Big Bang; and at least the restrictions imposed by SBBN would be relaxed somewhat.

However, there are many difficult problems associated with this theory, as well as a number of free parameters: the density contrast, the volume fractions occupied by the different density regions, their length scale and geometry. The diffusion hydrodynamics needs to be computed very carefully and most of the more exciting claims have turned out later to have resulted from excessively simplistic treatments; overproduction of ^{7}Li is still a severe constraint and it now appears that the constraints of SBBN are hardly relaxed at all (Fig. 4.12) and that any cosmological 'floors' or plateaux to abundances of elements not produced in appreciable quantities by SBBN will always be far too low to be detectable. However, the interest aroused by these ideas has been quite stimulating, notably in helping to lead to the discovery of beryllium and boron in Population II stars where these elements had not been thought to be readily detectable (cf. Chapter 9).

4.10 Conclusions

Thus SBBN theory still holds the field, and the resulting limits on η or Ω_b are more or less watertight subject to limitations in the knowledge of nuclear reaction rates and in the accuracy of primordial abundance estimates based on extrapolation of observations and measurements. The optimal region of concordance shown in Fig. 4.3 defines an acceptable range of baryonic densities

$$2.5 \leq \eta_{10} \leq 4, \quad \text{or } 0.009 \leq \Omega_{b0}h^2 \leq 0.015, \quad \text{or } 0.011 \leq \Omega_{b0} \leq 0.06, \tag{4.58}$$

which may be an underestimate of the true uncertainties, but probably not a very grave one; the whole range allowed by residual uncertainties in primordial deuterium would extend the above limits by at most a factor of 2 on each side.

Dark matter is known to exist from its gravitational effects on galactic rotation curves, the velocity dispersions in clusters and groups of galaxies and the departures from the smooth Hubble flow described by Eq. (4.1). This dark matter resides at least partly in the halos of galaxies such as our own, and there may be other dark matter that is more smoothly distributed and thus less easy to detect unambiguously. The smoothed-out cosmological density of luminous matter is deduced from galaxy red-shift surveys which give a luminosity density in blue light

$$l_B \simeq 2 \times 10^8 h L_{B\odot} \text{ Mpc}^{-3}, \tag{4.59}$$

where $L_{B\odot}$ is the blue-light luminosity of the Sun, which can be combined with an expression for the critical density in solar units

$$\rho_{\text{crit},0} = 2.85 \times 10^{11} h^2 M_\odot \text{ Mpc}^{-3} \tag{4.60}$$

to relate Ω_{lum} to a typical mass:light ratio (in solar units)

$$<M_{\text{lum}}/L_B> = 1500 \, h \, \Omega_{\text{lum}}. \tag{4.61}$$

A typical mass:light ratio of 5 solar units for stars and detected gas in galaxies thus contributes only $\Omega_{\text{lum}} = 0.003h^{-1}$ to the smoothed-out mass density corresponding to $\eta_{10} \leq 1$ as shown in Fig. 4.3. The remaining dark baryonic matter may reside in galactic halos, which contribute $\Omega_{\text{halos}} \simeq 0.014h^{-1}$ but need not be mainly baryonic; counts of micro-lensing objects reveal some dark objects with masses somewhere between about

0.1 and 1 times that of the Sun, but apparently not enough to account for the whole mass. It could also be in intergalactic ionized gas associated with the Lyman forest systems, in cold molecular gas or in black holes. The question of the nature (and chemical composition) of baryonic dark matter, raised by considerations of BBNS, is an on-going concern.

Similarly, the dynamical evidence that Ω_0 is at least 0.15 and could indeed be 1, combined with the BBNS arguments, gives a strong indication for the existence of non-baryonic dark matter (NDM). NDM is also invoked in theories of structure and galaxy formation from the gravitational effects of very small density perturbations in the early universe; a popular candidate for this is 'cold dark matter', i.e. some kind of particles which decoupled from radiation very early on and have been cooling non-relativistically ever since. These could be weakly interacting massive particles ('WIMPS') or very tiny particles known as axions. Massive neutrinos ('hot dark matter') may also exist and be involved in structure formation. BBNS considerations are a strong spur, though not the only one, to active laboratory searches for these exotic particles.

Notes to Chapter 4

Excellent descriptions of BBNS theory are given in the classic text:

S. Weinberg, *Gravitation and Cosmology*, Wiley 1972, and in

E.W. Kolb & M.S. Turner, *The Early Universe*, Addison-Wesley Press 1990.

H. Reeves, in F. Sánchez, M. Collados & R. Rebolo (eds.), *Observational and Physical Cosmology*, Cambridge University Press 1990, p. 73, includes a description of the physics of the quark–hadron phase transition.

The nuclear reaction rates and their uncertainties (at the time) are described in detail by

M.S. Smith, L.H. Kawano & R.A. Malaney, *Astrophys. J. Suppl.*, **85**, 219, 1993.

Observational data and their interpretation have been reviewed many times, notably by

A. Boesgaard & G. Steigman, *Ann. Rev. Astr. Astrophys.*, **23**, 319, 1985;

C.J. Copi, D.N. Schramm & M.S. Turner, *Science*, **267**, 192, 1995; and in

P. Crane (ed.), *The Light Element Abundances*, Springer-Verlag 1995.

More detailed discussions of dark matter can be found in:

V. Trimble, *Ann. Rev. Astr. Astrophys.*, **25**, 425, 1987; and in

D. Lynden-Bell & G. Gilmore (eds.), *Baryonic Dark Matter*, Kluwer, Dordrecht, 1990.

The formation of galaxies and large-scale structure according to the Cold Dark Matter theory is described by

G.R. Blumenthal, S.M. Faber, J.L. Primack & M.J. Rees, *Nature*, **311**, 517, 1984.

The value of Ω_0 derived from gravitational dynamics of galaxies inferred from red-shift surveys has been discussed by

M.A. Strauss & J.A. Willik, *Phys. Reports*, **261**, 272, 1995, who consider it to lie somewhere between 0.3 and 1.

Problems

1. Derive Eq. (4.9) by considering a unit mass at the surface of an expanding sphere of uniform density and radius $R(t)$ using Newton's laws supplemented by a repulsive force per unit mass $\Lambda R/3$ and taking kc^2 as an integration constant which can be identified with $-2\times$ the total energy per unit mass in the case when $\Lambda = 0$.

2. Estimate the present-day temperature of a population of WIMPS that decoupled from radiation at a temperature corresponding to their rest-mass of, say, 100 GeV and subsequently cooled like an adiabatic monatomic gas. ($1 \text{ GeV} \equiv 1.16 \times 10^{13} \text{ K}$).

3. Assuming $\eta = 3 \times 10^{-10}$, $T_{\gamma 0} = 2.73$ K, use Saha's equation to find the red-shift where hydrogen is 1 per cent ionized, corresponding to the epoch of last scattering. (The photon density is $20.3\, T^3$ cm^{-3}, the quantum concentration for electrons is $2.42 \times 10^{15}\, T^{3/2}$ cm^{-3} and the ionization potential is 13.6 eV.) Assess the validity of assuming Saha's equation given that the recombination coefficient is of the order of 10^{-12} cm^3 s^{-1}.

4. Find the red-shift of the epoch of mass-density equality between radiation and matter, assuming (a) $\Omega = 1$ (i.e. $\rho_0 = 1.88h^2 \times 10^{-29}$ gm cm^{-3}); (b) $\Omega_0 = 0.1$. Compare with the red-shift deduced in the previous example for the last scattering of the microwave background. The energy density of radiation is $7.56 \times 10^{-15}\, T^4$ erg cm^{-3}, which should be increased by a factor of up to 1.68 to allow for 3 species of massless neutrinos ($g_* = 3.36$ in Eq. 4.22).

5. Assuming a baryon:photon ratio $\eta = 3.16 \times 10^{-10}$ and a neutron:proton ratio $n/p = 0.15$, use Saha's equation (2.47) to calculate the D/H ratio in thermal equilibrium at temperatures of 10^9 and 2×10^9 K respectively, using the following information:

Neutron mass excess	8.07144 MeV	neutron spin 1/2
H atom mass excess	7.28899 MeV	proton spin 1/2
D atom mass excess	13.13591 MeV	deuteron spin 1

$kT = 0.0862\, T_9$ MeV, where T_9 is the temperature in units of 10^9 K;
photon number density is $2.4\pi^{-2}(kT/\hbar c)^3 = 2.0 \times 10^{28}\, T_9^3$ cm^{-3}.
The quantum concentration for protons is $(2\pi m_p kT/h^2)^{3/2} = 5.94 \times 10^{33}\, T_9^{3/2}$ cm^{-3}.
 Compare your answers with the graph in Fig. 4.2.

6. Given a value of η such that $Y_P = 0.24$ for 3 kinds of light neutrinos, find the corresponding value of Y_P for 4 kinds.

7. If the He/H ratio by number of atoms (e.g. as found from emission-line ratios) is y, show that

$$Y = \frac{4y(1-Z)}{1+4y}. \tag{4.62}$$

5 Outline of stellar structure and evolution

5.1 Introduction

A star is a ball of gas held in static or quasi-static equilibrium by the balance between gravity and a pressure gradient. The pressure can in general be supplied by one or more of a hot perfect ionized gas, radiation and a degenerate electron (or neutron) gas, depending on circumstances. For main-sequence stars like the Sun, nuclear reactions maintain stability over long periods and the pressure is predominantly that of a classical gas at water-like densities. For much larger masses (above about $100 M_\odot$), radiation pressure leads to instabilities, while for much smaller masses (below about $0.08 M_\odot$) the central temperature never becomes high enough to ignite hydrogen and the star slowly contracts releasing gravitational energy until halted by degeneracy pressure at densities $\sim 10^3$ gm cm^{-3}; such stars are called 'brown dwarfs'. Below about $10^{-3} M_\odot$ (the mass of Jupiter), ordinary solid-state forces combined with electron degeneracy pressure are able to balance gravity at a water-like density giving a planet rather than a star.

The formation of stars is a complicated process, many aspects of which are still poorly understood although it is observed to happen in dense, dusty molecular clouds. A basic concept is the Jeans instability in a uniform medium: gravitational collapse occurs on length scales $\lambda \geq \lambda_J$ (the 'Jeans length') such that the propagation time for pressure waves λ/c_s exceeds the free-fall time $(G\rho)^{-1/2}$, where c_s is the sound speed and ρ the density. This leads to a critical 'Jeans mass'

$$M_J \sim \left(\frac{\pi k T}{G \mu m_H} \right)^{3/2} \rho^{-1/2} \sim 1000 M_\odot \left(\frac{T}{10\,\mathrm{K}} \right)^{3/2} \left(\frac{\rho}{10^{-24}\,\mathrm{gm\,cm^{-3}}} \right)^{-1/2}, \qquad (5.1)$$

where μ is the mean molecular weight and m_H the mass of a hydrogen atom. The inverse dependence on density leads to the possibility of successive fragmentation in a collapsing cloud, as long as it can cool fast enough to remain isothermal by radiating away the gravitational energy released. Eventually a minimum fragment mass is formed that is optically thick and no longer able to cool within its collapse time scale (Hoyle 1953; Lynden-Bell 1973; Low & Lynden-Bell 1976; Rees 1976); this minimum stellar mass is estimated to be somewhere in the neighbourhood of $0.01 M_\odot$. However, collapse is resisted by magnetic fields, rotation and turbulence. For sufficiently high density or mass, the magnetic resistance may be overcome by gravity, or it may be circumvented more slowly by the process of ambipolar diffusion in which neutral molecules slip through the ions that are coupled to the magnetic field. Mm wave-length observations of molecular clouds show that most stars are formed in transient clusters in dense cloud cores with masses ranging from about 1000 to $1M_\odot$, e.g. the Trapezium cluster in Orion. This has a density of about 5×10^4 stars pc^{-3} or 4×10^{-18} gm cm^{-3}. Such clusters generally have a short lifetime because most of the mass of the cloud core is still in the form of gas which is blown away by radiation and winds from massive stars shortly after their formation. The resulting mass loss causes disruption of the cluster and dispersal of its constituent stars into a stellar association; if the efficiency of star formation is high, however, with less gas remaining to be lost, then a bound cluster can form.

Before a star settles into the more or less spherical configuration that is assumed to exist for most of its life, it has to lose angular momentum. In the early stages there is a disk-like configuration and the material of the disk loses angular momentum by dissipative processes. It is gradually accreted on to the central star, while presumably somewhere along the line and in at least some cases, planets are formed at locations within the disk. It appears that magnetic fields play an essential role in the process, linking the rotating star to the disk and to material expelled from the star in the form of winds and bipolar molecular outflows or jets, collimated along the rotation axis by the magnetic field. As the star grows in mass, it eventually reaches a point where residual circumstellar material is either exhausted or blown away, the proto-star emerges from its dusty cocoon and settles down into a state of quasi-static gravitational contraction; such stars are seen as variables with emission lines in their spectra known as T Tauri stars and related classes. The jets often impinge on nearby clouds of material and shock-excite them producing emission-line spectra of so-called Herbig–Haro objects.

Planets form by accretion of solid materials in the accreting disk and are thus distinguished from stars by their mode of formation as well as by their mass.

5.2 Time-scales and basic equations of stellar structure

5.2.1 Time-scales

Three time-scales are of importance in stellar evolution:

- Dynamical time-scale

$$\tau_{\text{dyn}} \simeq (G\bar{\rho})^{-1/2} \simeq 1 \text{ hour for the Sun,} \tag{5.2}$$

where $\bar{\rho}$ is the mean density. This is associated with gravitational collapse or radial pulsation.

- Thermal or Kelvin–Helmholtz time-scale

$$\tau_{\text{th}} \simeq \frac{GM^2}{RL} \simeq 10^7 \text{yr for the Sun,} \tag{5.3}$$

where R is the radius and L the luminosity. This is the time-scale for quasi-static gravitational contraction and for changes in the interior to propagate outwards to the surface.

- Nuclear time-scale

$$\tau_{\text{nuc}} \simeq \frac{0.007qXMc^2}{L} \simeq 10^{10} \text{ yr for the Sun,} \qquad (5.4)$$

where q is the fraction of the mass (about 10 per cent) occupied by the helium core when the star leaves the main sequence.

Usually

$$\tau_{\text{dyn}} \ll \tau_{\text{th}} \ll \tau_{\text{nuc}} \qquad (5.5)$$

and the star is effectively in hydrostatic equilibrium. Under somewhat less general conditions, the star is also in thermal equilibrium, i.e. the rate of nuclear energy generation in central regions is equal to the luminosity at the surface. When a departure from thermal equilibrium occurs, equilibrium is restored on the thermal time-scale.

5.2.2   Hydrostatic equilibrium

The equation of hydrostatic equilibrium is

$$\frac{dP}{dr} = -\frac{Gm(r)\rho(r)}{r^2}. \qquad (5.6)$$

Using the self-evident relation (in spherical symmetry)

$$\frac{dm}{dr} = 4\pi r^2 \rho, \qquad (5.7)$$

this can be written as

$$\frac{dP}{dm} = -\frac{Gm}{4\pi r^4}. \qquad (5.8)$$

It is convenient to treat all the internal variables as functions of the included mass $m(r)$, rather than of r itself.

From Eq. (5.8), one can get an estimate of the internal pressure of the Sun (mass $M_\odot = 2 \times 10^{33}$ gm; radius $R_\odot = 7 \times 10^{10}$ cm):

$$P_c \simeq P_c - Ps = -\int \frac{dP}{dm} dm > \frac{GM^2}{8\pi R^4} = 4 \times 10^{14} \text{ cgs or } 4 \times 10^8 \text{ atm.} \qquad (5.9)$$

5.2.3 Virial Theorem

Writing Eq. (5.8) in the form

$$4\pi r^3 dP = -\frac{Gm\,dm}{r}$$ (5.10)

and integrating over the star,

$$3\int_c^s V\,dP = -\int \frac{Gm\,dm}{r} = \Omega,$$ (5.11)

the total gravitational potential energy. Integrating by parts,

$$3[PV]_c^s - 3\int_0^{V_s} P\,dV = \Omega.$$ (5.12)

The first term on the left of Eq. (5.12) vanishes in the absence of external pressure, which leads to the Virial Theorem

$$3\int_0^M \left(\frac{P}{\rho}\right) dm + \Omega = 0,$$ (5.13)

which is the same as

$$3(\gamma - 1)U + \Omega = 0,$$ (5.14)

where U is the total internal energy and γ the ratio of specific heats. For a perfect gas, neglecting radiation pressure,

$$P = nkT = \frac{\rho}{\mu m_H} kT,$$ (5.15)

and, for a fully ionized gas, $\gamma = 5/3$. Since (with density increasing inwards)

$$-\Omega > \frac{0.6\,GM^2}{R},$$ (5.16)

$$k\overline{T} \equiv \frac{\int kT\,dm}{M} > \frac{m_H}{5}\frac{\mu M}{R},$$ (5.17)

which makes

$$\overline{T} > 5 \times 10^6 \mu\,\text{K} \quad \text{for the Sun.}$$ (5.18)

For a fully ionized gas,

$$\frac{1}{\mu} = 2X + \frac{3}{4}Y + \frac{1}{2}Z = \frac{1}{2} + \frac{3}{2}X + \frac{1}{4}Y < 2,$$ (5.19)

so the mean temperature inside the Sun is more than 2.5×10^6 K.

In the absence of nuclear energy sources, a star contracts on a thermal time-scale and radiates energy at the expense of gravitational potential energy. Since, by the Virial Theorem, the total energy

$$E = U + \Omega = -U(3\gamma - 4),$$ (5.20)

U, and hence the temperature, rises as the star contracts, as is also obvious from Eq. (5.17), as long as $\gamma > 4/3$. An ideal gaseous star can't cool!

5.2.4   Energy transport

There are three ways in which energy can be transported outwards in a star: radiative transfer, convection and conduction. The mean free path for radiation, $\lambda \equiv 1/\kappa\rho \simeq 1$ cm, greatly exceeds that for conduction ($\lambda \ll 1$ cm) except in degenerate material, so that radiative transfer dominates in those regions where the corresponding temperature gradient is stable to convection. The equation of radiative transfer can be treated in the plane-parallel diffusion approximation that was applied to stellar atmospheres in Chapter 3. From Eqs. (3.14) to (3.17), the radiative luminosity $l(r)$ at radius r, mass coordinate $m(r)$, is given by

$$\frac{l(r)}{4\pi r^2} = 4\pi H = -\frac{4}{3}\frac{1}{\kappa\rho}\frac{d}{dr}(\sigma T^4),$$ (5.21)

or, using the energy density constant $a \equiv 4\sigma/c$,

$$l(r) = -\frac{4}{3}\pi ac\frac{r^2}{\kappa\rho}\frac{dT^4}{dr}$$ (5.22)

or

$$\frac{dT}{dm} = -\frac{3\kappa l}{64\pi^2 acr^4 T^3}.$$ (5.23)

Convection occurs if $|dT/dr|$ from Eq. (5.23) is so steep that a rising/falling piece of gas expanding/contracting adiabatically under the ambient pressure continues to move owing to buoyancy forces, i.e. if (for constant chemical composition)

$$\nabla \equiv \frac{d\log T}{d\log P} > \frac{\gamma - 1}{\gamma}.$$ (5.24)

At sufficiently high densities (e.g. cores of upper main-sequence stars), the $>$ sign becomes an equality (adiabatic stratification), but at lower densities (e.g. envelopes of the Sun and cooler stars) an exact calculation is very difficult and in most models a crude approximation based on mixing-length theory is used. In a situation where the chemical composition changes with depth, Eq. (5.24) (known as the Schwarzschild criterion) needs to be replaced by more complicated considerations.

5.2.5   Opacity sources

The effective opacity is given by taking the wavelength-dependent opacity of the material at the relevant temperature and density and forming the (harmonic) Rosseland mean:

$$\frac{1}{\kappa} = \int \frac{1}{\kappa_\nu}\frac{dB_\nu}{dT}d\nu \bigg/ \int \frac{dB_\nu}{dT}d\nu.$$ (5.25)

This is a harmonic mean because it is really the mean free path that is relevant. At high temperatures ($> 10^5$ K), the main sources of opacity and approximate formulae for them (the first two originally due to H. Kramers) are:

- Bound–free transitions, due to photoionization of H-like ions of oxygen etc.:

$$\kappa_{BF} \simeq k_1 \rho Z (1 + X) T^{-3.5}; \tag{5.26}$$

- Free–free transitions (*Bremsstrahlung*):

$$\kappa_{FF} \simeq k_2 \rho (1 + X) T^{-3.5}; \quad \text{and} \tag{5.27}$$

- Thomson scattering from free electrons

$$\kappa_{es} = 0.2(1 + X) \text{ cm}^2\text{gm}^{-1}. \tag{5.28}$$

The $(1 + X)$ in all these expressions comes from the number of free electrons per atomic mass unit, usually expressed by the molecular weight per electron

$$\frac{1}{Y_e} \equiv \mu_e = \frac{2}{1 + X}. \tag{5.29}$$

At the lower temperatures of stellar envelopes and atmospheres, neutral hydrogen, H^- and bound–bound transitions from absorption lines make a significant contribution and the above formulae do not apply. In the Sun, the opacity is of the order of 1 cm^2 gm^{-1} in the photosphere and the deep interior, but rises to a peak of the order of 10^5 cm^2 gm^{-1} at a temperature of about 30 000 K in the outer envelope.

5.2.6 Energy generation

The final equation governing stellar structure is that for nuclear energy generation:

$$\frac{dl}{dm} = \epsilon - \epsilon_\nu - \left\{ \frac{\partial u}{\partial t} + P \frac{\partial}{\partial t} \left(\frac{1}{\rho} \right) \right\} \tag{5.30}$$

$$\simeq \epsilon - \epsilon_\nu - \frac{3}{2} \rho^{2/3} \frac{\partial}{\partial t} (P \rho^{-5/3}). \tag{5.31}$$

Here ϵ represents both nuclear energy generation and any changes in gravitational energy, while ϵ_ν represents neutrino losses. Neutrino losses arise either from nuclear reactions or from processes that take place anyway at sufficiently high densities and temperatures as a result of direct $e - \nu$ coupling via the weak interaction, and they act as a local energy sink because the neutrinos instantly escape from the star[1]. The last term, which is usually negligible except in gravitational contraction phases, represents changes in thermal energy and $P dV$ work. The second version of the equation applies to a perfect monatomic gas with no radiation pressure.

---

[1] Except in the dense collapsing core of a massive supernova.

5.3   **Homology transformation**

The differential equations (5.8), (5.21) and (5.30), together with the equation of state and suitable boundary conditions at the stellar centre ($l = r = m = 0$) and surface ($P_s \simeq T_s \simeq 0$), fix the structure of a star of given mass and homogeneous chemical composition in radiative equilibrium with negligible radiation pressure.[2] Over limited regions of parameter space, in which κ and ϵ can be represented by simple power laws, the structure of such stars is homologous, i.e. the dimensionless numbers $r(m)/R$, $l(m)/L$, $P(m)/P_c$ etc. become fixed functions of the dimensionless mass coordinate m/M. This in turn implies that the derivatives in the above equations can be replaced (apart from constants that remain the same at corresponding points in different stars) by P_c/M, T_c^4/M etc, and the equations used to estimate how global quantities such as luminosity and radius scale with the mass, mean molecular weight etc.

To begin with, we assume that the opacity scales as

$$\kappa = \kappa_0 \, \rho^r \, T^{-t}. \tag{5.32}$$

From radiative equilibrium, Eq. (5.23), and hydrostatic equilibrium with the ideal-gas equation of state Eq. (5.15),

$$T_c^{t+4} \propto \kappa_0 M^{r+1} L / R^{3r+4} \propto (\mu M / R)^{t+4}, \tag{5.33}$$

so that

$$L \propto \kappa_0^{-1} \mu^{t+4} M^{t-r+3} R^{3r-t} \tag{5.34}$$

$$\propto \kappa_0^{-1} \mu^4 M^3 \text{ (e.s.) or } \kappa_0^{-1} \mu^{15/2} M^{11/2} R^{-1/2} \text{ (Kramers)}. \tag{5.35}$$

Thus, without having yet made any appeal to the law of energy generation, one has a mass–luminosity–radius relation. This is independent of the radius for electron-scattering opacity (which dominates for stellar masses greater than about $3M_\odot$) and only slowly dependent on it for Kramers-type opacity. The system of equations is closed by the law of energy generation, which fixes the radius and hence the effective temperature. We assume (cf. Section 2.5)

$$L \propto \epsilon_0 M \rho \, T_c^\nu, \tag{5.36}$$

where $\nu \simeq 4$ for the proton-proton chain (dominating in low-mass stars $\leq 1.1 M_\odot$) and $\simeq 17$ for the CNO cycle (dominating in higher-mass stars and in later stages of evolution). Replacing T_c by $\mu M / R$, we have

$$L \propto \epsilon_0 \mu^\nu M^{\nu+2} / R^{\nu+3}. \tag{5.37}$$

L or M can then be eliminated from Eqs. (5.35) and (5.37), giving R, and hence T_{eff}, as functions of M or L respectively. The general equations become rather messy at this point, so only two selected sets of numerical results will be given, both assuming

---

[2] For an estimate of radiation pressure, see the description of Eddington's Standard Model in Appendix 4.

Table 5.1. *Homology relations for main-sequence stars*

	Case (a) $r = 1,\ t = 3.5,\ v = 17$	Case(b) $r = t = 0,\ v = 17$	
$R(M) \propto$	$(\epsilon_0 \kappa_0)^{2/39} \mu^{19/39} M^{9/13}$	$(\epsilon_0 \kappa_0)^{1/20} \mu^{13/20} M^{4/5}$	
$L(M) \propto$	$\epsilon_0^{-1/39} \kappa_0^{-40/39} \mu^{283/39} M^{67/13}$	$\kappa_0^{-1} \mu^4 M^3$	$(\forall v)$
$M(L) \propto$	$\epsilon_0^{1/201} \kappa_0^{40/201} \mu^{-283/201} L^{13/67}$	$\kappa_0^{1/3} \mu^{-4/3} L^{1/3}$	$(\forall v)$
$\tau_{\rm ms}(M) \propto XM/L \propto$	$\epsilon_0^{0.026} \kappa_0^{1.026} \mu^{-7.26} X M^{-4.15}$	$\kappa_0 \mu^{-4} X M^{-2}$	$(\forall v)$
$\tau_{\rm ms}(L) \propto XM/L \propto$	$\epsilon_0^{0.005} \kappa_0^{0.20} \mu^{-1.41} X L^{-0.81}$	$\kappa_0^{1/3} \mu^{-4/3} X L^{-2/3} (\forall v)$	
$L \propto R^2 T_{\rm eff}^{\,4} \propto$	$\epsilon_0^{0.15} \kappa_0^{0.52} \mu^{-1.33} T_{\rm eff}^{\,5.5}$	$\epsilon_0^{0.21} \kappa_0^{1.36} \mu^{-1.79} T_{\rm eff}^{\,8.6}$	

energy generation by the CNO cycle with $v = 17$, Case (a) for Kramers opacity and Case (b) for electron scattering (Table 5.1).

The homology relations give a rough guide to the dependence of luminosity, effective temperature and main-sequence lifetime $\tau_{\rm ms}$ on the star's mass and chemical composition. The luminosity increases as a high power of the mass and mean molecular weight, which latter depends on the helium (or hydrogen) content according to

$$\mu = \frac{4}{3 + 5X - Z} \simeq \frac{4}{3 + 5X} \tag{5.38}$$

since $Z \leq 0.02$ (cf. Eq. 5.19). X typically varies from about 0.7 in Population I stars like the Sun to about 0.76 ($= 1 - Y_P$) in Population II stars (*e.g.* metal-weak globular clusters), resulting in a change in μ from 0.62 to 0.59, i.e. a reduction of 5 per cent, which in itself leads to a reduction of about 30 per cent in the luminosity at a given mass. This, however, is more than compensated by the reduction in opacity, which is proportional to Z as long as bound–free processes dominate, *i.e.* as long as Z does not become too small. Thus the lifetime at a given mass is lower for Population II stars, but the lifetime at a given main-sequence luminosity is much less sensitive to chemical composition.

The last line of the table gives main-sequence relations. The effect of decreasing metallicity is to lower the luminosity at a given $T_{\rm eff}$, since $\kappa_0 \propto Z$ (if not too small) in Case (a) and ϵ_0 is so in Case (b). On the other hand, this is partly compensated by the aforementioned reduction in μ from the helium abundance, leading to a near-balance among Population I stars with metallicities differing by factors of a few. At very low metallicity ($< 0.1 Z_\odot$ or so), κ_0 becomes constant from free–free processes and electron scattering, and μ likewise settles down to its minimum value, leading to a subdwarf main sequence that lies below the Population I main sequence (cf. Fig. 4.6), although stars at a given mass are actually more luminous than their Population I counterparts. Numerical results for main-sequence stars with different masses are given in Chapter 7.

5.4   Degeneracy, white dwarfs and neutron stars

5.4.1   Introduction

Examination of the Fermi–Dirac distribution function Eq. (2.40) shows that the condition for applicability of the ideal-gas distribution to electron velocities is

$$(m_e c^2 - \mu)/kT \gg 0 \tag{5.39}$$

whereas under the opposite condition

$$(m_e c^2 - \mu)/kT \ll 0 \tag{5.40}$$

all possible momentum states are filled and electron energies are fixed by the Pauli exclusion principle rather than by the temperature; this is the state known as (complete) electron degeneracy. From Eq. (2.43) for the chemical potential, one can therefore define a degeneracy parameter

$$\psi \equiv \ln \frac{n_e}{2n_Q} = \ln \frac{1}{2} n_e \left(\frac{2\pi\hbar^2}{m_e kT} \right)^{3/2} = \ln \left(1.2 \times 10^{-4} \rho T_8^{-3/2}/\mu_e \right), \tag{5.41}$$

where T_8 is the temperature in units of 10^8 K. As a star consumes its hydrogen fuel in central regions, the core contracts, roughly following the homology law (cf. Eq. 5.17)

$$T \propto M^{2/3} \rho^{1/3}, \tag{5.42}$$

so that ψ tends to increase, roughly as $\frac{1}{2} \ln \rho$. At what stage degeneracy is actually reached depends mainly on the star's mass: high-mass stars may not reach it until the pre-supernova stage, whereas stars with mass of the order of $0.1 M_\odot$ or less do so even before hydrogen-burning can set in. Stars with masses in the range 1 to $8 M_\odot$ or so develop degenerate cores after leaving the main sequence; subsequent mass-loss episodes, especially along the AGB and in the planetary nebula stage, eventually leave this degenerate core exposed and it cools down and becomes a white dwarf.

5.4.2   Equations of state for degenerate matter

When thermal energy is negligible, cells in phase space are uniformly occupied up to the Fermi momentum p_F, given by

$$n_e = \int_0^{p_F} 8\pi p^2 dp/h^3, \text{ or } p_F = h (3n_e/8\pi)^{1/3}. \tag{5.43}$$

For a non-relativistic electron gas ($p_F \ll m_e c$), the pressure is given by

$$P \simeq P_e = \frac{1}{3} n_e <pv> = \frac{1}{3} \frac{n_e}{m_e} <p^2> \tag{5.44}$$

$$= \frac{1}{5} \frac{n_e}{m_e} p_F^2 = \frac{1}{20} \frac{h^2}{m_e} \left(\frac{3}{\pi}\right)^{2/3} n_e^{5/3}. \tag{5.45}$$

Because at this stage the material has usually been processed into helium and heavier elements, electrons outnumber nuclei and make the dominant contribution to the pressure, the more so when their energy from degeneracy greatly exceeds kT; it is the small electron mass that ensures that electrons become degenerate long before nucleons. The electron density is given by

$$n_e = \frac{\rho}{\mu_e m_H} = \frac{1}{2}(1+X)\rho/m_H = \frac{1}{2}\rho/m_H \tag{5.46}$$

(cf. Eq. 5.29), since $X \to 0$.

There is a critical density ρ_{crit} above which the electrons become relativistic, i.e.

$$p_F^2/2m_e = m_e c^2, \text{ or } p_F = m_e c \sqrt{2}, \tag{5.47}$$

whence

$$\rho_{\mathrm{crit}} = \frac{8\pi}{3} \mu_e m_H (m_e c \sqrt{2}/h)^3 = \frac{6 \times 10^6}{1+X} \text{ gm cm}^{-3}. \tag{5.48}$$

For fully relativistic degeneracy ($\rho \gg \rho_{\mathrm{crit}}$), the pressure is given by

$$P = P_e = \frac{1}{3} n_e c <p> = \frac{hc}{8} (3/\pi)^{1/3} n_e^{4/3}. \tag{5.49}$$

In each case, the pressure is mainly fixed by the density and quite insensitive to temperature. A consequence of this is that, when a new nuclear reaction (e.g. helium burning) sets in in degenerate material, the additional heat cannot be taken up (as it would be in an ideal gas) by expansion and there is a thermal runaway. The temperature rises very strongly until it becomes high enough for the degeneracy to be removed, and the star adjusts to a new structure with lower central density.

5.4.3 Structure of white dwarfs

Once again, using the idea of a homology transformation, we can approximate the hydrostatic equation (5.8) replacing dP/dm with P_c/M, leading to

$$P_c \simeq GM^{2/3} \bar{\rho}^{4/3}. \tag{5.50}$$

Expressing P_c in terms of $\bar{\rho}$ using the non-relativistic expression Eq. (5.45), together with Eq. (5.46),

$$M(\bar{\rho}) \simeq \left(\frac{2\hbar^2}{Gm_e}\right)^{3/2} (m_H \mu_e)^{-5/2} \bar{\rho}^{1/2} \tag{5.51}$$

$$\simeq 0.1 \left(\frac{m_e c^2}{Gm_H \mu_e}\right)^{3/2} \rho_{\text{crit}}^{-1} \bar{\rho}^{1/2}. \tag{5.52}$$

This leads to a mass-density or mass-radius relation

$$R \propto M^{-1/3} \mu_e^{-5/3} \simeq \text{ a few } \times 1000 \, \text{km}. \tag{5.53}$$

White dwarfs are formed hot and gradually cool at constant radius. As the mass is increased, the star becomes squashed down until it is highly relativistic and very small.

In the fully relativistic case, Eqs. (5.50), (5.49) and (5.46) lead to a unique mass:

$$M_{\text{Ch}} \simeq \left(\frac{\hbar c}{G}\right)^{3/2} \left(\frac{\rho_c}{\bar{\rho}}\right)^2 (\mu_e m_H)^{-2} \simeq \frac{2}{\mu_e^2} \left(\frac{\rho_c}{\bar{\rho}}\right)^2 M_\odot. \tag{5.54}$$

This is the Chandrasekhar–Landau limiting mass for white dwarfs, whose actual value (derivable from the theory of polytropic stars; see Appendix 4) is

$$M_{\text{Ch}} = 5.8 M_\odot / \mu_e^2 = 1.46 M_\odot \tag{5.55}$$

for a He or CO white dwarf. A white dwarf cannot have a mass greater than this: larger masses either collapse into a neutron star or black hole, or become incinerated, which is the favoured mechanism for type Ia supernovae.

An interesting feature of Eq. (5.54) is that it can be re-written

$$M_{\text{Ch}} \simeq \left(\frac{e^2}{\alpha Gm_H^2}\right)^{3/2} m_H / \mu_e^2 = (\alpha_G \alpha)^{-3/2} m_H / \mu_e^2, \tag{5.56}$$

where α is the well-known fine-structure constant $1/137$ and α_G is the corresponding 'gravitational fine-structure constant' $Gm_H^2 / e^2 = 8.0 \times 10^{-37}$, where e is the electronic charge. The number of nucleons in the Chandrasekhar limiting mass is thus essentially the $3/2$ power of the famous large cosmological numbers of order 2×10^{38} which fascinated Eddington and Dirac. Arguments can also be made that the number of nucleons in an ordinary main-sequence star, which avoids degeneracy from being too small and also avoids disruption by radiation pressure from being too large, must also be of this order, which is set by a few basic dimensionless physical constants (Carr & Rees 1979).

5.4.4   Neutron stars

The chemical potential of highly degenerate electrons is essentially equal to the Fermi energy cp_F, since virtually all levels below this are filled. At densities that are so high that the Fermi energy exceeds the mass-energy difference between protons and neutrons (i.e. $\rho > \sim 10^7 \mu_e$ gm cm^{-3} for a difference in binding energy of 1 MeV), the reaction

$$p + e^- \rightarrow n + \nu_e \qquad (5.57)$$

goes forward to the right, i.e. electrons are crushed on to protons to make neutron-rich nuclei by inverse β-decays, e.g. in the collapsing core of a massive supernova. Equilibrium (which may be rapidly brought about at high densities by pycnonuclear reactions) then favours increasingly neutron-rich nuclei, e.g. $^{118}_{36}$Kr$_{82}$ at a density of 4×10^{11} gm cm^{-3}. At this point, however, the binding energy for additional neutrons vanishes, so that at still higher densities there is a release of free neutrons (the effect known as neutron drip; cf. Fig. 6.9), which dominate when the density exceeds about 10^{13} gm cm^{-3}. The neutrons themselves become degenerate at densities exceeding 10^{15} gm cm^{-3} (a few times nuclear density), corresponding to a radius of about 10 km. Neutron stars have a maximum mass corresponding to the Chandrasekhar mass for white dwarfs and of the same order of magnitude, since Eq. (5.54) is independent of the mass of the degenerate particles and the replacement of $\mu_e = 2$ for white dwarfs by $\mu_n = 1$ for neutron stars is approximately compensated by general-relativistic effects. However, the equation of state, and consequently the exact limiting mass, are uncertain because of uncertainties in the properties of such extreme nuclei and the possible formation of meson condensates, quark matter or other exotic phases. The limiting mass is believed to lie somewhere in the neighbourhood of $2M_\odot$ or less. Neutron stars are observed in the form of radio pulsars and of components of binary stellar X-ray sources, and are thought to be the typical remnants of at least a subset of massive supernovae energised by core collapse (Type II and related classes).

5.5 Hayashi effect

In cool stars ($T_{\text{eff}} \leq 6000$ K), neutral hydrogen becomes positively ionised below the photosphere, reducing the adiabatic gradient (from low γ) and increasing the radiative gradient through the growth of opacity (mainly from H$^-$ formed from the extra free electrons). Thus the surface convection zone becomes deep and the outer boundary condition grows more significant: the approximations of Section 5.3 ($P_s \simeq T_s \simeq 0$) are no longer adequate. At some effective temperature T_H the star is fully convective down to the centre and for $T_{\text{eff}} < T_H$ stars cannot have a stable structure. This effect is important both in pre-main-sequence contraction stages (see Fig. 5.1) and in post-main-sequence evolution, since it limits the region of the HR diagram that a star can traverse, independently of its nuclear history.

The Hayashi track (Hayashi 1961, 1966) is the locus of fully convective stars in the HR diagram, dependent on the mass, luminosity and chemical composition. However, it is mainly the surface properties — luminosity and radius or luminosity and effective temperature — that govern the entire stellar structure by fixing the adiabatic constant. A sketch derivation of its properties is the following.

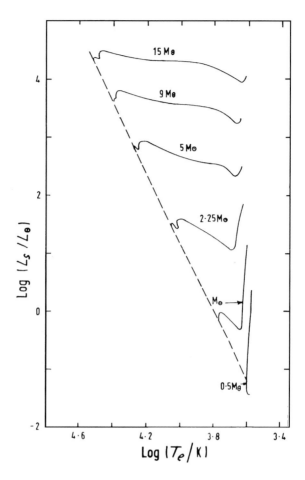

Fig. 5.1. Pre-main sequence contraction of stars with different masses. Adapted from Iben (1965).

Throughout the star, there is a single adiabat:

$$P = K\rho^{5/3} \propto K^{-3/2} T^{5/2}. \tag{5.58}$$

In the interior, we can apply the hydrostatic equation (5.8)

$$P_{\text{int}} \propto M^2/R^4, \tag{5.59}$$

which relates the adiabatic constant to the mass and radius:

$$K \propto \frac{M^2}{R^4} \left(\frac{M}{R^3}\right)^{-5/3} \propto M^{1/3}R, \tag{5.60}$$

whence throughout the star (and in particular at the surface where $T = T_{\text{eff}}$)

$$P \propto M^{-1/2}R^{-3/2} T^{5/2}. \tag{5.61}$$

At the surface, there is another hydrostatic condition:

$$dP = -g\rho dz \quad \text{and} \quad d\tau = -\kappa\rho dz, \tag{5.62}$$

whence

$$\frac{dP}{d\tau} = \frac{g}{\kappa}, \tag{5.63}$$

or

$$P_{\text{ph}} \simeq \frac{g}{\kappa} \propto \frac{M}{R^2} P_{\text{ph}}^{-a} T_{\text{eff}}^{-b}, \tag{5.64}$$

say, where $a \simeq 1$ and $b \simeq 3$ and we have taken $\tau \sim 1$. Equating (5.61) with $T = T_{\text{eff}}$ to (5.64) and replacing R by $L^{1/2}/T_{\text{eff}}^2$, one finally obtains

$$T_{\text{eff}} \propto M^{\frac{a+3}{11a+2b+3}} L^{\frac{3a-1}{22a+4b+6}} \quad \text{or} \quad M^{0.2} L^{0.05}, \tag{5.65}$$

which leads to an almost vertical line in the HR diagram with $T_{\text{eff}} \sim 4000$ K for solar metallicity (Fig. 5.1).

A proto-star resulting from condensation of a fragmented gas cloud thus first reaches quasi-hydrostatic equilibrium somewhere on the Hayashi track, along which it then contracts gravitationally at nearly constant effective temperature on the thermal timescale, Eq. (5.3). Meanwhile its internal temperature increases in accordance with the Virial Theorem and minor nuclear reactions like deuterium burning are ignited, but have little effect (apart from destruction of D and of ^{6}Li and ^{7}Li in cooler stars) because the relevant abundances are so low. Eventually, the interior ceases to be convective and the star then contracts radiatively, with its luminosity slightly rising again in accordance with the homology relation (5.35), until hydrogen-burning is ignited at the centre; this halts further contraction and the star is now on the zero-age main sequence (ZAMS).

5.6 Hydrogen-burning

5.6.1 *pp* chains and the solar neutrino problem

The first major set of nuclear reactions in stellar evolution involves hydrogen-burning through the *pp* chains and the CN cycle or CNO bi-cycle, which liberate 6.68 MeV of energy per proton minus neutrino losses (2 neutrinos are emitted for each ^{4}He nucleus synthesised). The first two reactions of the *pp* chains are

$$^1\text{H} + {}^1\text{H} \rightarrow {}^2\text{D} + e^+ + \nu + 0.42\,\text{MeV}, \tag{5.66}$$

with a further 1.42 MeV released by annihilation of the positron; and

$$^2\text{D} + {}^1\text{H} \rightarrow {}^3\text{He} + \gamma\,(5.49\,\text{MeV}). \tag{5.67}$$

Reaction (5.66) involves the weak interaction through the β^+ decay and is accordingly exceedingly slow. The condition for an allowed decay is that the two leptons (e^+ and ν) are emitted with zero orbital angular momentum ℓ and with spins that are either parallel (Gamow–Teller transition) or anti-parallel (Fermi transition). Since the nucleons in the

Table 5.2. *Reactions of the pp chains*

		$p(p, e^+v)d$		
		$d(p, \gamma)^3\mathrm{He}$		
$^3\mathrm{He}(^3\mathrm{He},2p)^4\mathrm{He}$			$^3\mathrm{He}(\alpha, \gamma)^7\mathrm{Be}$	
		$^7\mathrm{Be}(e^-, v)^7\mathrm{Li}$		$^7\mathrm{Be}(p, \gamma)^8\mathrm{B}$
		$^7\mathrm{Li}(p, \alpha)^4\mathrm{He}$		$^8\mathrm{B}(e^+, v)^8\mathrm{Be}^*$
				$^8\mathrm{Be}^*(\alpha)^4\mathrm{He}$
$pp - 1$ (86%)		$pp - 2$ (14%)		$pp - 3$ (0.02%)
$Q_{\mathrm{eff}} = 26.20$ Mev		$Q_{\mathrm{eff}} = 25.26$ Mev		$Q_{\mathrm{eff}} = 19.17$ Mev
(2.0% loss)		(4.0% loss)		(28.3% loss)

Data from C. Rolfs & W.H. Rodney, *Cauldrons in the Cosmos*, University of Chicago Press 1988, p. 354.

^{2}D nucleus have $\ell = 0$, and ℓ is unchanged by the allowed β-decay, the same applies to the protons, which thus have to scatter in an *s*-wave and have opposite spins by virtue of the Pauli Exclusion Principle. However, the deuteron spin is 1, so the proton that decays into a neutron has to flip its spin and the balancing spin is carried off by the two leptons. Their spins then have to be parallel, making this a pure Gamow–Teller transition, and the cross-section thus depends on the GT axial vector coupling constant C_A (deduced from laboratory measurements of related decays of neutrons, ^{3}H and ^{6}He) as well as on spin and spatial matrix elements and phase-space and barrier-penetration factors. The cross-section at $E_p(\mathrm{lab}) = 1$ MeV is only 10^{-47} cm^2, far too small to be measured directly. The calculated value of $S(E_0')$ in the interior of the Sun is about 5×10^{-22} keV barn, so that from Eq. (2.59) with $\tau = 13.7$ the rate coefficient is

$$<\sigma v> \simeq 7.6 \times 10^{-44} \, \mathrm{cm}^3 \, \mathrm{s}^{-1}. \tag{5.68}$$

At a density of 150 gm cm^{-3} and $X \simeq 0.5$, typical of the centre of the Sun, i.e. $n_p = 6 \times 10^{25}$ cm^{-3}, this leads to a hydrogen-burning lifetime

$$\tau_{pp} = (n_p <\sigma v>)^{-1} \simeq 10^{10} \, \mathrm{yr}. \tag{5.69}$$

The remaining reactions are much faster, e.g. the time for (5.67) is only 1.6 s, leading to a steady-state D/H ratio of the order of 10^{-18}.

After reaction (5.67), there are two alternatives. 86 per cent of the time (in the case of the Sun), there is the final link of the $pp - 1$ chain

$$^3\mathrm{He} + {}^3\mathrm{He} \rightarrow {}^4\mathrm{He} + 2^1\mathrm{H} + 12.86 \, \mathrm{MeV}, \tag{5.70}$$

which gives a steady-state ^{3}He/H ratio $\sim 10^{-5}$ at the centre, although a larger ^{3}He abundance is built up some way out. The remainder of the time one has the $pp - 2$ and $pp - 3$ chains summarised in Table 5.2, with associated neutrino losses.

In recent years, the energetic neutrinos from $pp - 3$, and the less energetic neutrinos from other reactions in the chains (see Fig. 5.2), have been detected in direct exper-

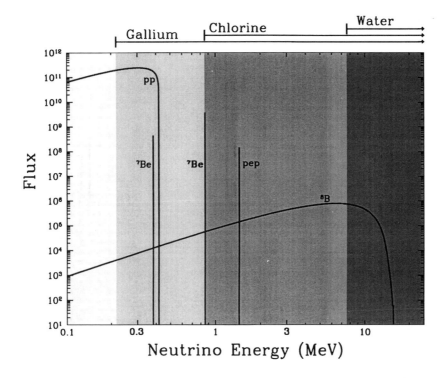

Fig. 5.2. Energy spectrum of solar neutrinos predicted from a standard solar model (e.g. Bahcall *et al.* 1982), omitting the undetectably small flux due to the CNO cycle. Fluxes are in units of cm^{-2} s^{-1} MeV^{-1} for continuum sources and cm^{-2} s^{-1} for line sources. Detectors appropriate in various energy ranges are shown above the graph. Courtesy J.N. Bahcall.

Table 5.3. *Solar neutrino fluxes expected and detected in different experiments*

Detector	Source	Exp. flux (snu)	Obs. flux (snu)
^{37}Cl	^7Be	1.2	
	^8B	5.5	
	Total	7.2 ± 1 :	2.2 ± 0.3
Kamiokande II	^8B	5.1 ± 1 :	2.7 ± 0.3
Gallex,	*pp*	71	
SAGE II	^7Be	33	
	^8B	12	
	Total	127 ± 5	$87 \pm 16 \pm 8$

After D.O. Gough, *Phil. Trans. R. Soc. London A*, **346**, 39 (1994). 1 snu ('solar neutrino unit') $\equiv 10^{-36}$ events per target atom per second.

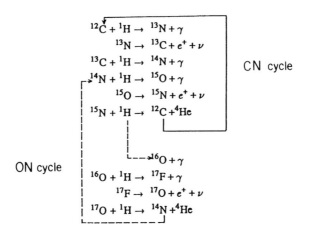

Fig. 5.3. The CNO bi-cycle. Adapted from R. Kippenhahn & A. Weigert, *Stellar Structure and Evolution*, Springer-Verlag 1990.

iments, but with somewhat lower fluxes than predicted (Table 5.3). The reason for the discrepancy could lie in the solar model, in nuclear reaction rates or in neutrino physics, e.g. if neutrinos were to have a small mass which enabled them to oscillate between ν_e and ν_μ, say. At the time of writing, the latter type of explanation is the most widely favoured.

5.6.2 The CNO cycles

At somewhat higher temperatures than inside the Sun, hydrogen burning takes place more rapidly by a catalytic cycle involving the CNO elements (provided, of course, that at least one of these is present, i.e. we are not dealing with a star of the first generation). The reactions involved are shown in Fig. 5.3. The main part of the cycle involves C and N, while the ON cycle usually contributes little energy, but becomes significant at high temperatures when it can lead to destruction of oxygen and enhancement of the ratio $^{17}O/^{16}O$ compared to the Solar-System value of 4×10^{-4}. Owing to the significant Coulomb barriers, the CN cycle is highly temperature-sensitive (see Section 2.5), its rate being fixed by the slowest reaction $^{14}N\,(p,\gamma)^{15}O$ (Coulomb barrier plus electromagnetic interaction). Fig. 5.4 shows the temperature dependence of the *pp* chain and CNO cycle.

At temperatures of several $\times 10^7$ K, the CNO cycle breaks out into third and fourth branches leading to destruction of ^{18}O relative to its solar-system abundance ($^{18}O/^{17}O$ = 5.5) and to higher H-burning cycles, the neon–sodium and magnesium–aluminium cycles (see Fig. 5.5). There are uncertainties in relevant nuclear reaction rates, but spectroscopic observations of enhanced features of sodium and aluminium in red giants in some globular clusters (though not in Population II field stars) provide evidence that these latter cycles can actually operate in later stages of hydrogen shell burning.

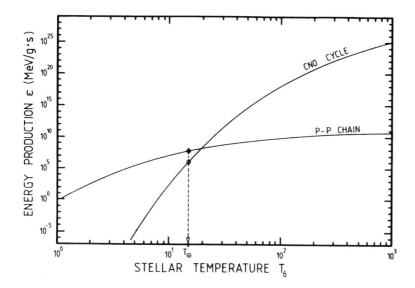

Fig. 5.4. Dependence on temperature of energy production rates by the *pp* chain and CNO cycle (for solar abundances). After Iben (1965) and C.E. Rolfs & W.S. Rodney (1988), *Cauldrons in the Cosmos*, University of Chicago Press, p. 337. ©1988 by the University of Chicago. Courtesy Claus Rolfs.

At still higher temperatures ($> 10^8$ K), e.g. in novae, nuclear reactions become so fast that there is not enough time for β-decays such as that of ^{13}N (which take times in the range 100 to 1000 s) and one then has a so-called fast CNO process leading into hot Ne-Na and Mg-Al cycles known as the rp process.

Because of the differing rates of the component reactions, the CNO cycle modifies the initial composition of the material, which typically has O > C > N, making more ^{14}N (and more ^{13}C relative to ^{12}C) at the expense of carbon and (at the higher temperatures) oxygen. Thus in H-burning zones the CN mixture is rapidly modified, and the CNO mixture somewhat more slowly modified, to reach steady-state ratios,[1] e.g. at $T = 3 \times 10^7$ K the steady-state percentages of CNO isotopes, compared to solar-system values, are as follows:

	^{12}C	^{13}C	^{14}N	^{15}N	^{16}O	^{17}O
CNO 'eq.' 3×10^7 K	1.6	0.39	97	3×10^{-3}	1.3	4.3×10^{-4}
Solar System	28	0.30	8.3	3×10^{-2}	64	8.3×10^{-3}

and virtually all the initial C and N (and at higher temperatures O) in the burning zone ends up as ^{14}N. Subsequent dredge-up episodes to the surface lead to minor or major

---

[1] These are often referred to as 'equilibrium' ratios, but this is not strictly correct because cyclic processes are involved.

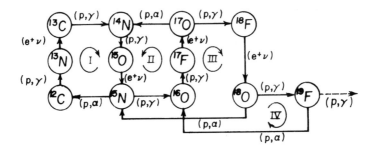

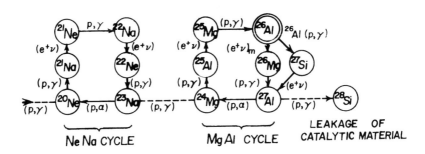

Fig. 5.5. CNO tri and quadricycles and Ne-Na and Mg-Al cycles. Adapted from C.E. Rolfs & W.S. Rodney, *Cauldrons in the Cosmos*, University of Chicago Press 1988.

corresponding changes in the surface composition of red giants. In some cases, e.g. WN stars, it appears that the exposed layers have reached complete CNO 'equilibrium'. The sum of C, N and O abundances remains unchanged, however.

Owing to its steep temperature dependence, the CNO cycle energy source in upper main-sequence stars is highly concentrated in a small central core which is convective because a finite surface area is required to carry radiation through a fixed temperature gradient (cf. Eq. 5.22). Consequently, in stars bigger than about $1.1M_\odot$, the chemical profile $X(m)$ resembles a step function, and when all the hydrogen in the core is used up one has an isothermal helium core surrounded by a hydrogen-burning shell. Less massive stars have a radiative core and a convective envelope (e.g. the outer 30 per cent of the solar radius), while stars with $\leq 0.25M_\odot$ are fully convective (apart from a thin radiative skin at the photosphere) in accordance with the Hayashi effect.

5.7   Evolution from the main sequence; the Schönberg–Chandrasekhar limit

As hydrogen becomes used up in the core of a star, a discontinuity in molecular weight develops, which causes the star to move upwards from the ZAMS in the HR

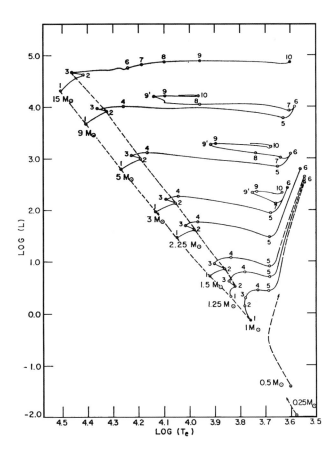

Fig. 5.6. Evolutionary tracks for solar-metallicity stars with different masses in the HR diagram. (Luminosities are in solar units.) Points labelled 1 define the ZAMS and points labelled 2 the terminal main sequence (TAMS), the point where central hydrogen is exhausted. Points marked 3 show the onset of shell hydrogen burning and the Schönberg–Chandrasekhar limit is reached somewhere near the points marked 4 (for $M \geq 1.4 M_\odot$). Few stars are found in the 'Hertzsprung gap' between points 4 and 5. Adapted from Iben (1967).

diagram (points marked '2' in Fig. 5.6). When central hydrogen is exhausted, the entire star undergoes homologous contraction, which heats the interior to the point where hydrogen is ignited in a shell (points marked '3' in the figure) and the star begins to expand again. One thus has a growing isothermal core supporting the surrounding envelope, and it was shown by Schönberg & Chandrasekhar (1942) that there is an upper limit to the mass fraction q of the star occupied by the core for which such a configuration can be in equilibrium. A sketch of the argument is the following: Apply the Virial Theorem, Eq. (5.14), to the isothermal core with radius R_c, but now with an external pressure P_0.

$$2U + \Omega = 4\pi R_c^3 P_0, \text{ or} \tag{5.71}$$

$$P_0 = \frac{2U + \Omega}{4\pi R_c^3} = \frac{C_V M_c T}{2\pi R_c^3} - \frac{\beta G M_c^2}{4\pi R_c^4} \tag{5.72}$$

$$= \frac{C_1}{\mu_c} \frac{M_c T_0}{R_c^3} - C_2 \frac{M_c^2}{R_c^4}, \tag{5.73}$$

where C_V is the specific heat, β is a number of order unity, T_0 is the temperature of the H-burning shell and C_1, C_2 are constants. Differentiation of Eq. (5.73) shows that $P_0(R_c)$ has a maximum given by

$$R_{max} = C_3 \frac{M_c \mu_c}{T_0}; \quad P_{0,max} = \frac{C_4}{\mu_c^4} \frac{T_0^4}{M_c^2}. \tag{5.74}$$

Applying the homology version of the hydrostatic equation (5.8) to the envelope, one obtains another condition for the pressure at the edge of the core:

$$P_{ext} \propto \frac{M^2}{R^4} = \frac{C_5}{\mu_{env}^4} \frac{T_0^4}{M^2}. \tag{5.75}$$

The fitting condition at the core boundary

$$P_{ext} < P_{0,max} \tag{5.76}$$

then implies

$$q_0 \equiv \frac{M_c}{M} \leq q_{SC} \simeq 0.37 \left(\frac{\mu_{env}}{\mu_c} \right)^2 \simeq 0.09. \tag{5.77}$$

When this limit is reached, the isothermal core is no longer capable of supporting the envelope, the core contracts and the envelope expands by virtue of a sort of 'mirror' effect: contraction increases the rate of energy generation in the hydrogen-burning shell and the envelope expands so as to transfer the additional energy outwards (Iben 1993). However, the situation is complicated by changes in the opacity of the envelope, which increases as a result of the concomitant cooling and drives faster expansion in a positive feedback loop (Renzini *et al.* 1992) until the outer convection zone penetrates deeply enough to restore thermal equilibrium. The star accordingly moves quickly (i.e. on the thermal time-scale) across the 'Hertzsprung gap' in the HR diagram to the neighbourhood of the Hayashi track (Fig. 5.6) and begins its ascent of the red-giant branch (RGB) where its evolutionary lifetime is of the order of 0.1 of that on the main-sequence. However, there is no SC limit for stars with total mass below $1.4 M_\odot$, because the cores are already becoming partially degenerate at this stage.

5.8 Helium-burning

During RGB evolution, the H-exhausted core continues to contract in radius (while expanding in the mass coordinate as the H-burning shell eats its way outwards in mass) and heat up, although this process is slowed down by degeneracy in the case of low-mass stars. Eventually (at points marked '6' in Fig. 5.6), helium-burning, which is

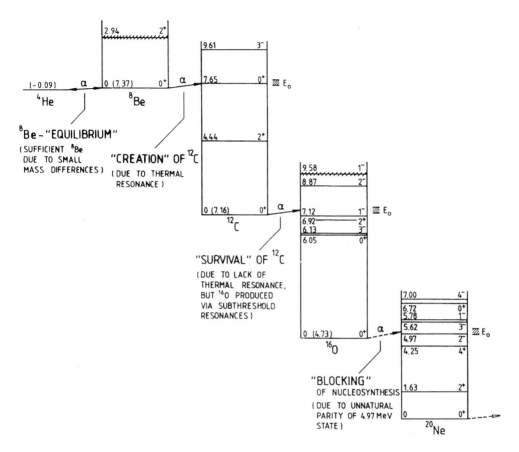

Fig. 5.7. Nuclear energy levels involved in the 3α reaction. After C.E. Rolfs & W.S. Rodney, *Cauldrons in the Cosmos*, University of Chicago Press 1988, Fig. 7.15, p. 411. ©1988 by the University of Chicago. Courtesy Claus Rolfs.

the start of nucleosynthesis proper, sets in at temperatures of the order of 10^8 K and densities $\sim 10^4$ gm cm^{-3} through the triple α reaction (Öpik 1951; Salpeter 1952)

$$2\alpha \rightleftharpoons {}^8\text{Be} - 0.09\,\text{MeV};\tag{5.78}$$

$$^8Be + \alpha \rightleftharpoons {}^{12}\text{C}^{**}(7.65\,\text{MeV})(\gamma\gamma)\,^{12}\text{C (ground state)}.\tag{5.79}$$

Fig. 5.7 shows the nuclear energy levels involved, which display a number of remarkable coincidences that have played an essential role in the creation of the elements needed for life. A Saha type equilibrium exists between 2α and ^{8}Be (which decays in a time of the order of 10^{-15} s), and also between ^{8}Be and the excited 7.65 MeV 0^+ state of ^{12}C, whose existence was predicted by Hoyle (1954) on the grounds that carbon would not otherwise exist in anything like its cosmic abundance. The 7.65 MeV state of ^{12}C mostly decays back into 3α, but has a small probability (corresponding to $\Gamma_\gamma = 3.6 \times 10^{-3}$ eV) of electromagnetic decay eventually leading to the ground state.

In massive and intermediate-mass stars, $M > 2M_\odot$, the interior is still not degenerate when helium-burning sets in; the new energy source causes the core to expand and the envelope to contract, and the star undergoes some loops in the HR diagram (Fig. 5.6). In less massive stars, the core is degenerate and a thermal runaway ensues, known as the (core) helium flash. This is followed by a readjustment of the structure, with helium burning in a non-degenerate convective core, and metal-rich stars end up back somewhat above points 5 in Fig. 5.6; old, metal-poor stars spread out along a horizontal branch (cf. Figs. 4.6, 5.14). The spread is largely caused by stochastic variations in mass loss during previous RGB evolution, whereas the luminosity is essentially fixed by the mass of the core when the helium flash set in (about $0.5M_\odot$).

Some other important reactions involving He-burning are the following:

- $^{12}C(\alpha, \gamma)^{16}O$ converts a substantial fraction of the carbon from 3α into oxygen; the rate is still somewhat uncertain because it is affected by the tails of sub-threshold resonances (Fig. 5.7; cf. also Fig. 2.12), and this uncertainty is significant for calculations of more advanced stages of nucleosynthesis.
- More advanced He-burning reactions $^{16}O(\alpha, \gamma)^{20}Ne$ $(\alpha, \gamma)^{24}Mg$ may take place at temperatures $\sim 10^9$ K under some conditions, *e.g.* shell burning in advanced stages and explosive nucleosynthesis.
- Helium-burning on CNO material previously converted into ^{14}N by H-burning leads to the series of reactions

$$^{14}N(\alpha, \gamma)^{18}F(\beta^+ v)^{18}O(\alpha, \gamma)^{22}Ne, \tag{5.80}$$

which lead to a neutron excess proportional to the inital $Z_{CNO} \simeq Z$. This is important in explosive nucleosynthesis (where the overall n/p ratio is unaltered because there is no time for β-decays) and as a source of free neutrons for the neutron capture processes in massive stars from $^{22}Ne(\alpha, n)^{25}Mg$. Another important neutron source, which operates also in low-mass stars, is $^{13}C(\alpha, n)^{16}O$.

5.9   Further burning stages: evolution of massive stars

In stars below a certain initial mass M_{up}, core burning is halted by formation of a degenerate CO core which later becomes a white dwarf. The value of M_{up} is rather uncertain, depending on mass loss in the AGB phase, but is thought to be around $8M_\odot$. Stars with initial mass about $10M_\odot$ or more ignite carbon in the core non-degenerately. Owing to neutrino (and anti-neutrino) emission at the high temperatures involved, due to $e^\pm$ annihilation and other processes, subsequent evolution is greatly accelerated, the nuclear time-scale becomes shorter than the thermal time-scale and nuclear evolution occurs before the envelope (blue or red supergiant) has time to adjust. The main reactions are

- Carbon burning ($T \simeq 10^9$ K, $t \simeq 600$ yr)

$$^{12}C + ^{12}C \quad \rightarrow \quad ^{20}Ne + \alpha + 4.6\,\mathrm{MeV} \tag{5.81}$$

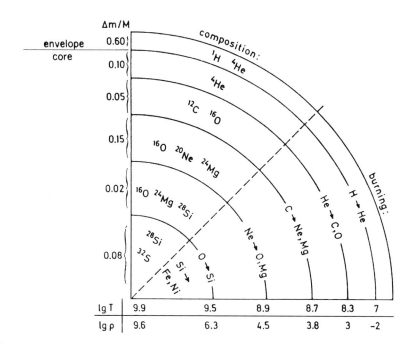

Fig. 5.8. Schematic illustration (not to scale) of the 'onion-skin' structure' in the interior of a highly evolved massive star ($25 M_\odot$). Numbers along the vertical axis show some typical values of the mass fraction, while those along the horizontal axis indicate temperatures and densities (gm cm^{-3}). Adapted from R. Kippenhahn & A. Weigert, *Stellar Structure and Evolution*, Springer-Verlag 1990.

$$\rightarrow \quad {}^{23}\text{Na} + p + 2.2\,\text{MeV} \tag{5.82}$$

$$\rightarrow \quad {}^{23}\text{Mg} + n - 2.6\,\text{MeV}, \tag{5.83}$$

followed by ${}^{23}\text{Na}(p,\alpha){}^{20}\text{Ne}$.

- Neon burning

$${}^{20}\text{Ne}(\gamma,\alpha){}^{16}\text{O}; \quad {}^{20}\text{Ne}(\alpha,\gamma){}^{24}\text{Mg}; \quad {}^{24}\text{Mg}(\alpha,\gamma){}^{28}\text{Si}. \tag{5.84}$$

- Oxygen burning ($T \simeq 2 \times 10^9$ K, $t \simeq \frac{1}{2}$ yr)

$$\text{O} + \text{O} \rightarrow \text{Si, S etc.} \tag{5.85}$$

in partially degenerate conditions. Core O-burning is accompanied by shell C-burning etc., leading to a concentric onion-like structure with increasingly heavy nuclei towards the interior (Fig. 5.8). However, convective zones caused by some of the shell sources will tend to complicate this picture by blurring composition boundaries and the extent of this effect is obscured by uncertainties in nuclear reaction rates and in the influence of convective overshooting and

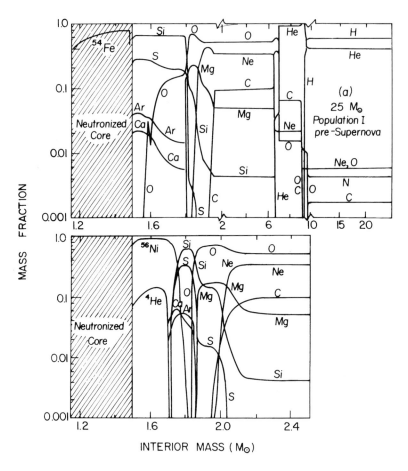

Fig. 5.9. Upper panel: Chemical profile of a $25M_\odot$ star immediately before core collapse. (Note change in horizontal scale at $2M_\odot$.) Lower panel: The same, after modification by explosive nucleosynthesis in a supernova outburst. The amount of ^{56}Ni (which later decays to ^{56}Fe) ejected depends on the mass cut, somewhere in the ^{28}Si$\rightarrow^{56}$Ni zone, and is uncertain by a factor of 2 or so. Adapted from Woosley & Weaver (1982).

semi-convection, which latter is a slow mixing process that can arise in regions with a gradient in molecular weight.

- Silicon burning ($T \geq 3 \times 10^9$ K, $t \simeq 1$ day) can be described as 'photo-disintegration-rearrangement', with

$$^{28}\text{Si}(\gamma, \alpha)^{24}\text{Mg} \qquad (5.86)$$

followed by

$$^{28}\text{Si} + \alpha + p + n \rightarrow \text{iron peak elements.} \qquad (5.87)$$

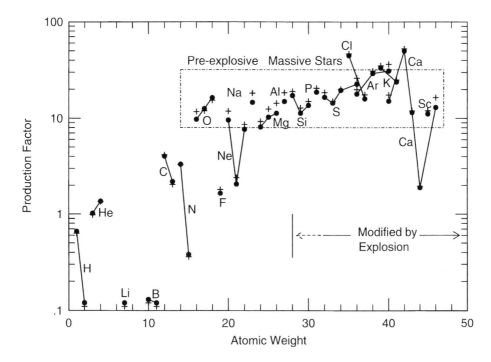

Fig. 5.10. Amounts, in units of relative solar-system abundances, of nuclear species resulting from hydrostatic evolution of an average pre-supernova. Filled circles represent an initial mass function with slope -2.3 and plus signs one with slope -1.5. The dashed box encloses 28 species co-produced within a factor of 2 of solar values, assuming a $^{12}C(\alpha,\gamma)^{16}O$ rate $1.7 \times$ that given by Caughlan & Fowler (1988). Reprinted from *Phys. Reports*, **227**, T.A. Weaver & S.E. Woosley, 'Nucleosynthesis in massive stars and the $^{12}C(\alpha,\gamma)^{16}O$ reaction rate', pp. 65–96, ©1993, with kind permission of Elsevier Science - NL, Sara Burgerhartstraat 25, 1055 KV Amsterdam, The Netherlands. Courtesy Tom Weaver.

Eq. (5.87) summarises many fast reactions which take place while ^{28}Si slowly 'melts'. The fast reactions in many cases are balanced by reverse reactions, so that there is an approximation to nuclear statistical equilibrium ('quasi-equilibrium'), favouring the more stable nuclei; the outcome is then a function of just three parameters, the temperature, the density and the neutron excess, resulting from previous nuclear reactions and inverse β-decays. At this stage, an inner mass of the order of M_{Ch} is degenerate and dominated by neutron-rich nuclei (mainly ^{54}Fe), owing to the high density (cf. Section 5.4.4), while further out there are 'onion skins' consisting successively of silicon, oxygen, helium and hydrogen with minor constituents (Fig. 5.8).

The next stage is dynamical collapse of the core, caused by electron capture and/or photodisintegration (and/or $e^{\pm}$ pair creation for $M \geq 100 M_{\odot}$) caused by the increase in temperature following gravitational contraction when the silicon fuel is exhausted.

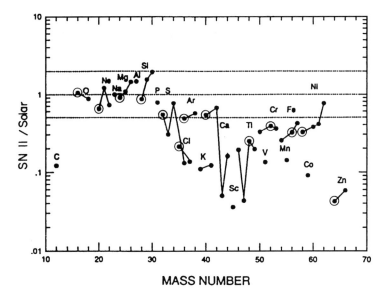

Fig. 5.11. Calculated abundances after decay of ^{56}Ni and other radioactive nuclei, relative to solar, in material ejected from a typical Type II supernova explosion, averaged over initial masses $10 - 50\ M_\odot$. Dominant isotopes of each element are circled. Adapted from Tsujimoto (1993).

According to the standard picture, the collapse leads to the formation of a neutron star, leading to copious neutrino emission and to a bounce at nuclear density which causes a shock to propagate outwards. The shock, reinforced by absorption of energetic neutrinos, leads to expulsion of the outer layers in an explosion identified with supernovae of Type II (and related types Ib etc.) and to modification of the inner layers by explosive synthesis due to sudden heating caused by the shock and subsequent freeze-out from the expansion (Fig. 5.9). The gravitational energy released is of the order of 10^{53} erg, 99 per cent of which is carried away by the neutrinos. About 1 per cent, i.e. 10^{51} erg, goes into the shock and only 1 per cent again of that appears as visible radiation, largely powered by radio-active decay in the stages following maximum light; around 10^{50} erg, or 10 per cent of the shock energy, are needed to eject the envelope against gravity. Fig. 5.10 shows the relative amounts of elements previously synthesized in hydrostatic stellar evolution that are ejected in a typical model of the explosion.

The shock takes the form of an expanding high-entropy bubble, dominated by radiation, so that the maximum temperature reached in each layer is essentially fixed by the shock energy E_0 (usually input into calculations as a parameter $\simeq 10^{51}$ erg) and the radius:

$$\frac{4}{3}\pi r^3 a T^4 = E_0,\qquad\qquad(5.88)$$

giving $kT_{\max} \geq 0.5$ MeV (adequate for complete explosive silicon burning) at a radius of 3700 km (corresponding to an included mass of $1.7 M_\odot$ in the silicon layer of a $20M_\odot$ star). The composition of the silicon and the inner part of the oxygen layer is then modified by explosive synthesis during the expansion time-scale of the order of seconds, which differs from previous 'hydrostatic' nucleosynthesis by higher temperatures and entropies and by allowing no time for β-decay. The nuclear statistical quasi-equilibrium is thus subject to the overall neutron–proton ratio previously established, which corresponds to a neutron excess

$$\eta \equiv \sum_i (N_i - Z_i)Y_i = 1 - 2Y_e = 1 - 2/\mu_e \sim 10^{-3}. \tag{5.89}$$

(Here N and Z are neutron and proton numbers in each nucleus present with a numerical abundance $Y_i \equiv X_i/A_i$ per amu.) The value of the neutron excess depends somewhat on the initial chemical composition (more initial CNO leading to more ^{22}Ne after helium burning), but this dependence is damped down by various nuclear reactions that have taken place in previous hydrostatic burning stages. With the small neutron excess of order 10^{-3} and the high entropy, the result of explosive synthesis favours α-particle nuclei (mainly ^{56}Ni; see Fig. 5.9) and α-particles themselves, resulting in a so-called α-rich freeze-out in the innermost ejected layers. (An α-rich freeze-out occurs generally at low densities where the 3α reaction is not fast enough to keep the helium abundance in equilibrium during the expansion and cooling in explosive events.) Resulting abundances calculated after explosion and radioactive decay are shown in Fig. 5.11, assuming the mass cut above which material is ejected to be such as to yield $0.07M_\odot$ of ^{56}Ni in accordance with observations of supernova 1987A in the Large Magellanic Cloud, i.e. at $1.6M_\odot$ for a $20M_\odot$ star. These massive stars are believed to supply most elements up to and including the iron group in stars of Population II and nuclides shown in the box in Fig. 5.10 in Population I, either in the explosions or in preceding mass loss episodes from stellar winds. However, there is a deficit by a factor of the order of 2 or 3 in the relative abundance of iron-group elements compared to the Sun, which is believed to be made up by supernovae of Type Ia.

Broadly speaking, O, Ne and Mg originate mainly from hydrostatic burning shells and the amount synthesized and ejected rises sharply with progenitor mass (or strictly speaking the mass of the helium core), whereas S, Ar, Ca and Fe are mostly due to explosive burning and their ejected mass is assumed to vary much less.

However, there are uncertainties in the range of initial stellar masses where this standard picture applies: more massive stars may leave behind a black hole instead of a neutron star, with or without a supernova explosion, depending on the equation of state of nuclear matter and how much initially outgoing material falls back, and mass loss in the pre-supernova stages can have drastic effects on the evolution of stars of $40M_\odot$ or more, especially at higher metallicities (Maeder 1992, 1993). Fig. 5.12 shows the chemical profile calculated for a solar-metallicity star of initial mass $40M_\odot$ (He core mass $M_\alpha = 19M_\odot$) in which the peeling away of the outer layers by stellar winds gradually exposes increasingly nuclear-processed material that appears at the surface of Wolf–Rayet stars and reduces the total mass to $12M_\odot$ before the supernova explosion. Fig. 5.13 shows schematically, as a function of initial stellar mass and for

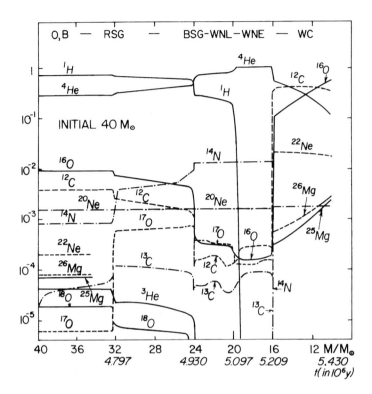

Fig. 5.12. Chemical profile of a $40M_\odot$ star that becomes a Wolf–Rayet star as a result of the outer layers peeling off in stellar winds. The spectrum evolves from type O to type B to a red supergiant (RSG) and then back to a blue supergiant (BSG) and towards increasing effective temperatures ending up well to the left of the main sequence. The chemically modified spectrum evolves from nitrogen-rich late, i.e. relatively cool (WNL), to nitrogen-rich early (WNE) to carbon-rich (WC); in some cases still hotter stars are observed that are oxygen-rich (WO). After A. Maeder & G. Meynet (1987), *A & A*, **182**, 243.

two initial chemical compositions, the amounts of different sorts of material ejected in winds and subsequent supernova explosions calculated under the assumption that strong mass loss normally occurs. The most massive stars (above $50M_\odot$, say) may lose their expected supernova ejecta in a black hole, but still contribute lighter elements like C, N, O and Ne (especially ^{22}Ne) in winds during their prior evolution.

5.10   Evolution of intermediate and low-mass stars

5.10.1   Introduction

Fig. 5.14 gives an overview of stellar evolution in the HR diagram. Both low and intermediate-mass stars end their lives as white dwarfs after having expelled a substantial amount of mass in winds and planetary nebulae, the basic reason being the

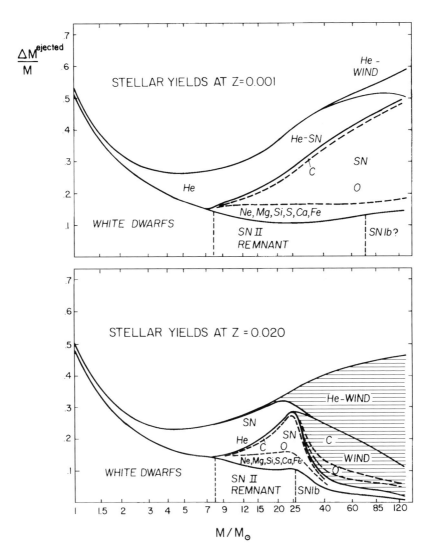

Fig. 5.13. Element production in winds and supernova ejecta from stars affected by strong mass loss, as a function of initial mass. Upper panel: stars with about 1/20 solar heavy-element abundance. Lower panel: stars with approximately solar composition, for which the effects of mass loss are believed to be more drastic. Horizontal shadings indicate outer layers that are expelled in winds prior to SN explosion. After A. Maeder (1992), *A & A*, **264**, 105.

formation of a degenerate CO core that is not massive enough to ignite carbon. They differ in that intermediate mass stars (2.5 to $8M_\odot$ or so) start with convective cores and ignite helium non-degenerately, whereas lower-mass stars start with small or non-existent convective cores, become degenerate during hydrogen shell burning and eventually ignite helium with a 'flash'.

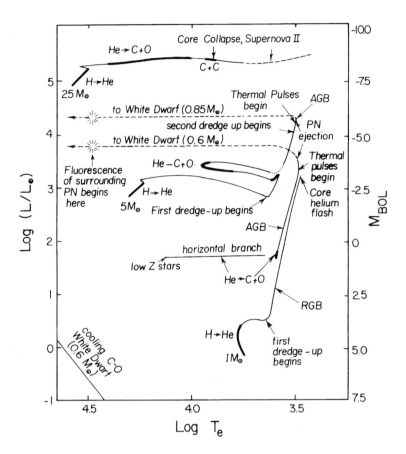

Fig. 5.14. Evolutionary tracks of stars with mass 1, 5 and $25M_\odot$ in the HR diagram. Adapted from Iben (1985, 1991).

5.10.2 Evolution of intermediate-mass stars

Fig. 5.15 shows a model for the evolution of an intermediate-mass star. At A it is on the ZAMS burning hydrogen in a convective core which gradually retreats, leaving behind a radiative region enriched in helium. At B the central hydrogen is reaching exhaustion and overall gravitational contraction sets in, leading to ignition of hydrogen in a thick shell at C. The star then rapidly moves across the Hertzsprung gap to D, where an outer convection zone develops, the burning shell becomes thin and the first dredge-up phase sets in: Li, Be and B in the atmosphere are destroyed by dilution and some ^{13}C and ^{14}N may be dredged up, depending on the depth of convective penetration plus overshoot. At E, core helium burning sets in, halting the contraction of the core and concomitant expansion of the envelope, and the outer convection zone retreats again. During core helium burning, from E to G, the star carries out a loop in the HR diagram, due to another phase of core gravitational contraction and envelope

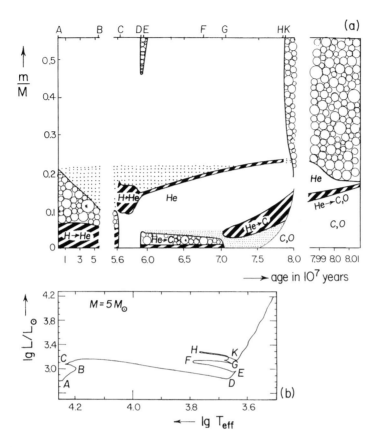

Fig. 5.15. Evolution of a $5M_\odot$ star of extreme Population I, shown in chemical profile and in the HR diagram. 'Cloudy' areas indicate convective regions; heavily hatched areas indicate significant nuclear energy generation ($> 10^3$ erg gm^{-1} s^{-1}); and dotted areas are regions of variable chemical composition. After R. Kippenhahn & A. Weigert, *Stellar Structure and Evolution*, Fig. 31.2, p. 294, ©Springer-Verlag Berlin Heidelberg 1990.

expansion as the helium fuel runs out, and at G it starts another loop when the helium shell source is ignited and eventually causes the hydrogen shell source to be quenched at H by cooling due to expansion. The whole helium core then contracts, leading to expansion of the envelope back to the Hayashi track at K, where the second dredge-up occurs (above a certain mass limit) and causes an increase in helium abundance and conversion of C and O into ^{14}N in the atmosphere. (These loops take the star several times through the cepheid instability strip.) At K, the hydrogen shell source is reignited by heat from the helium shell source and the star now evolves up the asymptotic giant branch (AGB) where extensive mass loss takes place culminating in PN ejection and the cooling of the degenerate CO core to form a white dwarf (see Fig. 5.14). Mixing processes during later stages of AGB evolution, to be described in more detail below,

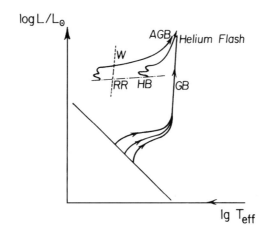

Fig. 5.16. Sketch of the evolution of low-mass stars in the HR diagram. For three slightly different initial masses, with a given chemical composition, the evolutionary tracks from the main sequence merge in the RGB, along which variable amounts of mass (between 0 and $\sim 0.2M_\odot$) are lost through stellar winds. After the helium flash, they appear on the ZAHB, evolve (after some loops) towards the upper right and merge in the asymptotic giant branch (AGB). The broken line indicates locations of pulsating variables, RR Lyrae (RR) and W Virginis (W) stars. After R. Kippenhahn & A. Weigert, *Stellar Structure and Evolution*, Fig. 33.1, p. 323, ©Springer-Verlag Berlin Heidelberg 1990.

lead to the production and expulsion in winds and PN of ^{7}Li, C, N and s-process isotopes and maybe others.

5.10.3 Evolution of low-mass stars to the helium flash

Low-mass stars ($\leq\sim 2.3M_\odot$) have small or vanishing convective cores while on the main sequence and hydrogen burning by the p-p chain leads to some enrichment in helium out to a fair fraction (20 to 50 per cent or so) of the total mass, well beyond the central core which eventually becomes exhausted of hydrogen and degenerate. They do not experience the Schönberg–Chandrasekhar limit and Hertzsprung gap, and their main-sequence configurations are in any case fairly close to the Hayashi track (cf. Fig. 5.6). The first dredge-up occurs at the base of the red-giant branch, point 5 in Fig. 5.6, and results in dilution of the helium from previous nuclear core burning in regions just above the deepest layer reached by the outer convection zone, as well as in some changes in surface abundances: that of ^{14}N is approximately doubled at the expense of ^{12}C which is reduced by about 30 per cent, the ^{12}C/^{13}C ratio is reduced to between 20 and 30, Li, Be and B are virtually wiped out by dilution, but ^{16}O is essentially unaffected. This phase is followed by a long period on the RGB where the degenerate helium core gradually grows by the slow combustion of hydrogen in its surrounding shell. The rate of growth of the core is governed by the luminosity, which itself is a strong function of the core mass M_c and nearly independent of the mass of the increasingly distended, tenuous envelope around it, basically because the core is highly concentrated, gravity at its surface is very large making a large pressure drop from $|dP/dm| \propto m/r^4$ (with core radius $R_c \sim 10^4$ km) and a much smaller one further out with typical radii orders of magnitude more and only a slightly greater included mass. Thus the pressure drops to a small fraction of its inner value in a small range of mass through the burning shell; the envelope is nearly weightless and has little influence on the shell.

Under these conditions, the dependence of physical parameters of the shell, and hence of the total luminosity, on the mass M_c of the core can be estimated by a homology argument due to Refsdal & Weigert (1970), described in the textbook by Kippenhahn & Weigert (1990). Assume negligible radiation pressure and negligible energy production from gravitational contraction, electron-scattering opacity, $\kappa = \kappa_0/\mu = $ const. (the proportionality to $1/\mu$ is a rough approximation), CNO burning, $\epsilon = \epsilon_0 \rho T^\nu/\mu$, ideal gas conditions $P \propto \rho T/\mu$ and a constant included mass M_c through the shell, in which we also assume there to exist homology relations (with $x \equiv R_c/r$ where R_c is the radius of the core):

$$\rho(x) = M_c^{\rho_1} R_c^{\rho_2} \mu^{\rho_3} f_\rho(x); \tag{5.90}$$

$$T(x) = M_c^{t_1} R_c^{t_2} \mu^{t_3} f_t(x); \tag{5.91}$$

$$l(x) = M_c^{l_1} R_c^{l_2} \mu^{l_3} f_l(x), \tag{5.92}$$

where the f's are universal functions. The exponents of M_c, R_c and μ in these equations can be found from the basic equations of stellar structure. The hydrostatic equilibrium equation (5.6) can be written

$$dP = G\rho M_c\, d(1/r) = G\rho M_c R_c^{-1} dx \tag{5.93}$$

which integrates up to

$$\rho(x)T(x) \propto \mu P(x) \simeq \frac{GM_c\mu}{R_c} \int_0^x \rho\, dx' = GM_c^{\rho_1+1} R_c^{\rho_2-1} \mu^{\rho_3+1} \int_0^x f_\rho(x')dx'. \tag{5.94}$$

This implies

$$T \propto \mu M_c/R_c, \quad \text{or} \quad t_1 = t_3 = 1 = -t_2. \tag{5.95}$$

The radiative transfer equation, Eq. (5.22), can be written

$$d(T^4) \propto \kappa \rho l R_c^{-1} dx \tag{5.96}$$

and integrates to

$$M_c^4 R_c^{-4} \mu^4 \propto T^4(x) \simeq \kappa_0 M_c^{\rho_1+l_1} R_c^{\rho_2+l_2-1} \mu^{\rho_3+l_3-1} \int_0^x f_\rho(x')f_l(x')dx', \tag{5.97}$$

which implies

$$\rho_1 + l_1 = 4; \quad \rho_2 + l_2 = -3; \quad \rho_3 + l_3 = 5. \tag{5.98}$$

Finally, the energy generation equation

$$dl \propto \epsilon \rho d(r^3) \propto \epsilon \rho R_c^3 x^{-4} dx \qquad (5.99)$$

integrates to

$$M_c^{l_1} R_c^{l_2} \mu^{l_3} \propto l(x) \simeq \epsilon_0 M_c^{2\rho_1+\nu} R_c^{2\rho_2+3-\nu} \mu^{2\rho_3+\nu-1} \int_0^x x'^{-4} f_\rho(x') f_t^{\ \nu}(x') dx', \qquad (5.100)$$

which implies

$$l_1 = 2\rho_1 + \nu; \quad l_2 = 2\rho_2 + 3 - \nu; \quad l_3 = 2\rho_3 + \nu - 1. \qquad (5.101)$$

The simultaneous equations (5.98) and (5.101) give the values of the exponents:

$$\rho_1 = (4 - \nu)/3 \simeq -3; \ \rho_2 = -\rho_3 = (\nu - 6)/3 \simeq 7/3; \qquad (5.102)$$

$$l_1 = (8 + \nu)/3 \simeq 7; \quad l_2 = -(3 + \nu)/3 \simeq -16/3; \quad l_3 = (9 + \nu)/3 \simeq 22/3, \qquad (5.103)$$

assuming $\nu \simeq 13$. We thus have

$$L \equiv l(0) \propto \mu^{22/3} M_c^7 R_c^{-16/3}, \qquad (5.104)$$

or, using as a very crude approximation the mass-radius relation for cold white dwarfs, Eq. (5.53),

$$L \propto \mu^{16} M_c^{8.8}, \qquad (5.105)$$

a very steep relation. The core mass then grows according to

$$\dot{M}_c = \frac{L}{X_H E_H} \propto M_c^{8.8}, \qquad (5.106)$$

where E_H is the energy gain per unit mass of hydrogen, except during a brief phase when the shell source penetrates into the slightly more hydrogen-rich region left behind by the retreating outer convection zone and the luminosity is temporarily reduced by the μ factor. In the meantime, the temperature (which is more or less uniform throughout the highly conducting core) grows according to

$$T(x \geq 1) \propto \mu M_c/R_c \sim \mu M_c^{4/3}, \qquad (5.107)$$

reaching the value of about 10^8 K, where core helium-burning sets in (somewhat off-centre as a result of neutrino losses), when M_c reaches a value close to $0.5 M_\odot$ and a corresponding luminosity $\sim 2000 L_\odot$; these values are quite insensitive to the total stellar mass, but dependent on chemical composition through the effects of helium on molecular weight and of heavy elements on opacity and energy generation. Because the degenerate material cannot take up the resulting extra heat energy by expansion, there is a thermal runaway known as the (core) helium flash, the core heats up suddenly to temperatures $\sim 2.5 \times 10^8$ K at which degeneracy is removed and then expands, leading to contraction of the envelope and a new structure on the zero-age

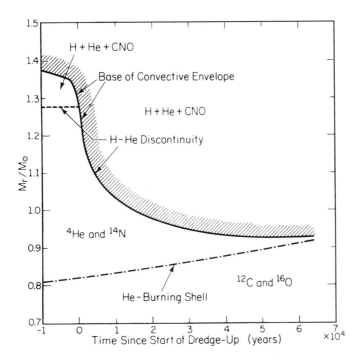

Fig. 5.17. Interior composition of a model $7M_\odot$ star during the second dredge-up phase. After I. Iben, Jr, *Ap. J. Suppl.*, **76**, 55, 1991.

horizontal branch (ZAHB). Here the luminosity is again mainly fixed by the core mass (leading to a virtually constant HB lifetime of about 10^8 yrs), but the core is no longer degenerate, its radius is larger and the luminosity (still coming mainly from the hydrogen shell) correspondingly lower (around $50L_\odot$). The core flash is not believed to lead to significant mixing that might affect the composition of the atmosphere. The location on the ZAHB depends on the chemical composition and the total mass (after any mass loss that may have occurred along the RGB). Metal-rich stars form a clump close to the RGB, whereas metal-poor stars, e.g. in globular clusters, have higher effective temperatures and a spread, the spread probably being largely due to stochastic variations in the mass loss, and the very bluest ZAHB stars have but a very thin envelope overlying the helium core, which has the 'helium main sequence' structure of a homogeneous helium star. While the average effective temperature along the ZAHB of globular clusters tends to increase with diminishing metallicity, there is not a one-to-one correlation, a fact often referred to as the 'second-parameter problem'. For a fixed core mass, the location on the ZAHB depends on the total mass, less massive stars being hotter and bluer. Thus a major factor must be some combination of the initial mass (and hence the age) with variable amounts of mass loss during the preceding passage up the RGB.

5.10.4   AGB evolution and the third dredge-up

ZAHB stars have a helium-burning convective core surrounded by unburned helium surrounded by a hydrogen-burning shell surrounded by a hydrogen-rich envelope. The helium core gradually increases in mass from hydrogen shell burning, leading to increasing luminosity, and the star evolves towards the asymptotic giant branch (AGB; cf. Fig. 5.16). In the meantime, helium burning at the centre leads to the formation of a growing CO inner core, which becomes degenerate and increases in mass and temperature up to a point where helium burning is ignited in a shell around the CO core. This causes expansion and cooling of the helium layer and the hydrogen shell source is temporarily switched off during early AGB (E-AGB) evolution at which point in the more massive intermediate-mass stars the surface convection zone penetrates deeply into the helium layer in the so-called second dredge-up phase (see Fig. 5.17). This causes substantial changes in the composition of the envelope and atmosphere: helium is enhanced and both carbon and oxygen are partly replaced by ^{14}N. (There is no second dredge-up for low-mass stars.)

As the surface convection zone deepens bringing down fresh hydrogen, it approaches closer and closer to the helium-burning shell coming up towards it (speaking in terms of the mass coordinate) from below, and this eventually supplies enough heat to reignite the hydrogen shell source. The E-AGB phase then gives way to a situation where there are two shell burning sources separated by a small interval in the mass coordinate; this is subject to an instability giving rise to thermal pulses (also called 'helium shell flashes' although in this case degeneracy is not directly involved) and is known as the TP-AGB stage (see Figs. 5.14, 5.18). The instability is a consequence of the high temperature sensitivity of the rate of helium burning combined with the thinness of the shell in which it occurs. Consider a thin shell of thickness D surrounding a sphere of much larger radius r. If the shell expands as a result of local heating, its change in density is magnified compared to that which results in the sphere as a whole:

$$-\frac{d\rho}{\rho} = \frac{dD}{D} = \frac{r}{D}\frac{dr}{r}. \tag{5.108}$$

But from the hydrostatic equation (5.8), for a fixed included mass m, there is a reduction in pressure,

$$-\frac{dP}{P} = 4\frac{dr}{r} = 4\frac{D}{r}\left(-\frac{d\rho}{\rho}\right) < -\frac{d\rho}{\rho} \text{ if } \frac{D}{r} < \frac{1}{4}. \tag{5.109}$$

For a perfect gas with $P \propto \rho T$, the temperature must therefore go up, which leads to instability, provided that the temperature sensitivity of the reaction rate is high enough to cover the losses. The result of this instability is a series of thermal pulses in which the helium burning shell greatly increases in luminosity, which causes expansion of the inter-shell region and cooling of the hydrogen-burning shell which is temporarily switched off, and the intense He-burning creates a convective layer ascending through the inter-shell region which is thereby enriched in ^{12}C and ^{22}Ne; this latter generates neutrons through (α, n) reactions and contributes to s-process nucleosynthesis, products

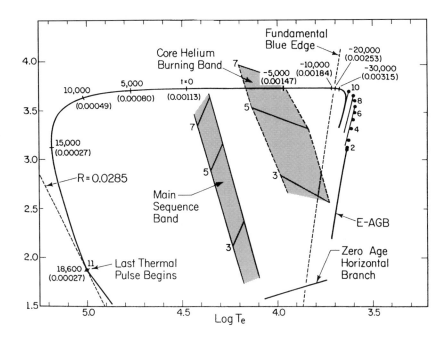

Fig. 5.18. Evolutionary track in the HR diagram of an AGB model of total mass $0.6M_\odot$, initial composition $(Y, Z) = (0.25, 0.001 \simeq Z_\odot/20)$. Heavy dots marked 2 to 11 indicate the start of a series of thermal pulses (cf. Fig. 5.19), which lead to excursions along the steep diagonal lines. Numbers along the horizontal and descending track indicate times in years relative to the moment when an ionized planetary nebula appears and (in parentheses) the mass of the envelope in units of $M_\odot$. $R = 0.0285$ indicates a line of constant radius (R in solar units) corresponding to the white-dwarf sequence. Shaded areas represent earlier evolutionary stages for stars with initial masses 3, 5 and $7M_\odot$ and the steep broken line marks the high-temperature boundary of the instability strip in which stars pulsate in their fundamental mode. Adapted from Iben & Renzini (1983).

of which are also mixed out into practically the whole of the helium layer. Eventually the helium-burning shell is sufficiently expanded and cooled to die down, the inter-shell region contracts and warms up again and the hydrogen-burning shell resumes, contributing the bulk of the star's luminosity, until the next pulse comes along (Fig. 5.19). Resulting changes in surface properties are shown for a low-mass star in Fig. 5.18. After the pulse has died down, the outer convection zone penetrates the former inter-shell region (see Fig. 5.19) and brings up carbon and products of the s-process to the surface (this is the third dredge-up process). Furthermore, protons and/or ^{13}C (an alternative neutron source that operates at lower temperatures and therefore in less massive stars than does the ^{22}Ne source) are probably ingested into the helium layer at some stage (cf. Chapter 6). This might occur during a pulse, if the inner convective zone reaches high enough to penetrate the H–He discontinuity; this is not predicted by conventional convection models, but could result from over-shooting or

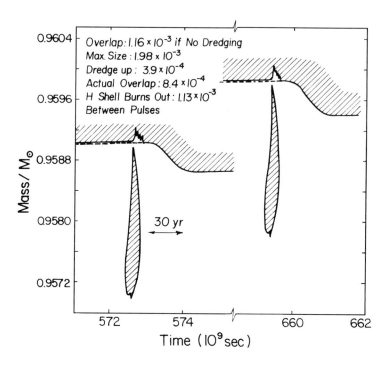

Fig. 5.19. Convective regions during the 15th and 16th pulses in a model of mass $7M_\odot$ and solar composition. Shading indicates where convection occurs, and the dashed lines indicate the location of the H–He discontinuity before dredge-up begins. After I. Iben, Jr. & A. Renzini (1983), with permission, from the *Annual Review of Astronomy and Astrophysics*, Volume **21**, p. 271, ©1983 by Annual Reviews, Inc.

plume convection. A more likely mechanism is thought to be semi-convective mixing of protons during a third dredge-up episode into the ^{12}C-rich region left behind by the preceding pulse, followed by their conversion into ^{13}C when hydrogen shell burning is reignited. When the next pulse occurs, the ^{13}C is ingested into the helium-burning shell and generates neutrons by ^{13}C$(\alpha, n)^{16}$O; this reaction may also occur during the inter-pulse period. For the process to work, the proton abundance should be comparable to, or less than, the ^{12}C abundance, since an excess of protons would lead to production of ^{14}N which is a neutron 'poison' that competes with ^{56}Fe as a neutron absorber. Finally, for sufficiently massive stars (details depend on the assumed mixing length, composition, opacities etc.), the base of the outer convection zone becomes hot enough in the intervals between pulses to convert the fresh ^{12}C into ^{14}N (and ^{13}C) by a process known as 'hot-bottom burning'; this is a so-called primary source of nitrogen nucleosynthesis, in which the amount produced is relatively insensitive to the initial chemical composition of the star, whereas the amount produced in the first and second dredge-ups is limited by the initial C and O abundances and these are known as secondary sources.

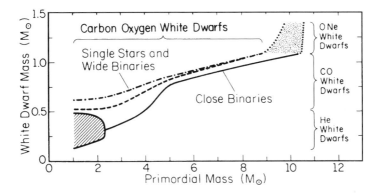

Fig. 5.20. Initial mass – final mass relation for intermediate-mass stars. After I. Iben, Jr, *Ap. J. Suppl.*, **76**, 55, 1991.

A further effect during evolution up the AGB is mass loss through stellar winds, at an increasing rate as the star increases in luminosity and radius and becomes unstable to pulsations which drive a 'super-wind' in the case of intermediate-mass stars. For stars with an initial mass below some limit, which may be of order $6M_\odot$, the wind 'evaporates' the hydrogen-rich envelope before the CO core has reached the Chandrasekhar limiting mass (see Section 5.4.3), the increase in luminosity ceases and the star contracts at constant luminosity eventually becoming a white dwarf (Figs. 5.14, 5.18). A computed relation between initial stellar mass and the final white-dwarf mass is shown in Fig. 5.20.

The ejected material forms a cool molecular and dusty envelope which initially veils the star from optical observations as it goes through the stages of Mira variable followed by OH-infra-red star or infra-red carbon star; later the star becomes hot enough to ionize part or all of the expanding gas-dust envelope forming a planetary nebula. Stars that are of too low mass to ignite carbon non-degenerately but of high enough mass to produce a degenerate CO core above the Chandrasekhar limit have been predicted to undergo a 'carbon flash' and explode in a supernova outburst referred to as a Type $1\frac{1}{2}$ supernova (sharing some characteristics of both Types I and II); there is no evidence that this ever happens, but they could be a subset of Type II. Stars at the upper end of the intermediate-mass range may ignite carbon non-degenerately but form a degenerate O,Ne,Mg core, which could either collapse and explode as Type II supernovae or end up as white dwarfs with that composition.

A group of planetary nebulae known as M. Peimbert's Type I is especially rich in nitrogen (N/O $\simeq$ 1.2), with more or less normal oxygen and modest enhancements of helium and carbon, and probably represents the combined effects of the first and third dredge-ups with hot-bottom burning, whereas in classical carbon stars and other types of planetary nebulae there seems to be more or less normal nitrogen and oxygen with an enhancement of ^{12}C (C/O $\simeq$ 1.3; Kingsburgh & Barlow 1994). Carbon stars tend to be more common in regions of lower metallicity (e.g. the Magellanic Clouds compared

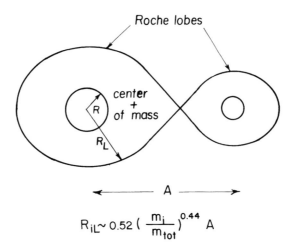

$$R_{iL} \sim 0.52 \left(\frac{m_i}{m_{tot}} \right)^{0.44} A$$

Fig. 5.21.   Roche-lobe geometry. Adapted from Iben (1991).

to the solar neighbourhood and the Galactic bulge), reflecting the fact that less carbon needs to be dredged up in order to make $C/O > 1$ when the oxygen abundance is lower in the first place.

Another consequence of the third dredge-up is the appearance of fresh lithium at the surface of luminous AGB stars (WZ Cas carbon stars and some giants in the Magellanic Clouds). This may result from the ^{3}He + ^{4}He $\to ^7$Be reaction (Cameron & Fowler 1971) at the base of a 'hot-bottom' convective envelope, followed by mixing up to the surface where it decays by K-electron capture to ^{7}Li (Sackmann & Boothroyd 1992).

5.11   Interacting binary stars

Among the equipotential surfaces in a system of coordinates rotating with the line joining the two components of a binary star system treated as two point masses, there is a critical equipotential consisting of two so-called Roche lobes which touch each other as shown in Fig. 5.21. At the contact point, the gravitational attractions of the two stars cancel each other out; thus if one star in the course of expansion from evolution or accretion, or as a result of orbital shrinkage, comes to fill its Roche lobe, the overflowing material is lost to that star and may end up on the other one, normally by way of an accretion disk in which orbital angular momentum is dissipated. In other cases, it may form a common envelope or leave the system altogether. The resulting mass transfer effects, which may lead to more than one exchange between the initially more massive and the initially less massive component, have rich consequences, including the production of cataclysmic variables, novae, Type Ia supernovae and binary X-ray sources from accretion on to a compact star (white dwarf, neutron star or black hole); the appearance of abundance peculiarities of extraneous origin in, e.g., Ba II stars and some varieties of carbon star, notably CH stars, owing to accretion of material from an AGB star on to a normal star; and enhancement of the mass-loss rates from evolving binary components compared to otherwise similar but isolated stars. Since a supergiant or AGB star can reach a radius of the order of an astronomical unit ($200R_\odot$), it is evident that effects of this kind can arise in binary stars with (initial) orbital periods of up to a few years.

Mass transfer in a red giant – white dwarf binary system can cause explosive burning processes giving rise to novae. If the accretion rate of hydrogen-rich material does not exceed $6 \times 10^{-8} M_\odot$ yr^{-1}, unburned hydrogen layers pile up on the surface of the white dwarf. After an envelope of 10^{-5} to $10^{-4} M_\odot$ has formed (depending on the mass of the white dwarf), pycnonuclear ignition under degenerate conditions triggers a thermonuclear runaway until the degeneracy is lifted and thermonuclear burning sets in reaching peak temperatures of maybe $2 - 3 \times 10^8$ K at densities of the order of 10^3 to 10^4 gm cm^{-3} and going on for $10 - 1000$ s before the partially burned hydrogen is ejected. During this time there is a sudden rise in luminosity, more or less reversing the previous evolution from AGB tip to white dwarf, followed by fading over weeks or months and a fresh outburst after a suitably long period. The typical energy release in such an outburst is between 10^{46} and 10^{47} erg. The main energy source of the thermal runaway is the fast CNO cycle (Section 5.6.2) in which ^{12}C, ^{14}N and ^{16}O are converted by proton captures in the first second into ^{14}O, ^{15}O and ^{17}F respectively, followed by ^{17}F$(p, \gamma)^{18}$Ne$(\beta^+, \nu)^{18}$F$(p, \alpha)^{15}$O; ^{14}O and ^{15}O subsequently β^+ decay after 1 or 2 minutes to ^{14}N and ^{15}N respectively, produced in comparable amounts. If ^{20}Ne and ^{24}Mg are initially abundant, as in an O,Ne,Mg white dwarf giving rise to a neon nova, the fast Ne-Na cycle begins to operate after 10s or so (owing to slowness of the triggering ^{20}Ne (p, γ) reaction) and may reach up to P and S isotopes (Van Wormer *et al.* 1994). The ejected envelope has a small mass, about $10^{-5} M_\odot$, so that the importance of this process in galactic chemical evolution is confined to rare species like ^{15}N. X-ray burst sources are also thought to result from a thermonuclear runaway, this time from accretion on to a neutron star, which leads to nuclear reactions with still higher temperatures and shorter time scales.

Supernovae of Type I are distinguished by having no hydrogen lines in their spectra, and it was already suggested by Hoyle & Fowler (1960) that they result from explosive carbon burning in a CO white dwarf. More recently they have been divided into subclasses Ia, Ib and Ic, with different light curves, spectra and progenitor masses (which can be estimated from the stellar populations with which they are associated). Type Ia supernovae occur in galaxies of all types and are not confined to regions of very recent star formation; their spectra at maximum light display absorption lines of Si, Ca, Fe and Ni and their light curves (which are so nearly homogeneous as to allow their use as standard candles) are energised after maximum by the decay ^{56}Ni(EC, half-life 6.6d)^{56}Co(EC,β^+, ν, half-life 77d)^{56}Fe, just as in the case of Type IIs, but involving a larger mass ($\sim 0.6 M_\odot$) of radioactive material. Type Ib have helium features in their spectra and are associated with young stellar populations; they are believed to result from core collapse in stars that have undergone severe mass loss (cf. Fig. 5.13), possibly as a result of binary interaction, and become helium or WR stars, so that their internal properties relate them more closely to Type II. The same

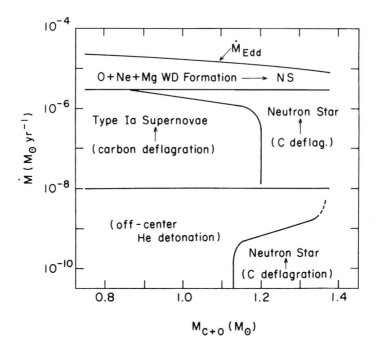

Fig. 5.22. Possible outcomes of accretion on to a CO white dwarf, according to its mass and accretion rate. $\dot{M}_{Edd}$ is the critical (Eddington) rate above which radiation pressure drives material out. After K. Nomoto & Y. Kondo, *Ap. J. Lett.*, **367**, L19, 1991. Courtesy Ken-ichi Nomoto.

probably applies to Type Ic, which, however, have much weaker helium lines and more rapidly declining light curves, and may result from an initally massive star that has been stripped down to its CO inner core by binary mass transfer.

SN Ia are believed to result from accretion of material from a companion on to a white dwarf until it reaches the Chandrasekhar limiting mass, at which point it collapses and explodes. Some possible fates of an accreting CO white dwarf, depending on the accretion rate and the initial mass, are shown in Fig. 5.22. The accreted material may come from a normal companion star, or, in the 'double-degenerate' model of Iben & Tutukov (1984a, 1987), there may be two white dwarfs that merge after orbital shrinkage by the formation of a common envelope followed by radiation of gravitational waves. Its arrival may have a variety of consequences, including the formation of a neutron star; the ignition of helium shell flashes already at masses well below M_{Ch}; resulting detonations of helium and carbon; and deflagration of carbon, which latter seems to give the best fit to observations, although a late detonation may also be involved. The problem with the straightforward detonation scenarios is that they lead to incineration of virtually all the C and O to ^{56}Ni, whereas observations reveal the presence of substantial amounts of lighter elements like Si and Ca. It turns out that, for a reasonably large rate of accretion, hydrogen and helium shell flashes

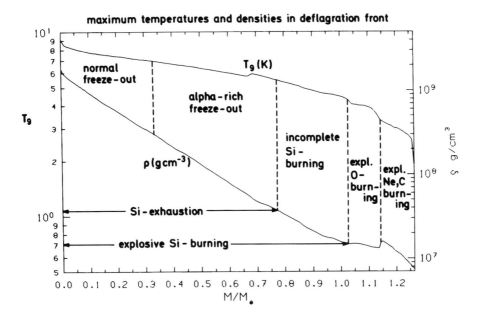

Fig. 5.23. Maximum temperatures and densities calculated for an outward propagating deflagration front in a model (model W7) of an SN Ia explosion from accretion on to a CO white dwarf with initial mass $1M_\odot$ at a rate of $4 \times 10^{-8} M_\odot \text{ yr}^{-1}$. Zones of different burning conditions are indicated. After F.-K. Thielemann, K. Nomoto & K. Yokoi, *A & A*, **158**, 17, 1986. Courtesy Ken-ichi Nomoto.

are relatively weak so that the main effect of accretion is a steady growth in the mass and temperature of the CO core similar to that in an AGB star. When the mass comes close to M_{Ch}, carbon is explosively ignited at the centre and a deflagration wave propagates outwards at a subsonic speed and causes explosive burning (see Fig. 5.23) and disruption of the entire star with an energy of the order of 10^{51} erg. The inner $0.7M_\odot$ or so is incinerated to nuclear statistical equilibrium, but the diminution of peak temperature and density with time and distance from the centre of the star allows reasonable quantities of elements from oxygen upwards to survive. The outcome of one of the more successful models is shown in Fig. 5.24, where it can be seen to more or less compensate for the deficiency of iron-group elements expected from Type II supernovae (Fig. 5.11). Fig. 5.25 shows the outcome when the two types of supernovae are combined in suitable proportions; the overall fit is fair, but some obvious discrepancies remain.

Notes to Chapter 5

An excellent account of stellar structure and evolution is given in the classic text:
R. Kippenhahn & A. Weigert, *Stellar Structure and Evolution*, Springer-Verlag 1990.

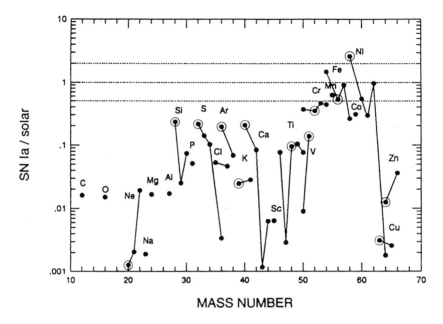

Fig. 5.24. Nucleosynthetic outcome (after radioactive decay) of model W7 for Type Ia supernovae (Nomoto, Thielemann & Yokoi, ApJ, 286, 644, 1984, and Thielemann, Nomoto & Yokoi, *A & A*, **158**, 17, 1986), compared to solar-system abundances. Dominant isotopes of multi-isotope elements are circled. Adapted from Tsujimoto (1993).

This is usefully supplemented (especially for the evolution of close binaries) by the following review article:

I. Iben, Jr, 'Single and Binary Star Evolution', *Astrophys. J. Suppl.*, **76**, 55 (1991).

Basic physical principles are given, and the consequences of nuclear reactions worked out, in much more detail in the older, but still very useful classic text:

D.D. Clayton, *Principles of Stellar Evolution and Nucleosynthesis*, McGraw-Hill 1968 and University of Chicago Press 1984.

A more up-to-date treatment of many aspects of stellar evolution and nuclear astrophysics is available in the book by

David Arnett, *Supernovae and Nucleosynthesis*, Princeton University Press, 1996.

Detailed information about specific nuclear reactions at different stages of stellar evolution is given (together with an overview of the whole of astrophysics) in

C.E. Rolfs & W.S. Rodney, *Cauldrons in the Cosmos*, University of Chicago Press 1988.

An excellent readable elementary introduction to the subject is given by

R.J. Tayler, *The Stars: Their Structure and Evolution*, Cambridge University Press 1994,

and a useful elementary introduction to some of the physical principles is given in

A.C. Phillips, *The Physics of Stars*, Wiley 1994.

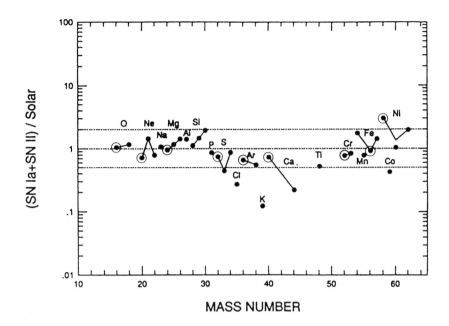

Fig. 5.25. Nucleosynthesis products from SN Ia (Fig. 5.24) and SN II (Fig. 5.11) combined in a ratio of 1:10, compared to solar-system abundances. (A slightly higher ratio of 1:7 gives optimal fit to elemental, as opposed to individual nuclidic abundances.) Dominant isotopes of multi-isotope elements are circled. Adapted from Tsujimoto (1993).

An overview of the reaction sequences in all stages from helium to silicon burning is given by Arnett & Thielemann (1985). Evolution of massive stars is reviewed by Nomoto & Hashimoto (1988) and by Weaver & Woosley (1993). Physics of supernovae is discussed by Woosley & Weaver (1986). Evolution of intermediate mass stars is discussed by Iben & Renzini (1983, 1984) as well as in the article by Iben (1991) quoted above.

Tables of stellar opacities computed at Los Alamos are given by Hübner *et al.* (1977) and Weiss, Keady & McGee (1990). More up-to-date and accurate opacities for temperatures up to 10^8 K are given by Rogers & Iglesias (1992) and up to 10^7 K by Seaton *et al.* (1994).

Abundance anomalies displayed by red giants in globular clusters, and the difficulties that they raise for canonical stellar evolution theory, are discussed, e.g., by Kraft (1994). Much information on the theoretical and observational aspects of AGB evolution can be found in

H.R. Johnson & B. Zuckerman (eds.), *Evolution of Peculiar Red Giant Stars*, Cambridge University Press 1989.

Problems

1. A homogeneous sphere of cold (i.e. pressure-free) gas with density ρ collapses homologously starting from rest. Show that it collapses to a point after a time $\pi/2\sqrt{2G\rho}$.

2. Compare the Fermi energy, the electrostatic energy due to nearby ions and the thermal energy per electron,
   (a) in a lump of iron (singly ionised) at room temperature, $\rho \simeq 10$ gm cm^{-3};
   (b) in a carbon white dwarf, $T = 10^8$ K, $\rho \simeq 10^5$ gm cm^{-3};
   (c) in a hydrogen 'brown dwarf', $T = 5 \times 10^6$ K, $\rho \simeq 10^3$ gm cm^{-3}.
   What does this mean for the relative importance of degeneracy pressure and electrostatic forces in the three cases? Make a rough estimate of the mass of the largest body that can be held up against gravity by electrostatic forces.

3. Verify using Saha's equation (2.47) that the most populated form of the oxygen in the centre of the Sun is fully stripped of electrons (O^{8+}) and estimate the proportion of oxygen having one bound electron (O^{7+}), using the following information:
   $T_6 = 15$, $X = 0.5$, $\rho = 100$ gm cm^{-3}, O^{7+} ionization potential $= 0.87$ keV, $kT \equiv 0.0862\, T_6$ keV, $(m_e kT/2\pi\hbar^2)^{3/2} \equiv 2.4 \times 10^{27}\, T_6^{3/2}$ cm^{-3}, $n_e \equiv \rho/(\mu_e m_H)$.
   Given that the absorption cross-section of O^{7+} at the photo-ionization threshold is 10^{-19} cm^2 and that the oxygen abundance is 10^{-2} by mass, find the bound-free opacity (cm^2 gm^{-1}) due to oxygen at that frequency, and compare it to the electron scattering opacity.
   Show also that hydrogen has to be fully ionized under these conditions.

4. Derive the formula (5.25) for the Rosseland mean opacity by comparing the transfer equation for K, Eq. (3.15), for monochromatic and integrated radiation.

5. Show that, if the monochromatic opacity varies as $v^{-3}T^{-1/2}$ over relevant frequency ranges, the Rosseland mean varies as $T^{-3.5}$, as in Eq. (5.26).

6. Show that the Eddington limiting luminosity, at which the gradient of radiation pressure balances gravity near the surface, is given by

$$L_{\text{Edd}} = 4\pi c G M/\kappa \tag{5.110}$$

and estimate its value in units of the solar luminosity $L_\odot = 4 \times 10^{33}$ erg s^{-1} when κ is given by electron scattering.

7. From Eq. (5.8), derive the inequality

$$P_c < \left(\frac{\pi}{6}\right)^{1/3} GM^{2/3}\rho_c^{4/3} \tag{5.111}$$

for the central pressure of a star of mass M. Supposing that a fraction β of the pressure comes from an ideal gas and the remainder $(1-\beta)P_c$ from radiation pressure, find an upper limit to the value of $(1-\beta)/\beta^4$ and estimate numerical values for stars of masses 1, 20 and $40M_\odot$. (The radiation density constant $a = 7.56 \times 10^{-15}$ erg cm^{-3} deg^{-4}. $G = 6.67 \times 10^{-8}$ dyn cm^2 gm^{-2}.)

8. Using the parameters for the centre of the Sun given in Problem 3, estimate the degeneracy parameter from Eq. (5.41).

9. Supposing that at the surface of a star ($m = M$, $l = L$), the opacity is given by a formula of the form

$$\kappa = \kappa_0 P^p T^{-t}, \tag{5.112}$$

use the hydrostatic equation (5.6) and the radiative transfer equation (5.22) to show that the pressure is related to the temperature by

$$P^{p+1} - P_s^{p+1} = \frac{16\pi ac}{3\kappa_0} \frac{GM}{L} \frac{p+1}{4+t} (T^{4+t} - T_{\text{eff}}^{4+t}), \tag{5.113}$$

where P_s is the pressure at the photosphere where the temperature is T_{eff}. Thus, as long as $t \geq -2$ or so, which holds while hydrogen is mainly positively ionized and neutral hydrogen the main opacity source, the pressure-temperature relation below the surface is quite insensitive to the values at the surface itself, justifying the 'zero boundary condition' $P_s \simeq T_s \simeq 0$. A different situation arises when hydrogen is mainly neutral and opacity comes from H$^-$.

10. Supposing that the atmospheric opacity of a 'fully' convective star is proportional to the abundance of easily ionized metals (which at low temperatures replace hydrogen as the source of free electrons to make H$^-$), correct Eq. (5.65) for the Hayashi track by finding an appropriate power of Z to multiply by.

11. Use the Saha type equilibrium that exists between α-particles, ^{8}Be and the 7.654 MeV excited state of ^{12}C to estimate the production rate per unit volume of ground-state ^{12}C nuclei in helium with a density of 10^5 gm cm^{-3} and a temperature of 10^8 K. The mass excesses are

^{4}He 2.425 MeV
^{8}Be x
^{12}C 0.000

and the spins are zero in all cases. The contribution of the width of the ^{12}C** excited state from γ-ray emission is $\Gamma_\gamma = 0.00367$ eV $\ll \Gamma_\alpha$. The frequency associated with 1 eV is 2.42×10^{14} s^{-1}.

12. Use the formula Eq. (2.43) for chemical potential to show that the equilibrium number abundance of a nucleus i with mass number $A_i = Z_i + N_i$, partition function

u_i and binding energy B_i with respect to free protons and neutrons is given by

$$Y_i = \left(\frac{\rho}{m_H}\right)^{A_i-1} \frac{u_i}{2^{A_i}} A_i^{3/2} \left(\frac{2\pi\hbar^2}{m_H k T}\right)^{3(A_i-1)/2} e^{B_i/kT} Y_p^{Z_i} Y_n^{N_i}. \tag{5.114}$$

13. When the core of a massive star exceeds the Chandrasekhar limit, it collapses and energy is absorbed by photodisintegration of ^{4}He through the reaction

$$^4He + \gamma \,(28.3\,\text{MeV}) \rightarrow 2p + 2n. \tag{5.115}$$

Assuming thermal equilibrium at a density of 10^8 gm cm^{-3}, find the temperature at which half the helium has been dissociated.

14. Using Eqs. (5.105) and (5.106), show that the time spent by a low-mass star evolving up the RGB in a given magnitude interval (and hence the relative number of stars in that interval) is predicted to vary approximately as $L^{-0.9}$. (In practice, the variation in globular clusters is more like $L^{-0.5}$.)

15. Estimate the amount of energy generated when a carbon white dwarf with the mass of the Sun is incinerated to ^{56}Ni (mass excess -53.9 MeV) and compare it with the energy needed to disrupt the star if its radius is 5000 km and its density assumed uniform. (1 MeV $\equiv 1.6 \times 10^{-6}$ erg; $G = 6.67 \times 10^{-8}$ dyn cm^2 gm^{-2}.)

6 Neutron capture processes

6.1 Introduction

As was mentioned in Chapter 1, elements above the iron group are not effectively produced by reactions between charged particles. Instead — apart from the rare 'p-process' nuclides — they result from a succession of neutron captures on a seed nucleus — predominantly ^{56}Fe. Broadly speaking, these neutron capture processes are of two distinct types: the slow or s-process, where neutrons are added on a long time-scale compared to that of most β-decays, and nuclides are built ascending the β-stability valley up to ^{209}Bi (cf. Figs 1.6, 6.1); and the rapid or r-process, in which very neutron-rich unstable nuclei are built up under extreme conditions, followed by a series of β^--decays (accompanied by fission of the heaviest nuclei) which continue after freeze-out leading to nuclides on the neutron-rich side of the stability valley (cf. Figs 1.6, 6.9). The s-process occurs in cool giant stars, especially during mixing episodes in the course of AGB evolution, whereas the r-process may occur during SN II outbursts, or just possibly in other high-energy events related to neutron stars.

6.2 The s-process

6.2.1 The physical environment

A significant clue to the order of magnitude of neutron fluxes, temperatures and densities relevant to the s-process comes from the outcome of branchings such as those shown in Fig. 6.1. The shortest lifetimes (e.g. 42 s for ^{104}Rh, 54 hr for ^{115}Cd) lead to β-decay preceding n-capture in all cases, whereas the longest (*e.g.* 7 Myr for ^{107}Pd) make the nucleus effectively stable. Branchings occur when

$$(t_{1/2})^{-1} \sim \sigma\phi, \tag{6.1}$$

where $t_{1/2}$ is the β-decay half-life (possibly less than under laboratory conditions if the nucleus has low-lying excited states), σ is the neutron capture cross-section and ϕ the neutron flux $n_n<v>$. B^{2}FH noticed that ^{151}Sm (90 yr) decays along the s-process path, whereas ^{99}Tc (2×10^5 yr) does not, leading to time scales of the order of hundreds or thousands of years. However, the argument is complicated by the temperature dependence of both neutron-capture and some β-decay rates and consideration of these factors enables permitted regions of the (n_n, kT) plane to be identified. A crude

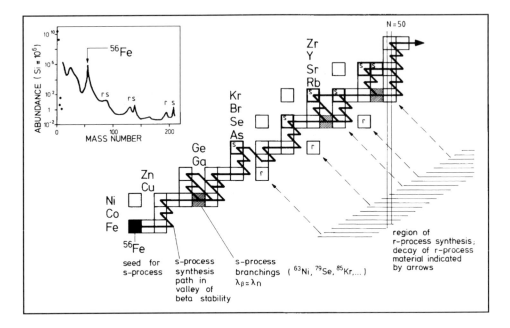

Fig. 6.1. Part of the s-process path, showing some s-only nuclei (marked 's') and some branchings between *n*-capture and β^- decay (shaded boxes), which give an idea of relevant neutron densities and temperatures. After F. Käppeler, H. Beer & K. Wisshak, 's-process nucleosynthesis — nuclear physics and the classical model', *Rep. Prog. Phys.*, **52**, 945, ©1989 by IOP Publishing Ltd. Courtesy Franz Käppeler.

argument, based on a typical temperature of 10^8 K (characteristic of helium burning), giving $<v> \simeq 10^8$ cm s^{-1}, a typical cross-section of 100 mb (cf. Fig. 6.2) and a lifetime of 300 yr gives

$$n_n \sim 10^7 \text{ cm}^{-3}. \tag{6.2}$$

More detailed considerations (Käppeler, Beer & Wisshak 1989) lead to a temperature range of up to 4×10^8 K and neutron densities up to about 10^8 cm^{-3}. Electron-density effects on β-decay lifetimes also enable the total density to be placed in the range 2500 to 13 000 gm cm^{-3}; all these parameters are characteristic of helium shell-burning zones as expected.

6.2.2   Analytical theory of the s-process

The classical analysis of the s-process assumes a simple chain starting from ^{56}Fe as the seed. The independent variable is the neutron 'irradiation', 'exposure' or 'fluence'

$$\tau \equiv \int \phi \, dt = \int n_n <v> \, dt \tag{6.3}$$

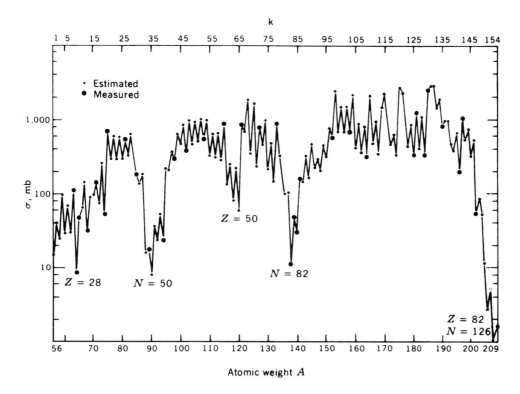

Fig. 6.2. Neutron-capture cross-sections at energies near 25 keV. Very large dips occur at the magic numbers. After D.D. Clayton, *Principles of Stellar Evolution and Nucleosynthesis*, McGraw-Hill, 1968 and University of Chicago Press 1984, p. 556. ©1984 by the University of Chicago. Courtesy Don Clayton.

and the abundances N_i are governed by the differential equation

$$\frac{dN_i}{d\tau} = \sigma_{i-1}N_{i-1} - \sigma_i N_i \; ; \quad 56 \le i \le k \le 209 \tag{6.4}$$

with the initial conditions

$$N_{56}(0) = 1 \, ; \; N_{i \ne 56}(0) = 0. \tag{6.5}$$

Thus, in a steady state, $dN_i/dt = 0$, one has $\sigma N = \text{const.}$, which actually holds over considerable ranges of mass number (thus being usable as a local approximation), but not globally (except if $\tau \to \infty$); in particular, it breaks down near the magic numbers (see Fig. 6.3).

The general solution of Eq. (6.4) is

$$N_k(\tau) = \sum_{i=56}^{k} C_{ki} e^{-\sigma_i \tau}, \tag{6.6}$$

where

$$C_{56\,56} = 1; \ C_{57\,56} = -C_{57\,57} = \frac{\sigma_{56}}{\sigma_{57} - \sigma_{56}}; \ C_{ki} = \prod_{j=56}^{k} \left(\frac{\sigma_j}{\sigma_j - \sigma_i} \right); \ j \neq i. \qquad (6.7)$$

This solution is of little practical use, because small inaccuracies in $\sigma_j - \sigma_i$ lead to large errors in large terms with alternating signs, swamping the true differences. Clayton *et al.* (1961) derived an approximation, based on the Poisson-like distribution obtained when all the cross-sections are equal:

$$\sigma_k N_k(\tau) = \frac{\lambda_k^{m_k} \tau^{m_k - 1}}{\Gamma(m_k)} e^{-\lambda_k \tau}; \ \lambda_k \equiv \frac{<\sigma^{-1}>_k}{<\sigma^{-2}>_k}; \ m_k \equiv \frac{<\sigma^{-1}>_k^2}{<\sigma^{-2}>_k} (k - 55). \qquad (6.8)$$

This formula for a single exposure was found not to give a good fit to the solar-system σN curve for s-only isotopes. Seeger, Fowler & Clayton (1965) found, however, that a much better fit is obtained by assuming an exponential distribution of exposures given by the probability distribution

$$p(\tau)\, d\tau = e^{-\tau/\tau_0}\, d\tau/\tau_0. \qquad (6.9)$$

The resulting abundances are given by

$$\tilde{N}(\tau_0) = \int_0^\infty N(\tau)\, p(\tau)\, d\tau, \qquad (6.10)$$

which is a Laplace transform of the solution for a single exposure (Clayton & Ward 1974).

The Laplace transformation makes the solution of the differential equation (6.4) very easy. Operating on that equation as in Eq. (6.10), one obtains

$$\int_0^\infty \frac{dN_{56}}{d\tau} e^{-\tau/\tau_0} \frac{d\tau}{\tau_0} = -\sigma_{56}\tilde{N}_{56}(\tau_0). \qquad (6.11)$$

Integrating the lhs of Eq. (6.11) by parts,

$$\tilde{N}_{56}(\tau_0) - 1 = -\tau_0\,\sigma_{56}\tilde{N}_{56}(\tau_0). \qquad (6.12)$$

Similarly,

$$\frac{1}{\tau_0}\tilde{N}_{57}(\tau_0) = -\sigma_{57}\tilde{N}_{57}(\tau_0) + \sigma_{56}\tilde{N}_{56}(\tau_0) \qquad (6.13)$$

etc. This results in a series of *algebraic* equations, easily solved from the top down. Dropping the 'tildes',

$$\sigma_{56}N_{56} = \frac{1}{\tau_0}\left(1 + \frac{1}{\tau_0\,\sigma_{56}}\right)^{-1}; \qquad (6.14)$$

$$\sigma_{57}N_{57} = \frac{1}{\tau_0}\left(1 + \frac{1}{\tau_0\,\sigma_{56}}\right)^{-1}\left(1 + \frac{1}{\tau_0\,\sigma_{57}}\right)^{-1}; \qquad (6.15)$$

$$\sigma_k N_k = \frac{1}{\tau_0}\prod_{i=56}^{k}\left(1 + \frac{1}{\tau_0\,\sigma_i}\right)^{-1}. \qquad (6.16)$$

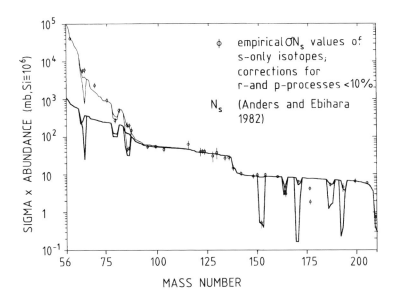

Fig. 6.3. Product σN of abundance and neutron capture cross-section for s-only nuclides in the Solar System. The main and weak s-process components are shown by the heavy and light curves respectively. Units are mb per 10^6 Si atoms. After F. Käppeler, H. Beer & K. Wisshak, 's-process nucleosynthesis — nuclear physics and the classical model,' *Rep. Prog. Phys.*, **52**, ©1989 by IOP Publishing Ltd. Courtesy Franz Käppeler.

Thus, as long as $\tau_0 \, \sigma_i$ is large, σN declines slowly with increasing atomic mass number k, but when σ becomes very small at the magic neutron numbers, there is a sudden drop (Fig. 6.3). (Such behaviour is characteristic of any distribution of neutron exposures that declines gradually with τ; the exponential formula is mathematically convenient and has a physical interpretation that will be described below.) Eq. (6.16) with $\tau_0 \simeq 0.3$ mb^{-1} (for cross-sections at $kT = 30$ keV) gives an excellent fit to solar-system abundances above $A = 80$ (Kr) to 90 (Zr), apart from the dips caused by branchings, and this exponential distribution of exposures is known as the main component of the s-process; the quality of the fit obtained from such simple considerations is arguably the most elegant result in the whole of nucleosynthesis theory. For lower mass numbers, an additional component, known as the weak component, is required, corresponding to either a single exposure with $\tau = (4$ mb$)^{-1}$ or an exponential distribution with $\tau_0 = (16$ mb$)^{-1}$; of these, the single exposure (perhaps arising in helium and carbon-burning zones of a massive star) seems to be more successful in fitting certain details (Raiteri *et al.* 1993).

Table 6.1 shows some other best-fit parameters to solar-system s-process abundances. The seed nucleus is basically ^{56}Fe; light nuclei have low cross-sections (but can act as neutron 'poisons', e.g. ^{14}N for the ^{13}C(α, n) neutron source), whereas heavier nuclei are not abundant enough to have a major influence. Certain nuclidic ratios, e.g.

Table 6.1. *Parameters of the two s-process components*

| | Fraction of Fe seeds,% | | Exposure τ or τ_0 | | n_c |
	Single	Exponential	Single	Exponential	Exponential
Main	—	0.057 ± 0.004	—	0.30 ± 0.01	10.7 ± 0.7
Weak	0.32	0.26	0.23	0.06 ± 0.01	1.4 ± 0.4

Adapted from Gallino (1989) and Käppeler *et al.* (1990).

^{37}Cl/^{36}Ar and ^{41}K/^{40}Ca, indicate that under 1 per cent of solar-system material has been s-processed.

Another parameter is the mean number n_c of neutrons captured by an iron seed, given by

$$n_c = \frac{\sum_{57}^{209}(A-56)\,N(A)}{N(56)} = \frac{1}{\tau_0}\sum_{57}^{209}\frac{A-56}{\sigma_A}\prod_{i=56}^{A}\left(1+\frac{1}{\tau_0\,\sigma_i}\right)^{-1}. \qquad (6.17)$$

The mean exposure $\tau_0 = (3\ \mathrm{mb})^{-1}$ implies $n_c \simeq 10$, which is quite a strong constraint on the physical environment, including in particular the abundance of the neutron source relative to iron. Thus the ^{22}Ne$(\alpha, n)^{25}$Mg source, which arises naturally in helium burning at temperatures of 3 to 4×10^8 K, is unlikely to contribute the bulk of the main component: the temperature and fluxes are too high and n_c is too low (because about 3/4 of the neutrons generated are captured by Ne and its neighbours acting as neutron 'poisons'). Furthermore, the Mg isotope ratios are observed to be unchanged in carbon stars, whereas one could expect an enhancement of the heavy isotopes ^{25}Mg, ^{26}Mg by a factor of the same order as that of the heavier s-process elements, i.e. an order of magnitude. The ^{22}Ne source could provide the weak component, however, and it is thought to make a minor contribution to the main one (cf. Fig. 6.5). The ^{13}C(α, n) source, which works at lower temperatures (10^8 K) and can fit the constraints under certain conditions, requires a non-classical mixing process (cf. Section 5.10.4) leading to a need for some arbitrary assumptions in the computational treatment.

6.2.3   Meaning of the exponential exposure distribution

Originally, the exponential distribution of exposures was thought to be an accident of Galactic reprocessing, but it was shown by Ulrich (1973) that it arises naturally as a result of successive thermal pulses in a single AGB star (Fig. 6.4; cf. Fig. 5.19). During an interpulse period, the helium core increases in mass as a consequence of hydrogen shell burning by an amount ΔM_α; almost all of this is mixed into the intershell convective region of mass M_{isc} generated by the next pulse, giving rise to an exposure $\Delta\tau$ to slow neutrons (the source of the neutrons is discussed below). Thus, a fraction $\Delta M_\alpha / M_{\mathrm{isc}} \equiv 1 - r$ receives an exposure $\Delta\tau$ for the first time, while the remaining

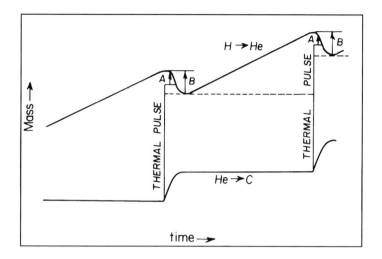

Fig. 6.4. Schematic view of ingestion during and after a pulse. Pulses appear as vertical lines owing to their short duration (~ 300 yr) compared to the interpulse period ($\sim 10^4$ to 10^5 yr). After G.J. Wasserburg *et al.* (1994), *Ap. J.*, **424**, 412. Courtesy G.J. Wasserburg.

fraction r overlaps with the previous pulse and has undergone an exposure of $2\Delta\tau$ (or more). A fraction r^2 overlaps 2 previous pulses, etc. Consequently, when a steady state is reached after a sufficient number of pulses, the probability of an exposure $n\Delta\tau$ is

$$(1-r)\,r^{n-1} = \frac{1-r}{r}\,r^{\tau/\Delta\tau} \propto e^{-\tau/\tau_0}, \tag{6.18}$$

where

$$\tau_0 = \Delta\tau/\ln(1/r). \tag{6.19}$$

Fig. 6.4 also illustrates the mechanism whereby ^{13}C may be introduced into the intershell convection zone of a low or intermediate mass star. When that zone reaches its maximum extension, it covers almost the whole intershell region apart from a thin interval marked A. The mass of this region decreases from about $2 \times 10^{-3} M_\odot$ in the first cycles to about $10^{-4} M_\odot$ in the final ones, as the mass of the helium core increases from about $0.5 M_\odot$ to about $0.7 M_\odot$. Immediately after the pulse, the outer layers of the star expand and cool, so that the H-burning shell is temporarily extinguished and the envelope convection may extend downward, dredging up H and He burning products to the surface. The material thus dredged up, labelled B, has two parts with different compositions: layer A contains only H-burning ashes, whereas the inner part has He-burning products like carbon and s-process nuclei. A small semi-convective region, driven by partial carbon recombination in the expanding material, may develop below the border of envelope convection at its maximum downward extension, causing a small amount of protons from the envelope to enter the carbon-rich zone and generate ^{13}C and a little ^{14}N in a tiny pocket (represented by the broken line in Fig. 6.4) when

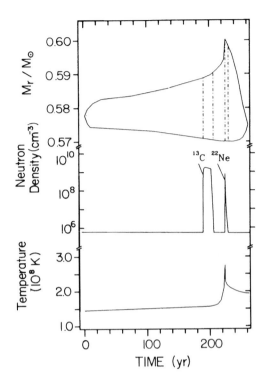

Fig. 6.5. Development of the convective region, neutron density from ^{13}C and ^{22}Ne sources and maximum temperature as functions of time during a thermal pulse in a low-mass star with $Z \simeq Z_\odot/3$, which seems to give the best fit to solar-system abundances from the main s-process. After F. Käppeler, R. Gallino, M. Busso, G. Picchio & C.M. Raiteri, *Ap. J*, **354**, 630, 1990. Courtesy Maurizio Busso and Claudia Raiteri.

the H-burning shell reignites. This pocket is then ingested by the next thermal pulse and generates neutrons. The amount of ^{13}C involved is a matter of guesswork; values in the range 10^{-6} to $10^{-5}M_\odot$ are considered in some current models (Wasserburg *et al.* 1994).

Fig. 6.5 shows a computation of the development of physical conditions during a pulse. The ^{13}C source is activated at a relatively low temperature at the moment when the intershell convection zone reaches up to the level of the pocket. Later, after exhaustion of ^{13}C, the temperature rises and the ^{22}Ne source is briefly activated, contributing significantly to a few isotopes that depend on temperature-sensitive branchings. In general, the results from this stellar evolutionary model agree well with the classical analytical treatment of Section 6.2.2, when the latter is corrected for branchings (with a neutron density of $(3.4 \pm 1.1) \times 10^8$ cm^{-3} and a temperature $kT = 29 \pm 5$ keV). The model leads to a mean neutron exposure $\tau_0 = 0.3$ mb^{-1} (for 30 keV), resulting from $\Delta\tau(30$ keV$) = 0.17$ mb^{-1} per pulse and an overlap ratio $r = 0.6$ (cf. Eq. 6.19), in excellent agreement with the classical analysis of the main component. However, it now becomes clear that at least two very different time-scales are involved, the short

Table 6.2. *Spectral types of red giants with carbon and/or s-process anomalies*

RGB		AGB	
	$\leftarrow T_{\mathrm{eff}}$		
	$L \rightarrow$		
G, K, M	M	Mira	OH-IR
	MS	"	"
BaII	S	"	$C \leq O$
BaII	SC	"	$C \simeq O$
CH, BaII	CS	"	$C > O$
(Early) R	N	"	C-stars

time of each neutron exposure $\Delta\tau$ (in this case 12 yr) and the long time between pulses (2×10^5 yr). Most nuclei have total lifetimes $\tau = 1/(\lambda_n + \lambda_\beta)$ that are either so long or so short that the pulsed nature of the exposure makes no difference, but an exception among the main s-process branch points is ^{85}Kr (shown by a shaded box in Fig. 6.1) with $\tau = 4$ yr at a temperature of 29 keV, so that there should be a noticeable effect on the abundance of its descendants ^{86}Kr and ^{87}Rb; in the detailed stellar model, however, this goes away because the temperature is only 1.5×10^8 K (13 keV) at the relevant times (Käppeler *et al.* 1990). In some still more recent models, which follow the temperature evolution after the decay of a pulse in considerable detail, the ^{13}C is already consumed in the inter-pulse period, leading to a relatively low neutron density of the order of 10^7 cm^{-3}, which helps to solve any problem with the ^{85}Kr branching (Lambert *et al.* 1995); in this case the ^{13}C neutron pulse shown in Fig. 6.5 goes away, although the ^{22}Ne pulse is still present (Straniero *et al.* 1995).

6.2.4 Surface abundances in red giants

Table 6.2 gives an overview of some of the stages of stellar evolution where carbon and/or s-process anomalies occur (cf. Fig. 3.30). The C/O ratio increases down the series. In addition to the types listed there, there are infra-red carbon stars such as IRC +10216[1], proto-planetary nebulae and a whole zoo of peculiar carbon stars including J stars (strong ^{13}C as in the case of HD 52432 shown in Fig. 1.7) and hydrogen-deficient carbon stars which can be cool (*e.g.* R Cor Bor, RY Sag and HD 137613 shown in Fig. 1.7) or hot (when they look like extreme helium stars); such stars may have lost their envelopes by binary mass transfer.

Planetary nebulae are often even more rich in carbon than cool carbon stars, and those classified by M. Peimbert as Type I are rich in nitrogen, indicating effects of hot-bottom burning in intermediate-mass progenitor stars. s-process elements are not

---

[1] IRC stands for the CalTech infra-red (2μm) catalogue.

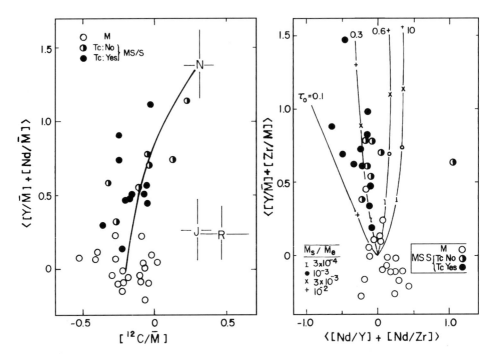

Fig. 6.6. Left panel: Correlation of s-process enhancement with carbon abundance in carbon stars. Right panel: Enhancement of lighter (main) s-process elements plotted against excess enhancement of a heavier element (Nd) relative to the lighter ones (Y and Zr), compared to predictions for different values of τ_0. m_s and m_e refer to mass of s-processed material dredged up and envelope mass respectively. After V.V. Smith & D.L.Lambert (1990), *Ap. J. Suppl.*, **72**, 387. Courtesy David Lambert.

normally detectable in PN or their central stars, but a remarkable case is that of FG Sagittae, the central star of a fossil planetary nebula, which has cooled in the course of the twentieth century from around 25 000 K to around 5000 K at constant bolometric luminosity. This star suddenly underwent an enhancement of s-process elements in its atmosphere between 1965 and 1972.

Observations of carbon stars in the Magellanic Clouds indicate luminosities between $10^{3.5}$ and $10^{4.3} L_\odot$, in agreement with theoretical estimates of where the TP–AGB phase of evolution sets in for stars with initial mass 1 to $5M_\odot$ (cf. Fig. 5.14). The warmer and less luminous Ba II and CH stars, with about $100 L_\odot$ (cf. Fig. 3.30), therefore presumably have never reached this stage of evolution, but rather have acquired processed material by binary mass transfer from an initially more massive companion which is now a white dwarf. These stars, being warmer, have spectra that are much easier to analyse than are those of AGB stars, and a Ba II star played an important role in early investigations of s-process nucleosynthesis (cf. Fig. 1.8). The same phenomenon of mass transfer may also be responsible for carbon and s-process enhancements in more luminous stars, e.g. some S stars which have similar composition to Ba II stars;

a plausible criterion for this is the absence of technetium (present in some S stars such as R And shown in Fig. 1.8, but absent in others), which does not decay very much between pulses but would have done so over the much longer periods that have elapsed since the termination of the mass transfer episode. CH stars, with strong G-bands of CH and s-process enhancement, occur at various luminosities (see Fig. 3.30) and in different population groups, and are also believed to result predominantly from mass transfer. Certain CH stars of Population II, with low metallicity, display an unusually high ratio of heavy s-process elements (e.g. Ba) to lighter ones (e.g. Sr), indicating high values of τ_0 and n_c; this is an expected result from a fixed amount of ^{13}C injection combined with a low abundance of iron seeds.

Fig. 6.6 shows some observational correlations between s-process and carbon enhancements and between lighter and heavier (main) s-process products. There is a steady progression from M to MS to S to N stars, giving a fair correlation between s-elements and ^{12}C as one might expect from the thermal pulsing model, but this is not shared by J and R stars which have enhanced ^{13}C. The relation between lighter and heavier (main) s-process products is consistent in most cases with a mean exposure $\tau_0 \simeq 0.3$ mb^{-1} (for $kT = 30$ keV), about the same as in the Solar System, which suggests that these low to intermediate mass carbon stars are typical of the progenitors of the solar-system distribution. Jura (1989) has estimated the rate at which carbon and s-processed material are ejected into the ISM near the Sun by winds from these stars. From the Cal Tech 2μm infra-red survey, there are about 40 carbon stars per kpc^2 with a typical scale height of 200 pc, corresponding to an initial mass of order 1.5 $M_\odot$, and the typical mass loss rate is around $2 \times 10^{-7} M_\odot$ yr^{-1}, but dominated by a few stars with a much higher mass loss rate of over $10^{-5} M_\odot$ yr^{-1}. The total mass loss rate from carbon stars, estimated from the IRAS 60μm survey, is within a factor of 2 or so $2 \times 10^{-4} M_\odot$ kpc^{-2} yr^{-1}, leading over 10^{10} yr to an output of 2 $M_\odot$ pc^{-2}. Enriched by an order of magnitude and mixed with the total surface density of about 50 $M_\odot$ pc^{-2}, this would be enough to supply an average s-process abundance of 0.4 solar, which is a reasonable average for Pop I stars.

Thus the ^{13}C neutron source (with a little assistance from ^{22}Ne) in thermally pulsing low and intermediate-mass stars is well established as the chief source of the main component of s-process nuclides in the Solar System. It is not quite clear, however, whether the τ_0 parameter is something unique, or just the average over a more-or-less broad distribution of values; nor is it clear why a similar s-process pattern is seen in stars that are metal-deficient by factors of up to 100 (cf. Pagel & Tautvaišienė 1997).

6.3 **The r-process**

6.3.1 Introduction

A separate neutron capture process is needed for neutron-rich nuclides by-passed by the s-process and for species above ^{209}Bi. A possible path for this 'rapid' or r-process is shown in Fig. 6.9.

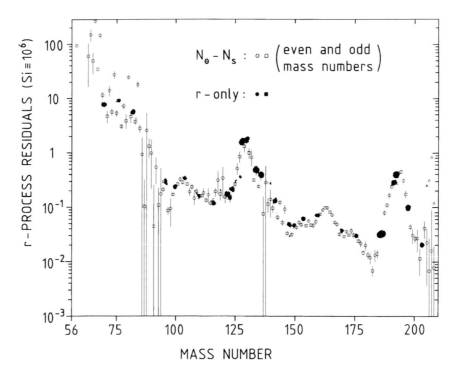

Fig. 6.7. r-process abundances in the Solar System. Filled circles represent r-only nuclides, while open circles with error bars show the result of subtraction of a calculated s-process contribution. After F. Käppeler, H. Beer & K. Wisshak, 's-process nucleosynthesis — nuclear physics and the classical model,' *Rep. Prog. Phys.*, **52**, ©1989 by IOP Publishing Ltd. Courtesy Franz Käppeler.

The abundance curve for r-process products in the Solar System is obtained from 27 r-only nuclides, combined with numbers obtained by subtracting predicted abundances of an s-process contribution in the other cases (Fig. 6.7); the latter are correspondingly less accurate, particularly when the s-process contribution dominates, as is shown by the error bars in the figure. The result is a fairly smooth curve, with broad peaks corresponding to the progenitor magic numbers $N = 82$, 126, and a third peak centred on $A = 165$. Fig. 6.8 shows the r-process curve in its relation to the corresponding curves for the s- and p-processes. It is presumably just a coincidence that the s- and r-process curves have about the same general level.

For purposes of comparison with stellar abundances, it is useful to have the relative contributions of s- and r-processes to the various elements (as opposed to nuclides) in the Solar System, because in most cases only element abundances without isotopic ratios are available from stellar spectroscopy. At the same time, elements formed in one process may often be expected to vary by similar factors in the course of stellar and galactic evolution, but to be found in differing ratios to elements formed in another process. Relative contributions are listed for some key elements in Table 6.3.

Table 6.3. *Relative r and s contributions to some elements in the Solar System*

Element	log N	s-weak	s-main	r
37 Rb	2.40	.14	.39	.47 ±.10
38 Sr	2.93	.09	.77	.14 ±.07
39 Y	2.22	.04	.85	.11 ± .06 :
40 Zr	2.61	.02	.78	.20 ± .03
56 Ba	2.21	.01	.88	.11 ± .02
57 La	1.20	.01	.75	.25 ± .08
58 Ce	1.61	.01	.77	.23 ± .01
59 Pr	0.71	.01	.45	.54 ± .09
60 Nd	1.47	.00	.46	.53 ± .03
62 Sm	0.97	.00	.30	.70 ± .03
63 Eu	0.54	.00	.03	.97 ± .06
66 Dy	1.15	.00	.12	.88 ± .15

Numbers in the first 4 rows (apart from the error estimates) are taken from Raiteri, Gallino & Busso (1992). The others are based on data for individual isotopes given by Käppeler, Beer & Wisshak (1989).

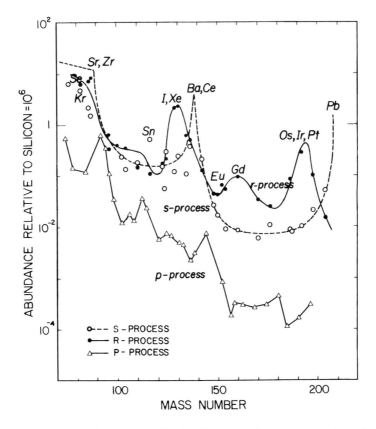

Fig. 6.8. Abundance curves for s-process (broken line), r- and p-process products (solid lines) in the Solar System. Adapted from Cameron (1982a).

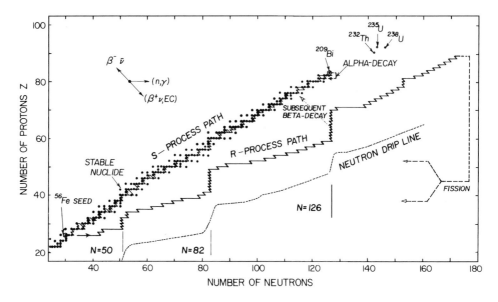

Fig. 6.9. Neutron capture paths in the N, Z plane. The r-process path was calculated for a temperature of 10^9 K and a neutron density of 10^{24} cm^{-3} (Seeger, Fowler & Clayton 1965). The dotted curve shows a possible location of the neutron drip line after Uno, Tachibana & Yamada (1992). Adapted from Rolfs & Rodney (1988).

6.3.2   Physical environment of the r-process

A possible path for the r-process is shown in Fig. 6.9. The iron seed nuclei need to capture many neutrons leading to unstable nuclei with very short β-decay half-lives, demanding a high neutron density:

$$t(n, \gamma) \ll t_\beta \simeq 10^{-3} \text{ to } 1 \text{ s} \Rightarrow n_n > 10^{19} \text{ cm}^{-3}. \tag{6.20}$$

Such neutron densities are presumably associated with very high temperatures, leading to reverse (γ, n) reactions. At each Z, neutrons are added up to a so-called waiting point (which defines the r-process path); in the 'waiting-point approximation', (n, γ) and (γ, n) reactions balance, so that for given Z, there is a Saha-type equilibrium

$$\frac{n(A+1, Z)}{n(A, Z)} \simeq 10^{-34} \, n_n \, T_9^{-3/2} e^{Q/kT}, \tag{6.21}$$

where Q is the neutron separation energy. Increasing n_n thus drives the abundance peak where $N(A+1, Z) = N(A, Z)$ towards lower values of Q, i.e. closer to the neutron drip line where $Q = 0$; the path shown in Fig. 6.9 was computed taking $Q = 2$ MeV for the abundance peak, corresponding to $T = 10^9$ K; $n_n = 10^{24}$ cm^{-3}. However, the estimation of $Q(A+1, Z)$ — or in other words the determination of where in the (N, Z) plane one has $Q = 2$ MeV — depends on understanding the interaction between nucleons sufficiently well that one can estimate nuclear masses in unknown territory;

the tentative neutron drip line shown in Fig. 6.9 is not the one used in computing the r-process path shown there, and in any case it is possible that the rises at the magic neutron numbers have been overestimated because nuclear shell effects are suppressed at the neutron drip line (Chen *et al.* 1995). The abundance peak will actually be a few units wide in A or N because of odd-even effects in the neutron separation energy; these effects also have the result that neutron capture always terminates at an even value of N.

From time to time, a β-decay occurs increasing Z by 1 unit; this leads to an increase in Q (corresponding to the increased distance above the neutron drip line) and consequently to further neutron captures until Q is again reduced to the appropriate value and a further β-decay occurs. At the magic numbers, this leads to a vertical zig-zag track, parallelling the rise in the neutron drip line. Along this track, each neutron capture is followed by a β-decay until the nucleus comes so close to the β-stability valley that its β-decay half life becomes long enough for the chain of neutron captures to resume. In the 'steady flow approximation', the chain of β-decays is assumed to reach a steady state, so that the abundance of a given element (fixed Z) is proportional to its β-decay lifetime at the waiting point[1], and this is preserved in a freeze-out. After freeze-out, each nucleus decays at leisure to the nearest stable isobar. The final abundances for given T, n_n therefore depend essentially only on nuclear masses and β-decay rates (which themselves depend on nuclear masses, *inter alia*); the abundance peaks are due to the relatively long β-decay lifetimes of the magic-numbered progenitors (especially those nearest to the β-stability valley) and neutron capture cross-sections do not come in.

The r-process path is terminated by (neutron-induced or β-delayed) fission near $A_{\max} = 270$, feeding matter back into the process at around $A_{\max}/2$, followed by recycling as long as the neutron supply lasts, assuming sufficient seed nuclei to start the process off. The number of heavy nuclei is thus doubled at each cycle, which could take place in a period of a few seconds. β-delayed fission also occurs after freeze-out, when the β-decay leaves nuclei with $A \geq 256$ or so with an excessive positive charge (cf. Eq. 2.89).

In modern computations, the waiting-point approximation is replaced by detailed dynamical calculations in which neutron capture cross-sections are significant. Alternatively, improved fits to abundances can be obtained by approximating the dynamical process with some combination of waiting-point calculations with different sets of parameters. However, all the parameters depend on estimates for nuclei that are far from stability and for which measurements do not yet exist. Some may eventually arrive in the future from the use of radioactive beams and there are prospects for an improved theoretical understanding of the structure of exotic nuclei, depending on pairing effects, shell effects and the like. In the meantime, there are major uncertainties

---

[1] More precisely, the inverse of the average decay constant of the various isotopes, weighted by their abundance according to the Saha equation expressed roughly by Eq. (6.21).

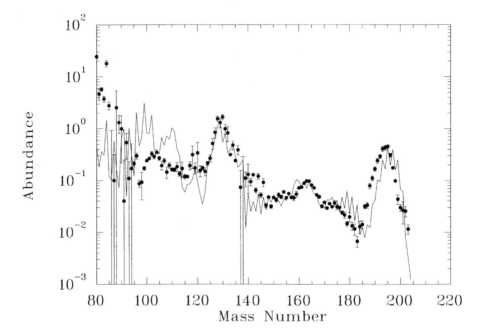

Fig. 6.10. Results of a dynamical calculation of the r-process in the hot neutrino bubble inside a $20M_\odot$ supernova (continuous curve) compared to the observed solar-system abundance distribution (filled circles). After S.E. Woosley *et al.*, *Ap. J.*, **433**, 229 (1994). Courtesy Brad Meyer.

in the nuclear physics, and r-process abundances may actually help in improving the understanding of nuclei (cf. Chen *et al.* 1995).

The site of the r-process is also not clear, but it seems that the conditions needed to reproduce solar-system r-process abundances may hold in the hot bubble caused by neutrino winds in the immediate surroundings of a nascent neutron star in the early stages of a supernova explosion (cf. Fig. 6.10). Circumstantial evidence from Galactic chemical evolution supports an origin in low-mass type II supernovae, maybe around $10M_\odot$ (Mathews, Bazan & Cowan 1992; Pagel & Tautvaišienė 1995). In stars with extreme metal-deficiency, the heavy elements sometimes display an abundance pattern characteristic of the r-process with little or no contribution from the s-process, and the pattern is remarkably close to that found for r-process elements in the Solar System (cf. Section 10.3.4).

Notes to **Chapter 6**

D.D. Clayton, *Principles of Stellar Evolution and Nucleosynthesis*, McGraw-Hill 1968, University of Chicago Press 1984, gives a very instructive account of both s and r-processes. Some more recent developments are described in

C.E. Rolfs & W.S. Rodney, *Cauldrons in the Cosmos*, University of Chicago Press 1988.
   An excellent review of the s-process is given in the article by
F. Käppeler, H. Beer & K. Wisshak, *Rep. Prog. Phys.*, **52**, 945 (1989).
   Modern dynamical calculations of the r-process are described by Cowan, Thielemann
& Truran (1991ab), Takahashi, Witti & Janka (1994) and Woosley *et al.* (1994). Chen
et al. (1995) discuss the nuclear physics issues.
   The origin of the *p*-nuclei is discussed by
D.L. Lambert, 'The p-nuclei: Abundances and Origins', *A & A Reviews*, **3**, 201 (1992).

Problems

1. Solve the differential equation (6.4) for the case when all neutron capture cross-sections are equal and show that it leads to the Poisson distribution for a single exposure τ. Use this to derive an equivalent to Eq. (6.16) for this case when there is an exponential distribution of exposures as in Eq. (6.9).

2. Given that seed nuclei are exposed to a flux of neutrons at $T = 1.5 \times 10^8$ K, $n_n = 10^9$ cm^{-3} for 20 years during a pulse, and $\tau_0 = 0.3$ mb^{-1}, estimate the overlap fraction r between successive intershell convective zones.

3. Show that, for β-decay with highly relativistic electrons, the decay rate is proportional to $Q_\beta{}^5$ (the Sargent rule).

7 Galactic chemical evolution: basic concepts and issues

7.1 Introduction

Different aspects of the evolution of galaxies include **dynamical**, involving diffuse material ('gas', which will be understood to include dust), stars and dark matter; **thermal** (mainly affecting the gas); **photometric** + **spectrophotometric** (involving stars and gas); and so-called **galactic chemical evolution** (GCE) which is not really about chemistry (an important topic in its own right) but concerns the origin and distribution of nuclear species (loosely referred to as elements) in stars and gas. True insights into the origin and evolution of galaxies need studies of all these different aspects and their inter-relations, but there are several results of GCE that can be at least partially understood using only the very broadest ideas about other poorly understood aspects (e.g. the physics of galaxy and star formation) and this makes GCE a topic that is worth studying in its own right. At the same time, there are still many uncertainties, both in the underlying theory of stellar evolution, nucleosynthesis and mass loss and in the galactic context in which these basic element-forming processes take place. These limit the extent to which safe deductions can be made and motivate one to use the simplest possible models. The present chapter describes some basic principles, while the following one will discuss some specific models and related observational results.

7.2 The overall picture

The general picture has been shown (very schematically!) in Fig. 1.1. The Big Bang leads to significant initial abundances of H, D, ^{3}He, ^{4}He and ^{7}Li in the primordial intergalactic medium (Chapter 4). The IGM is assumed to condense into galaxies consisting initially of gas (probably enclosed in dark-matter halos) which then cools, collapses and makes stars. The star formation may take place either on a short time-scale (as appears to have happened in the formation of spheroidal systems like elliptical galaxies and the bulges and halos of spirals) or on a long time scale (in the formation of disk components in spiral and irregular galaxies in which substantial quantities of gas are still present). This picture may, however, be seriously over-simplified, as there is evidence that sudden violent star formation is often triggered by interactions and mergers, and such effects may be involved in the formation of many or even all spheroidal systems (cf. Chapter 11). In any case, galaxy formation is followed and/or accompanied by star formation leading to synthesis of heavier elements and modification of primordial abundances: D is destroyed by stellar activity, ^{4}He is mildly topped up and ^{3}He and ^{7}Li are both created and destroyed by stars.

When stars are formed, there are broadly three things that can happen to the interstellar medium (ISM) as a result of their evolution (cf. Chapter 5). Small stars with less than about 1 solar mass ($M_\odot$) have such long lifetimes that essentially nothing happens (except perhaps minor effects of flares and dwarf novae): they simply serve to lock up part of the gas and remove it from circulation, but they remain visible today as archaeological tracers of the composition of the ISM at the time and place where they were formed. Middle-sized stars (say 1 to $9M_\odot$) undergo various dredge-up processes in which first some ^{12}C is changed into ^{13}C and ^{14}N and later (in the third dredge-up) ^{3}He, ^{4}He, ^{7}Li, ^{12}C, ^{14}N and s-process elements are synthesised to varying degrees. The products can be seen in carbon stars and some planetary nebulae (PN), and they are expelled in the form of stellar winds and PN, leaving behind a degenerate core (usually a white dwarf) which serves to lock up additional material in the course of time. Close binaries in this mass range can undergo still more dramatic effects as a result of mass transfer; in particular SN Ia, which contribute substantially to the nucleosynthesis of iron, are believed to result from a white dwarf which accretes enough mass from a companion (possibly even merging with a companion white dwarf) to exceed the Chandrasekhar limiting mass, ignite the degenerate CO core and disintegrate completely. Above about $9M_\odot$ we have the third category of stars, the big ones, which complete their evolution in well under 10^8 years and ignite carbon non-degenerately, whereafter (owing largely to neutrino losses) they rapidly evolve to core collapse. In the standard picture, the core becomes a neutron star and the rest is expelled in a supernova explosion (Type II and some others) in which earlier nucleosynthesis products are ejected after partial modification through explosive nucleosynthesis in the inner layers. There are, however, considerable uncertainties in the range of initial stellar masses in which this picture holds, and the more massive stars could collapse into black holes with or without a previous SN outburst. Before completing their evolution, massive stars also undergo mixing and mass-loss episodes of many different kinds, manifested by Wolf–Rayet (WR) stars, luminous blue variables like η Carinae and the circumstellar shells (presumably resulting from winds in the red supergiant phase) of certain WR stars and supernovae like the radio source Cassiopeia A and SN 1987A in the Large Magellanic Cloud (LMC). These all show evidence of CNO processing.

The net result of all these processes is to produce a 'standard' or 'local Galactic' abundance distribution in our vicinity with typical mass fractions $X \simeq 0.7$ for hydrogen, $Y \simeq 0.28$ for helium and $Z \simeq 0.02$ for heavier elements consisting chiefly of O, C, N, Ne, Mg, Si and Fe (cf. Fig. 1.4).

7.3　Ingredients of GCE models

To put together a model for the chemical evolution of galaxies, one needs the following ingredients:

(i) Initial conditions.

(ii) A picture of the end-products of stellar evolution, i.e. which stars eject how much of their mass in the form of various heavy elements after how much time?

(iii) The initial mass function (IMF) which gives the relative birthrates of stars with different initial masses.

(iv) A model of the total star formation rate as a function of time, gas mass, gas density and/or many other parameters including stochastic variables.

(v) Assumptions about all other relevant processes in galactic evolution besides the birth and death of stars.

7.3.1　Initial conditions

Mostly one assumes that we begin with pure gas with primordial abundances from the Big Bang. But there are also models with a 'prompt initial enrichment' or 'initial nucleosynthesis spike'. The latter could represent either a hypothetical pre-galactic or proto-galactic process involving preferentially high-mass objects (Truran & Cameron 1971) or prior enrichment of the system by products from some neighbouring more evolved system, e.g. of the disk of the Galaxy by the bulge or the halo.

7.3.2　End-products of stellar evolution

In principle these should be predictable from theory, but in practice there are many grey areas such as the effects of mixing and mass loss, the mechanism of stellar explosions, nuclear reaction rates such as $^{12}C(\alpha,\gamma)^{16}O$, the evolution of close binaries and the corresponding mass limits between which various things happen for differing initial chemical compositions. Fig. 5.13 shows a version of what may happen in single stars with different initial masses and two metallicities, $Z \simeq Z_\odot$ and $Z \simeq Z_\odot/20$.

Other representative data for stars with different initial masses are given in Table 7.1. Numbers in brackets refer to $Z = 0.001$, $Y = 0.24$, others to $Z = 0.02$, $Y = 0.28$, i.e. near solar. The luminosities of the most massive stars are quite insensitive to Z.

The contributions of different stars to nucleosynthesis depend on their initial mass and chemical composition, their mass-loss history in the course of evolution and effects of close binaries (especially SN Ia). When mass loss is small, as is believed to be the case for low metallicities, the distribution of 'primary' elements (those synthesised directly from hydrogen and helium) in ejecta from massive stars is mainly the result of hydrostatic evolution with some modifications to deeper layers resulting from explosive nucleosynthesis in the final SN outburst, classically associated with SN II (see Chapter 5). The resulting production of He, C and O is then quite insensitive to initial chemical composition as such, although it will be affected by the tendency of mass loss to

Table 7.1. *Representative stellar data*

$M_{init.}$ ($M_\odot$)	On zero-age main sequence (ZAMS):						τ_{ms} Myr or Gyr[a]	τ_{tot} Myr or Gyr[b]
	log $L/L_\odot$	T_{eff} (10^3K)	Sp	M_V	$B - V$			
120	6.2	53(59)					3.0(2.9)	3.9(3.2)
60	5.7	48(52)					3.5(3.9)	4.0(4.3)
40	5.4	44(48)	O5	−5.6	−0.32		4.4(5.1)	5.0(5.5)
20	4.6	35(39)	O8	−5.0	−0.31		8.2(9.4)	9.0(10.2)
12	4.0	28(32)	B0.5	−4.0	−0.28		16(18)	18(20)
7	3.3	21(25)	B2	−1.5	−0.22		43(45)	48(50)
5	2.8	17(21)	B4	−0.8	−0.19		94(88)	107(100)
3	1.9(2.1)	12.2(16.1)	B7	−0.2	−0.12		350(290)	440(340)
2	1.2(1.4)	9.1(12.2)	A2	1.4	0.05		1.16(.86)	1.36(1.03)
1.5	.68(.92)	7.1(9.6)	F3	3.0	0.40		2.7(1.84)	(2.0)
1.0	−.16(.15)	5.64(6.71)	G5	5.2	0.65		10.0(7.3)	
0.9	−.39(−.07)	5.30(6.31)	K0	5.9	0.89		15.5(10.7)	
0.8	−.61(−.31)	4.86(5.86)	K2	6.4	0.94		25(15)	

[a] Time to end of core H burning.
[b] Time to end of C or He burning.
Sources: Schaller *et al.* (1992); Meynet *et al.* (1994); Tinsley (1980); Allen (1973).

increase with metallicity, especially for the biggest stars. There is also the possibility that stars above some initial mass limit — which could itself depend on metallicity — may collapse into black holes instead of leaving a neutron star remnant. Production of other elements, often referred to as "secondary", such as ^{14}N and ^{22}Ne, will depend on the initial abundance of their progenitors (mainly CNO), while that of many odd-numbered elements, such as Na and Al, may depend on the neutron excess and hence again on the initial chemical composition. Some data on the expected net production of primary elements in the case of modest mass loss are given in Table 7.2, where all numbers are in units of $M_\odot$.

Comparison of Table 7.2 with similar computations by other authors (e.g. Arnett 1978, Thielemann *et al.* 1993, 1996, Woosley & Weaver 1995, Timmes *et al.* 1995) indicates that the predictions of helium, oxygen and 'Z' are fairly robust, especially as a function of M_α, as is the tendency for oxygen to account for an increasing proportion of 'Z' with increasing mass. Carbon abundances, on the other hand, are quite discrepant between different authors, varying by factors of up to 2 or 3 on either side of those quoted. (In any case, carbon could have a significant contribution from intermediate-mass stars.) According to detailed modelling of nuclear evolution in massive stars by Weaver & Woosley (1993), solar-system relative abundances of nearly all isotopes with $16 \leq A \leq 46$ are reproduced within a factor of 2 or better after purely hydrostatic evolution, when certain adjustments are made to the assumed ^{12}C$(\alpha, \gamma)^{16}$O rate (cf. Fig. 5.10). The subsequent explosion affects only rare species in this mass range (which will also depend on progenitor metallicity), as well as affecting those with $A \geq 44$.

Table 7.2. *Primary element production from massive stars with modest mass loss*

M_{init}	$M_{fin}{}^a$	$M_\alpha{}^b$	$M_{CO}{}^c$	He	C	O	Z
120	81	81	59	9.8	0.88	35	42
85	62	62	38	8.1	0.72	23	27
60	47	28	25	6.0	0.70	14	17
40	38	17	14	4.2	0.55	6.8	10
25	25	9	7	3.5	0.40	2.4	4.4
20	19	7	5	2.1	0.30	1.3	2.9
15	15	5	3	1.6	0.20	0.46	1.5
12	12	4	2	1.4	0.10	0.15	0.8
9	9	3	2	1.0	0.06	0.004	0.3
5	5	1	1	0.45			
3	3			0.09			

^{a} Final mass at end of carbon burning (or helium burning for lower masses).
^{b} Mass of He core at end of carbon or helium burning.
^{c} Mass of CO core at end of carbon or helium burning.
Source: Maeder (1992) for the case $Z = 0.001$, $Y = 0.24$.

For stars of higher metallicity, e.g. $Z = 0.02$ (approx. solar) and high mass ($M_{init} > 35 M_\odot$ or so), significant mass loss changes the outcome considerably and the uncertainties are correspondingly greater. Nucleosynthesis of helium and carbon leads to the ejection of substantial amounts of these elements in winds before the onset of core collapse, while at the same time the final masses (and hence those of the He and CO cores) are drastically reduced, leading to corresponding reductions in the amounts of heavy elements that are finally ejected. Further complications arise if stars above a certain upper mass limit collapse into black holes rather than leaving a neutron-star remnant, because such stars will still contribute the elements ejected in winds without contributing those in the SN ejecta. This can cause drastic changes in certain element:element ratios, e.g. dY/dZ or C/O (Maeder 1992, 1993). Table 7.3 gives Maeder's calculations of nucleosynthesis from massive stars undergoing strong mass loss.

Intermediate-mass stars with $M_{init} > 3$ to $5 M_\odot$, depending on chemical composition, undergo three dredge-up episodes in the course of evolution, the first one during the first ascent of the red-giant branch (RGB) and the second and third on the asymptotic giant branch (AGB) (cf. Section 5.10). Less massive stars experience only the first and third dredge-ups. In the first dredge-up, the outer convection zone reaches into regions where ^{12}C was partly converted into ^{13}C and ^{14}N during the main-sequence and post main-sequence phases. Correspondingly ^{14}N and ^{13}C are enhanced at the surface at the expense of ^{12}C while ^{16}O is substantially unchanged and a little ^{4}He is produced. In the second episode, following ignition of the He-burning shell, the convective envelope penetrates into the helium core, dredging up helium and nitrogen (see Fig. 5.17). Once again the surface abundances of ^{14}N and ^{4}He are enhanced, while C and O

Table 7.3. *Primary element production from stars with drastic mass loss*

M_{init}	M_{fin}	He		C		O		Z	
		a	b	a	b	a	b	a	b
120	2.4	42.7	−0.1	8.0	0.3	0.0	0.2	10.1	0.7
85	3.5	16.7	−0.4	13.5	0.4	4.0	0.6	19.3	1.6
60	3.0	13.5	−0.3	7.2	0.3	1.4	0.4	9.8	1.2
40	3.6	6.1	−0.4	4.9	0.4	2.1	0.6	8.0	1.6
25	11.3	1.5	0.6	0.30	0.32	2.6		4.5	
20	14.0	1.6	1.5	0.22		1.3		2.7	
15	13.6	1.4	1.3	0.14		0.4		1.3	
12	11.5	1.2		0.07		0.1		0.7	
9	8.6	0.9		0.03		0.0		0.2	
5	4.9	0.40							
3	3.0	0.07							

^{a} Sum of amounts freshly produced and expelled in wind and in final ejecta (SN or PN).
^{b} Total amount freshly produced and expelled in final ejecta (SN or PN).
Source: Maeder (1992) for the case $Z = 0.02$, $Y = 0.28$, high mass-loss rates.

are correspondingly reduced. The third phase consists of several individual mixing episodes: following each helium shell flash the base of the convective envelope may penetrate through the H–He discontinuity bringing up helium, carbon and s-process elements. If the temperature at the base of the convective zone is high enough, some or all of this carbon is changed into primary ^{14}N by so-called 'hot-bottom burning', which takes place above some mass limit which depends on chemical composition and the assumed model of convection. While all this is happening, the star continuously loses mass through a stellar wind, which gradually increases in intensity as the star expands, cools and becomes vibrationally unstable; eventually the remaining amount of mass above that of the final remnant is expelled in the form of a planetary nebula and the remnant becomes a white dwarf. The wind and the planetary nebula are the means by which nucleosynthesis in these stars affects galactic chemical evolution.

Table 7.4 gives some very tentative but still quite widely used estimates of the outcome of these processes, taken from Renzini & Voli (1981). All quantities are in units of $M_\odot$ and numbers in brackets indicate primary components of ^{13}C and ^{14}N, the rest being secondary. It is notable that, with these parameter values, the net production of ^{12}C is considerably reduced by transformation into primary nitrogen; more ^{12}C would survive for smaller values of the mixing-length parameter α. The helium production is a little less than for the same masses in Tables 7.2 and 7.3, which gives some indication of the uncertainties. Intermediate-mass stars contribute a large part of the enrichment in ^{4}He, they may contribute significantly to ^{12}C and they dominate the production of ^{13}C, ^{14}N and the main s-process elements. However, details of the actual yields from the third dredge-up episode are quite uncertain. Since the work of Renzini & Voli, there have been substantial advances in numerical models of AGB evolution

Table 7.4. *He, C and N production from intermediate-mass stars:Version 1*

M_{init}	M_{rem}	^{4}He		^{12}C$\times 100$		^{13}C$\times 1000$		^{14}N$\times 1000$	
	a	a	b	a	b	a	b	a	b
8.0	1.4	0.79	.86	−1.3	.09	2.1(0.6)	1.5(1.4)	130(94)	102(92)
6.0	1.4	0.40	.53	−.28	.27	3.1(2.8)	1.7(1.6)	118(97)	107(101)
5.0	1.4	0.18	.34	3.9	.51	3.9(3.4)	2.2(2.1)	63(50)	104(99)
4.3	1.3	0.09		4.5		3.9(2.4)		20(13)	
4.0	1.2	0.08	.15	4.7	2.2	1.6(1.0)	3.1(3.0)	3(0)	56(54)
3.5	1.1	0.05		3.0	2.2	0.14	2.6(2.4)	2.4	26(24)
3.25	1.1				2.8		3.0(2.8)		4.9(4.3)
3.0	1.0	0.03	.05	1.65	2.6	0.11	0.02	2.1	0.4
2.5	0.9	0.02	.03	0.65	1.14	0.09	0.02	1.8	0.3
2.0	0.8	0.01	.01	0.12	.36	0.07	0.01	1.4	0.3
1.0	0.6	0.00	.00	−.03		0.03	0.01	0.3	0.1

[a] Initial composition: $Z = 0.02$, $Y = 0.28$.
[b] Initial composition: $Z = 0.004$, $Y = 0.232$.
Source: Renzini & Voli (1981), Case A, convection parameter $\alpha = 1.5$, mass-loss parameter $\eta = 0.33$.

and it is not clear that stars with $M > 5M_\odot$ develop degenerate CO cores (a necessary condition for the third dredge-up) anyway; the luminosity function of carbon stars in the LMC suggests that they may not. Thus the nucleosynthesis contributions of stars between 5 and $8M_\odot$, if any, are quite unclear (apart from helium, perhaps).

A more recent set of estimates for production by low and intermediate-mass stars is given in Table 7.5, again in units of $M_\odot$. These also are to be taken as indicative only. They were calculated assuming values for certain parameters that were chosen to fit the luminosity function of carbon stars in the LMC. These include a choice of a minimum core mass of $0.58M_\odot$ for the onset of thermal pulses and a dredge-up efficiency (defined as the ratio of the mass shown as B in Fig. 6.5 to the mass ΔM_α added to the core by H-burning during the previous interpulse period) of 0.65, which is larger than expected from numerical studies of stellar evolution. The trends shown in Table 7.5 are in qualitative, but not quantitative, agreement with those in Table 7.4. According to Table 7.5, dredge-up begins above the horizontal line for $Z = 0.008$ (corresponding to LMC composition) and at a higher initial mass for solar composition. At lower masses, only the first dredge-up occurs. Thus for these masses, the nitrogen yield shows 'secondary' behaviour (increasing with Z), whereas at higher masses there is less sensitivity to Z because of primary carbon production and hot-bottom burning; this transition sets in at substantially higher masses in the table taken from Renzini & Voli. The production of ^{13}C in the model by Marigo *et al.* is sensitive in addition to the arbitrarily assumed mass of the ^{13}C pocket that enters the intershell zone after a pulse, and is not quoted here; nor are their yields of ^{15}N (< 0), ^{16}O ($\simeq 0$), ^{17}O, ^{18}O (< 0), ^{22}Ne and ^{25}Mg. There is no ^{20}Ne production in these models.

Table 7.5. *He, C and N production from intermediate and low-mass stars: Version 2*

M_{init}	M_{rem}		^4He		$^3He \times 1000$		$^{12}C \times 100$		$^{14}N \times 1000$	
	a	b	a	b	a	b	a	b	a	b
4.000	0.77	0.87	0.084	0.130	0.16	1.11	0.26	1.53	6.26	3.25
3.000	0.66	0.71	0.075	0.097	0.24	0.09	0.55	1.93	4.77	2.06
2.500	0.64	0.69	0.080	0.129	0.29	0.31	1.67	3.98	3.33	1.42
2.000	0.60	0.65	0.036	0.077	2.23	0.40	0.58	2.40	0.37	0.89
1.837	0.59		0.024		0.60		0.11		0.39	
1.831		0.63		0.058		0.46		1.70		0.72
1.665	0.59		0.022		1.65		0.15		0.36	
1.570		0.61		0.029		0.55		0.64		0.48
1.318	0.58		0.012		1.08		−0.30		0.29	
1.305		0.59		0.016		0.64		0.23		0.20
1.066	0.56		0.008		0.73		−0.20		0.19	
1.046		0.58		0.008		0.72		−0.07		0.065
0.915	0.56		0.006		0.57		−0.12		0.11	
0.819		0.58		0.004		0.53		−0.01		0.012

[a] Initial composition: $Z = 0.02$, $Y = 0.28$.
[b] Initial composition: $Z = 0.008$, $Y = 0.25$.
Source: Marigo, Bressan & Chiosi (1996).

Further important contributions to GCE come from Type Ia supernovae and possibly other interacting binary systems. In the canonical SN Ia model, a CO white dwarf accretes matter at a rate between 4×10^{-8} and 10^{-6} $M_\odot$ yr^{-1} from a binary companion, processes the accreted matter to C and O via hydrogen and helium shell flashes, approaches the Chandrasekhar limiting mass and ignites degenerate carbon at its centre. A subsonic nuclear burning front (a deflagration) propagates outwards, incinerating an innermost $0.6M_\odot$ of the star to iron-peak isotopes, principally radioactive ^{56}Ni, and burning an outer portion to elements of intermediate mass from O to Ca (cf. Fig. 5.23), corresponding to features observed in the light curve and spectrum development. The entire star is disrupted in the process. Indirect evidence (the behaviour of the O/Fe ratio as a function of metallicity in normal stars) suggests that about 2/3 of the iron in the Solar System comes from this process (which involves a significant time delay of the order of a Gyr), the remainder coming from massive stars associated with SN II. Table 7.6 gives estimated yields from a $1M_\odot$ CO white dwarf accreting hydrogen-rich material from a red-giant companion.

The table suggests that SN Ia may contribute noticeably to Si, S, Ar and Ca as well as iron. In addition, there might possibly be contributions to the s-, r- and especially the p- (or γ-) process (Howard, Meyer & Woosley 1991). Other types of SN, e.g. Ib, could also be binaries and make a distinctive contribution, but their nature is not well known at present. Roughly speaking, a mean SN Ia rate of 1 per century over 10 Gyr produces $5 \times 10^7 M_\odot$ of iron, equivalent to an overall abundance $Z(Fe) \simeq 10^{-3} \simeq Z_\odot(Fe)$, while a

Table 7.6. *Element production from SN Ia*

Species	Mass/$M_\odot$	$[X_i/X_{56}]^a$
^{24}Mg	.09	-1.1
^{28}Si	.16	-0.3
^{32}S	.08	-0.4
^{36}Ar	.02	-0.3
^{40}Ca	.04	0.1
^{54}Fe	.14	0.6
^{56}Fe	.61	0.0
^{58}Ni	.06	0.4
Cr–Ni	.86	

^{a} Logarithmic element:^{56}Fe ratio relative to solar.
Source: Nomoto *et al.* (1984), model W7; cf. also Thielemann *et al.* (1986).

similar SN II rate (with a typical oxygen production of $2M_\odot$) produces the equivalent of an overall abundance $Z(O) \simeq 3 \times 10^{-3} \simeq 0.3Z_\odot(O)$. Thus reasonable SN rates of 0.33 per century for Type Ia and 1.5 per century for Type II give the right orders of magnitude.

7.3.3   Initial mass function (IMF)

The IMF $\phi(m) \propto dN/dm$ or $\xi(m) \propto dN/d\log m$ describes the relative birth-rates of stars with different initial mass $mM_\odot$ in a given interval dm or $d\log m$, at a given time and in a given region such as the solar cylinder, i.e. a cylinder perpendicular to the Galactic plane through the Sun and presumably representing the population of the Galactic disk at the Sun's distance from the centre. For the solar neighbourhood the IMF is obtained in a series of steps, first applied by Salpeter (1955), beginning with the luminosity function dN/dM_V for the local volume, the mass–luminosity relation $dM_V/d\log m$ and the appropriate scale heights $h(M_V)$ (all referring to average main-sequence stars) to give the present-day mass function (PDMF) for the solar cylinder:

$$\text{PDMF}(\log m) = \frac{dN}{dM_V} \times 2h(M_V) \times \frac{dM_V}{d\log m}. \tag{7.1}$$

The relation between the PDMF and the IMF then depends on the lifetimes of the stars. For long-lived stars ($m \leq 0.9$),

$$\text{IMF} = \text{PDMF}/(T <\psi(t)>), \tag{7.2}$$

where T is the age of the Galactic disk and $<\psi(t)>$ is the mean rate of star formation (SFR) over that time. For short-lived stars, i.e. those that have only been able to live for such a short time that there has been no change in $\psi(t)$, say $m > 2$,

$$\text{IMF} = \text{PDMF}/\{\tau_{\text{ms}}(m)\psi(T)\}. \tag{7.3}$$

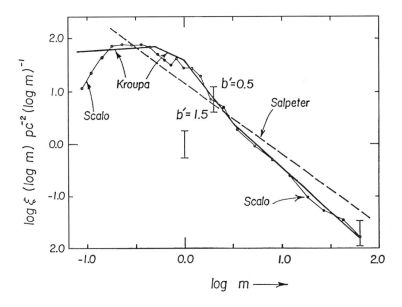

Fig. 7.1. Local IMF after Scalo (1986) with $b' = 1$ (points joined by thin lines), Kroupa *et al.* (1991) including 'distance effect' (thick lines below $m = 1$) and two additional power-law segments approximating Scalo's IMF. Salpeter's law is shown by a broken line. The IMFs are normalized to a total mass of stars ever born of $37 M_\odot \text{pc}^{-2}$ between mass limits $m = 0.1$ and $m = 100$.

For intermediate masses,

$$\text{IMF} = \text{PDMF} / \int_{T-\tau_{\text{ms}}(m)}^{T} \psi(t)dt. \tag{7.4}$$

Determination of the local IMF thus involves several non-trivial steps, including corrections for evolved stars and unresolved binary or multiple systems, knowledge of scale heights and the mass-luminosity relation and the procedure whereby the three different pieces are fitted together, which itself requires assumptions about the age of the Galaxy and the past history of the SFR, as well as an assumption that the IMF itself either has not changed (the more usual assumption, built-in to the above equations) or has changed in some specified manner. Furthermore, the rarity of very massive stars can introduce a statistical bias if the sample is not sufficiently large.

A number of arguments (mainly based on the observed distribution of stellar ages, together with the assumption that there are no large discontinuities in the IMF since there are none in Galactic clusters) suggest that the present SFR in the disk of the Galaxy is not very different from its average value in the past; specifically

$$0.5 \le b(T) \equiv \frac{\psi(T)}{<\psi(t)>} \le 1.5, \tag{7.5}$$

and this enables the three pieces to be patched together. Some resulting IMFs are shown in Fig. 7.1, together with the effect of the parameter $b' \equiv 12\,b(T)/T(\mathrm{Gyr})$. The normalization here is to the PDMF of long-lived stars pc^{-2}, so that the level above $m = 2$ or so is inversely proportional to b' as shown by the vertical error bars at $m = 2$ and $m = 60$. Additional typical errors are shown by the error bar near the centre of the figure.

For GCE purposes, the appropriate normalization is such that

$$\int_{m_L}^{m_U} m\phi(m)dm = 1 \tag{7.6}$$

where m_L and m_U are lower and upper limits of integration which will be taken as 0.1 and 100 respectively. Taking the Kroupa *et al.* IMF, which consists of two power-law segments, below $m = 1$, and approximating the Scalo IMF above $m = 1$ with two more power-law segments as shown in Fig. 7.1, and with the above normalization, we have (with f giving the mass fraction within the given mass interval)

$$m\phi(m) = 0.93m^{0.15}; \quad 0.1 \le m \le 0.5. \ f = 0.31. \tag{7.7}$$

$$m\phi(m) = 0.46m^{-.85}; \quad 0.5 \le m \le 1.0. \ f = 0.31. \tag{7.8}$$

$$m\phi(m) = 0.46m^{-2.4}; \quad 1.0 \le m \le 3.16. \ f = 0.26. \tag{7.9}$$

$$m\phi(m) = 0.21m^{-1.7}; \quad 3.16 \le m \le 100. \ f = 0.12. \tag{7.10}$$

The above exponents are usually referred to by the symbol $-x$, so that ϕ itself varies as $m^{-(1+x)}$. Other IMFs commonly assumed, especially for external galaxies, are single power laws, notably Salpeter's law which has $x = 1.35$ and gives a better fit to many properties of disk galaxies (Kennicutt 1983). With the above mass limits for normalization,

$$m\phi_{SAL}(m) = 0.17m^{-1.35}, \tag{7.11}$$

but in this case the coefficient (together with other significant parameters) is somewhat dependent on the adopted mass limits, whereas with the piece-wise function (or Scalo's function) changing m_L from 0.1 to zero only makes a difference of 7 per cent (or less) and changing m_U from 100 to 40 or infinity makes a difference of only 1 per cent in the coefficients above. The choice of 0.1 as the lower limit for the Salpeter function gives a similar balance between low and high-mass stars as do the Scalo or Kroupa functions which are more realistic at low masses.

The IMF has a number of important integral properties including the following:

(i) The mass fraction $F_M(> m)$ of a generation of stars that is born with stellar masses above some limit:

$$F_M(> m) = \int_m^{m_U} m'\phi(m')dm'. \tag{7.12}$$

In particular, one is interested in the mass fraction $\zeta \equiv F_M(> m_\tau)$ of stars with a lifetime $\tau(m)$ less than the age of the system. E.g. if $m_\tau = 1$, a characteristic value for the Galaxy, then ζ is close to 0.45 for all three functions; but the acknowledged uncertainties imply a corresponding uncertainty in ζ of at least ± 0.1.

(ii) The return fraction R or the lock-up fraction $\alpha \equiv (1 - R)$. R is the mass fraction of a generation of stars that is returned to the interstellar medium, and it increases with time as progressively smaller stars complete their evolution, while α correspondingly decreases, but by a proportionately smaller amount as long as $R < 0.5$. α is the fraction that remains locked up in long-lived stars and compact stellar remnants and is given by

$$\alpha = 1 - F_M(> m_\tau) + \int_{m_\tau}^{m_U} m_{\text{rem}}(m)\, \phi(m) dm, \tag{7.13}$$

where m_{rem} can be approximated by

$$m_{\text{rem}} = 0.11m + 0.45; \quad m \leq 6.8; \tag{7.14}$$
$$= 1.5; \quad m > 6.8 \tag{7.15}$$

(Iben & Tutukov 1984b). For $m_\tau = 1$, $\alpha \simeq 0.7$ to 0.8 with these IMFs, again with an uncertainty of the order of ± 0.1, which is comparable to the amount by which it decreases after the first few hundred million years.

(iii) The yield p of primary elements produced by massive stars, e.g. oxygen, which is defined as the mass of element freshly produced and ejected by a generation of stars in units of the mass that remains locked in long-lived stars and compact remnants. (This will be referred to as the *true yield*, as opposed to an *effective yield* deduced from observed abundances with the aid of GCE models.) The mass produced relative to the initial mass of the generation is $q \equiv \alpha p$. The true yield is given by

$$p_i = \alpha^{-1} \int_{m_\tau}^{m_U} m\, q_i(m)\, \phi(m)\, dm \tag{7.16}$$

where $q_i(m)$ is the fraction of the initial mass of a star ejected in the form of freshly synthesised element i and can be deduced (or summed over elements to find p_Z) from the kind of data given as $m\, q_i(m)$ in Tables 7.2 to 7.6. The oxygen yield for the Scalo IMF, using the modest mass-loss results in Table 7.2 and $m_\tau = 1$, is 0.006, or about 2/3 solar, but this is uncertain by about a factor of 2 because of the uncertainty in b' and by further factors related to mass loss and the upper mass limit for SN II.

7.3.4 Star formation rates (SFR)

It would be convenient if the SFR could be described, as has often been attempted, by some simple laws, e.g. as a given function of time or of the mass of gas present, possibly involving other parameters like the total surface density or Galactic rotation constants, all of which presumably play a role. Unfortunately there is no clear basis for such simple laws. In disk galaxies, star formation is quite sporadic, in gas-rich dwarf galaxies it may be dominated by a small number of bursts separated by extremely long intervals of time, and in large starburst and luminous IRAS galaxies one has a very violent burst of star formation triggered by an interaction or merger. Other effects that have been considered include stochastic self-propagating star formation, self-regulated star formation and the presence of (surface) density thresholds. In our own Galaxy, the highest SFR occurs in the 4 kpc ring, roughly half-way between here and the Galactic centre, where total surface density, gas density, gas pressure, temperature, chemical composition, gravitational potential, Galactic rotation effects, spiral shocks, magnetic fields and the frequency of collisions between clouds all may play a role. There is also strong star formation activity around the centre itself, but little in the region between, the main factor probably being a shortage of gas.

From the point of view of GCE, one is interested primarily in effects averaged over long periods of time of the order of Gyr; but in dwarf galaxies which may have experienced only a few star formation bursts over a Hubble time the sporadic character may have appreciable effects, especially when one bears in mind that much of the abundance data for such objects comes from H II regions which are intrinsically the result of a current burst, and there is indeed evidence for a cosmic dispersion in certain element abundance ratios such as N/O and $\Delta Y/\Delta Z$ in such objects (cf. Chapter 11).

No attempt will be made here to enter into a detailed discussion of the physics of the SFR, but a few comments will be made on the basis of some very simple parameterizations. Considering for example an exponential time-dependence

$$\psi(t) = \psi_0\, e^{-\lambda t}, \tag{7.17}$$

the restriction on $b(t)$ in Eq. (7.5) implies that $\lambda^{-1} > 0.8\, T \simeq 10$ Gyr. A related argument can be made about the past evolution of the luminosity of the disk — closely related to both the SFR and the IMF — from considerations of nuclear fuel consumption. All energy radiated corresponds to the synthesis of helium and/or heavy elements from hydrogen at a rate

$$L \simeq -0.007\, Mc^2\, dX/dt \tag{7.18}$$

which translates into an overall present-day abundance in the whole system (including stellar interiors and white dwarfs)

$$<Z + \Delta Y> \simeq 0.005\, \frac{L_1}{M}\, \frac{e^{\lambda T} - 1}{\lambda} \tag{7.19}$$

where L_1 is the present-day bolometric luminosity and the mass is assumed to have remained unchanged. Since about 10% of luminous matter in the solar neighbourhood is in the form of white dwarfs (Fleming, Liebert & Green 1986) and in the rest of the material $Z \simeq 0.02$ and $dY/dZ \simeq 4$ (Pagel *et al.* 1992), one can estimate that

$$<Z + \Delta Y> \simeq 0.2. \tag{7.20}$$

Since $L_1/M \simeq 1$ (in cgs units), it follows from a comparison of equations (7.19) and (7.20) that, for exponential decay in bolometric luminosity, $\lambda^{-1} \gtrsim \sim 5$ Gyr. Thus a single strong initial burst seems to be ruled out for the solar neighbourhood.

Table 7.7. *Some properties of the Galaxy and the Solar cylinder*

	Galaxy	Solar cylinder
Age	10 to 15 Gyr	
Mass now in stars[a]	$7 \times 10^{10} M_\odot$	$45\ M_\odot \mathrm{pc}^{-2}$
Mass now in gas[b]	$\sim 7 \times 10^9 M_\odot$	7 to 14 $M_\odot \mathrm{pc}^{-2}$
Gas fraction	~ 0.1	0.14 to 0.25
Surface brightness		$23 m_V, m_{bol}\ \mathrm{arcsec}^{-2}$
$(M/L_V)/(M/L_V)_\odot$ [a]	5	3
Processes tending to deplete the gas:		
Average past SFR	$(5\ \mathrm{to}\ 7)\alpha^{-1}\ M_\odot\ \mathrm{yr}^{-1}$	$(3\ \mathrm{to}\ 4.5)\alpha^{-1} M_\odot\ \mathrm{pc}^{-2}\ \mathrm{Gyr}^{-1}$
Gas consumption time	~ 1 Gyr	1.5 to 5 Gyr
Processes tending to restore the gas:		
Mass ej. from AGB+PN[c]		$0.8\ M_\odot\ \mathrm{pc}^{-2}\ \mathrm{Gyr}^{-1}$
Mass ej. from O stars[d]		$\sim 0.05 M_\odot\ \mathrm{pc}^{-2}\ \mathrm{Gyr}^{-1}$
Mass ej. from SN[d]	$\sim 0.15 M_\odot\ \mathrm{yr}^{-1}$	$\sim 0.05 M_\odot\ \mathrm{pc}^{-2}\ \mathrm{Gyr}^{-1}$
(Total mass ejection from stars		$\sim 1\ M_\odot\ \mathrm{pc}^{-2}\ \mathrm{Gyr}^{-1})$
Net inflow from IGM[e]	$\leq 2 M_\odot\ \mathrm{yr}^{-1}$	$\leq 1\ M_\odot\ \mathrm{pc}^{-2}\ \mathrm{Gyr}^{-1}$

Sources: [a] Binney & Tremaine (1987); [b] Kulkarni & Heiles (1987); [c] Jura (1989);
[d] Pottasch (1984); [e] Lacey & Fall (1985).

The simplest possible parameterization of the SFR is to assume that it is just proportional to the surface density of gas (a special case of what is known as Schmidt's (1959) law, with an exponent of 1). This at least has the merit of simplicity and of taking into account the necessity of having gas as the raw material from which stars are formed, and will be used widely here as it has been elsewhere, while always bearing in mind that the coefficient may vary with ambient conditions or stochastically. A number of the results of GCE theory are insensitive to the SFR, while others are affected by many other difficulties anyway.

7.3.5 The Galactic context

Fig. 3.31 shows a schematic cross-sectional view of the Galaxy, indicating the main stellar population groups (disk, bulge, halo and solar cylinder) that will figure in subsequent discussions and Table 7.7 gives some relevant statistics.

The final ingredient of GCE models is a set of assumptions about the conditions in which stellar births and deaths take place. In particular, it makes a considerable difference whether we are allowed to consider an isolated, well-mixed system or whether inflows and/or outflows of gas and inhomogeneities are important. For example, in 'chemo-dynamical' models (Burkert & Hensler 1989; Burkert, Truran & Hensler 1992), massive stars energise and enrich the hot component of the ISM, some of which may escape in the form of a metal-enhanced wind, whereas fresh stars are formed from cold material where the enrichment lags behind that of the hot component. It may also be necessary to contemplate changes in the yields (with or without variations in IMF) as a function of chemical composition or other variables. Changes in the slope or upper or lower mass limits of the IMF have all been proposed for particular situations, such as violent star formation bursts, but the evidence is not very clear; at the same time, Maeder (1992, 1993) has predicted substantial changes in the yields even with a constant IMF.

7.4 The GCE equations

7.4.1 Basic equations

In the standard case there are four variables to be calculated: the total system mass M, the mass of 'gas' g, the mass existing in the form of stars (including compact remnants) s and the abundance Z of the element(s) of interest, assuming certain initial conditions and laws governing the SFR and flows of material into and out of the system.

For the system mass, we have

$$M = g + s \tag{7.21}$$

and

$$dM/dt = F - E \tag{7.22}$$

where F is the rate of accretion of material from outside the system and E the rate of ejection in a galactic wind.

The mass of gas is governed by

$$dg/dt = F - E + e - \psi \tag{7.23}$$

where ψ is the star formation rate (by mass) and e the ejection rate of matter from stars, and the mass of stars by

$$ds/dt = \psi - e, \tag{7.24}$$

where

$$e(t) = \int_{m_{\tau=t}}^{m_U} (m - m_{\text{rem}}) \, \psi(t - \tau(m)) \, \phi(m) \, dm. \tag{7.25}$$

The abundance of a stable (i.e. non-radioactive) element in the gas (and in newly formed stars if the ISM is homogeneous) is governed by

$$\frac{d}{dt}(gZ) = e_Z - Z\psi + Z_F F - Z_E E, \tag{7.26}$$

where the first term on the rhs represents the total amount of the element ejected from stars, the second one the loss to the ISM by star formation, the third one the addition from any of the element that may exist in inflowing material and the fourth one the loss by a galactic wind (if any). Generally two possibilities are considered for a wind: if it is *homogeneous* then $Z_E = Z$, while if it is *metal-enhanced*, $Z_E > Z$ for the elements affected. In the latter case it may sometimes be necessary to distinguish between the abundances in different phases of the ISM, the hot one being metal-enhanced and contributing to the wind and the cool one fixing the abundances in stars (Burkert & Hensler 1989; Pagel 1994a).

The term e_Z is given by

$$e_Z(t) = \int_{m_{\tau=t}}^{m_U} [(m - m_{\mathrm{rem}})Z(t - \tau(m)) + mq_Z(m)]\, \psi(t - \tau(m))\, \phi(m)dm \qquad (7.27)$$

where the first term in square brackets represents recycling without change in abundance (this does not apply to fragile elements like D, Li, Be, B which are wholly or partly destroyed by recycling in stars or 'astration') and the second represents fresh production by nuclear processes in stellar evolution followed by ejection. The return fraction R, or rather $\alpha = 1 - R$, and the true yield p_Z are given by equations (7.13) and (7.16) respectively.

Full numerical solution of the above equations involves many detailed assumptions and it is not always easy to visualise the effects of these assumptions on the outcome. Therefore it is useful to consider the much simpler approach of the next subsection.

7.4.2 The instantaneous recycling approximation

In the instantaneous recycling approximation, first explicitly used by Schmidt (1963), one assumes that all processes involving stellar evolution, nucleosynthesis and recycling take place instantaneously on the time-scale of galactic evolution, so that quantities such as α and p do not depend explicitly on time, although they may of course be affected by changes in chemical composition or other parameters that themselves vary with time. This enables a variety of results to be derived analytically in a straightforward manner. The approximation is often quite good for products of massive-star evolution, like oxygen, if the SFR does not vary violently on a short time-scale, if the return fraction is not too dependent on long-lived, low-mass stars and the residual gas fraction not too small. It is poor in any case for elements like iron, nitrogen, s-process and possibly carbon, which come partly or wholly from lower-mass stars with significant evolutionary time-scales of the order of a Gyr. These can actually be handled to an extent by a "delayed production" approximation (Section 7.4.3).

The corresponding equations in instantaneous recycling are:

$$S(t) = \int_0^t \psi(t')dt', \qquad (7.28)$$

where $S(t)$ is the mass of all stars that have been born up to time t;

$$s(t) = \alpha S(t),\tag{7.29}$$

where $s(t)$ is the mass still in form of stars (or compact remnants);

$$M(t) = s(t) + g(t) = M_0 - M_{\mathrm{ej}} + M_{\mathrm{accr}}\tag{7.30}$$

and

$$dg/dt = F - E - ds/dt\tag{7.31}$$

or (eliminating time)

$$\frac{dg}{ds} = \frac{F - E}{\alpha\psi} - 1.\tag{7.32}$$

Assuming a homogeneous ISM, the abundance in the gas (and in newly formed stars) of a stable robust element is governed by

$$\frac{d}{dS}(gZ) = q + RZ - Z - Z_E\frac{E}{\psi} + Z_F\frac{F}{\psi}\tag{7.33}$$

where the suffix z has been dropped and the terms on the rhs represent respectively new production, recycling, lock-up in stars, loss in galactic wind and gain from accretion. (For a fragile element, the RZ term is absent or reduced.) For a homogeneous wind with $Z_E = Z$, combining $RZ - Z \equiv -\alpha Z$ and dividing by α, we have

$$\frac{d}{ds}(gZ) = p - Z\left(1 + \frac{E}{\alpha\psi}\right) + Z_F\frac{F}{\alpha\psi}\tag{7.34}$$

or, using Eq. (7.32),

$$g\frac{dZ}{ds} = p + (Z_F - Z)\frac{F}{\alpha\psi}.\tag{7.35}$$

Eqs. (7.34) and (7.35) can also be written in a form making no explicit mention of ψ or α as follows:

$$\frac{d}{ds}(gZ) = p - Z(1 + \frac{dM_{\mathrm{ej}}}{ds}) + Z_F\frac{dM_{\mathrm{accr}}}{ds}\tag{7.36}$$

or

$$g\frac{dZ}{ds} = p + (Z_F - Z)\frac{dM_{\mathrm{accr}}}{ds}.\tag{7.37}$$

7.4.3   The delayed production approximation

Instantaneous recycling cannot be used to describe the formation of elements, such as iron, to which there is a significant contribution from stars that take a non-negligible time to complete their evolution. This case can still be handled within the framework of analytical models by a 'delayed production approximation' (Pagel 1989a), which works by the simple device of assuming that the delayed element or component thereof starts to be released at a single time Δ after the onset of star formation and is then instantaneously recycled. Equation (7.34), with the the time variable re-introduced, is then modified as follows:

$$\frac{d}{dt}(gZ) = 0; \quad t < \Delta \tag{7.38}$$

$$= p\left(\frac{ds}{dt}\right)_{t-\Delta} - Z(\alpha\psi + E); \quad t \geq \Delta \tag{7.39}$$

(assuming $Z_F = 0$) and in general p is simply replaced by $p\,\psi(t-\Delta)/\psi(t)$. The models now have two additional parameters: the division between instantaneous and time-delayed contributions to the yield, and the dimensionless product $\omega\Delta$, where $\omega \equiv \alpha\psi/g$ (see Eq. 8.15 in the next chapter).

Notes to Chapter 7

A fine account of the structure and content of galaxies is given in Mihalas & Binney (1981) and excellent introductions to many questions in GCE are to be found in Truran & Cameron (1971) and Tinsley (1980). A good modern account of the kinematics and abundance distribution of the Galaxy is given by Gilmore, Wyse & Kuijken (1989).

Early GCE papers (Talbot & Arnett 1971, 1973a) used a slightly different definition of the yields q and p from that given here, which included the amounts of ejected elements already present in processed zones of the star before nucleosynthesis, so that the resulting GCE equations are slightly different; the difference is only significant in the case of helium, however. The definition here is identical to that introduced by Searle & Sargent (1972) and used by Renzini & Voli (1981), Maeder (1992) and other authors, although many of these use the symbol y for the yield whereas I have preferred p because it looks less like a variable. Note that there can be negative yields in some cases.

For a very thorough discussion of the IMF, see Scalo (1986), and tables of return fractions and oxygen yields for a number of different IMFs with various values of $m_{\tau=t}$ have been published by Köppen & Arimoto (1991). However, some authors consider more 'top-heavy' IMFs than the ones discussed here, with larger yields and larger return fractions (e.g. Kennicutt, Tamblyn & Congdon 1994); in such cases, the instantaneous recycling and delayed production approximations can become very poor because progressive gas consumption and enrichment are diluted or even reversed by mass ejection from older stars.

For critical discussions of supposed evidence for variations in the IMF, see Scalo (1986) and McGaugh (1991); star counts in the Large Magellanic Cloud between about 1 and 100 $M_\odot$ seem to be consistent with either the Salpeter law (Melnick 1987; Richtler, de Boer & Sagar 1991) or the steeper Scalo law (Parker & Garmany 1993), but observations of Galactic and extragalactic H II regions using the Hubble Space Telescope indicate close agreement with the Salpeter law (above $4M_\odot$ at any rate) over a broad range in metallicities; the upper mass limit is mainly governed by the age of the stars (Waller, Parker & Malamuth 1996). Information on the relative numbers of older and younger stars in the solar neighbourhood comes from the study of chromospheric ages based on the strength of the Ca^+ K_2 emission line component (Soderblom, Duncan & Johnson 1991), while in external galaxies models of past star formation rates (assuming a fixed IMF) are constrained by colours (e.g. Larson & Tinsley 1978) and spectra (e.g. Bica 1988) of the integrated light.

The question of upper mass limits to stars which explode as SN II and leave neutron-star remnants is discussed by Maeder (1992, 1993) and by Brown, Bruenn & Wheeler (1992); it is highly controversial. (Note that Köppen & Arimoto when referring to the Scalo IMF use the version with $b'(T) = 1$, as I have done, whereas Maeder (1993) uses the version with $b'(T) = 0.48$, corresponding to yields that are 3 times higher!)

Interesting variants on the simplest star formation laws include stochastic self-propagating star formation (Gerola & Seiden 1978; Dopita 1985), self-regulating star formation (Arimoto 1989; Hensler & Burkert 1990), stochastic star-formation bursts (Matteucci & Tosi 1985) and the existence of a threshold surface gas density for star formation (Kennicutt 1989; Chamcham, Pitts & Tayler 1993).

The estimates of mass densities and mass:light ratios are meant to include only gas, living stars and stellar remnants, not counting dark matter which may be detected by purely gravitational means, and this distinction is not so clear in the older literature. Furthermore, there is controversy about the amount of such dark matter in the solar neighbourhood; see Bahcall, Flynn & Gould (1992). This has a direct bearing on considerations of the past luminosity, since Larson (1986) suggested that there might be a very large population of undetected white dwarfs resulting from an early IMF strongly weighted towards massive stars and accounting for the 'missing mass'. Following the work of Bienaimé, Robin & Crézé (1987), Kuijken & Gilmore (1989) and Flynn & Fuchs (1994), it seems that there is no real evidence for considerable amounts of such dark matter in the solar neighbourhood, however, and so I do not count it.

Problems

1. Use the IMF, eqns (7.7) to (7.10), to calculate the relative numbers of stars born with less and more mass than that of the Sun.

2. Assuming the star formation rate for the Galaxy given in Table 7.6 and that all stars between 10 and $100M_\odot$ explode as Type II supernovae, estimate the corresponding supernova rate (a) for the Salpeter IMF with lower mass limit $0.1M_\odot$ and (b) for the IMF given by Eqs. (7.7) to (7.10). How much difference does it make if the upper mass limit for SN is $50M_\odot$? (The observed rate for SN II in galaxies like our own is of the order of 2 to 3 per century.)

3. Approximating the oxygen masses in Table 7.2 by power laws

$$m\,q_O(m) \;=\; 2.4(m/25)^3; \;\; 10 \le m \le 25 \tag{7.40}$$
$$=\; 2.4(m/25)^2; \;\; 25 \le m \le 100, \tag{7.41}$$

and using the IMF in Eq. (7.10), show that an average supernova ejects $2M_\odot$ of oxygen (or $2.6M_\odot$ for the Salpeter function). Now use the results of the previous problem to estimate the oxygen yield from a generation of stars and compare with the solar oxygen mass fraction $Z_\odot^O = 0.009$. (Note that the star formation rate cancels out in this calculation.)

8 Some specific GCE models and related observational data

8.1 The 'Simple' (1-zone) model

The 'Simple' model of galactic chemical evolution, which goes back to early papers by van den Bergh (1962) and Schmidt (1963), is still widely used despite its failure, pointed out in those papers, to account for the metallicity distribution function of long-lived stars in the solar neighbourhood (the 'G-dwarf problem', which will be discussed below). Its popularity is due to its simplicity, and to the fact that it may have at least a limited applicability in other contexts such as the Galactic halo, dwarf galaxies and elliptical galaxies and bulges. Furthermore, its simplicity makes it a useful standard of comparison for more sophisticated models.

The assumptions of the Simple model (with a capital 'S') are:

(1) The system is isolated with a constant total mass, i.e. no inflows or outflows, so

$$g(t) + s(t) = M = \text{const.} \tag{8.1}$$

(2) The system is well mixed at all times, i.e.

$$Z = Z(t) \tag{8.2}$$

is the abundance of any element(s) in the gas and in newly born stars.

(3) The system starts as pure gas with primordial abundances, i.e.

$$g(0) = M; \quad Z(0) = S(0) = 0. \tag{8.3}$$

(4) The IMF and nucleosynthetic yields of stars with given initial mass are unchanging as far as 'primary' elements (such as oxygen) are concerned.

8.2 The Simple model with instantaneous recycling

In the Simple model with instantaneous recycling (to be referred to as the instantaneous Simple model), we have from Eq. (8.1)

$$dg = -ds \tag{8.4}$$

and hence from Eq. (7.37) (applying to stable, robust elements)

$$g\frac{dZ}{ds} = -g\frac{dZ}{dg} = p \tag{8.5}$$

whence

$$Z = p\ln\frac{M}{g} = p\ln\frac{s+g}{g} = p\ln\frac{1}{\mu}, \tag{8.6}$$

say, where μ is the 'gas fraction'. Equation (8.6), first derived explicitly by Searle & Sargent (1972) although it was implicit in earlier work, is the fundamental equation of the instantaneous Simple model. Since the yield is assumed constant, and is not very well determined anyway, it is often convenient to use $z \equiv Z/p$ in place of Z. If there is a finite initial abundance, e.g. in the case of helium, Z in Eq. (8.6) is replaced by $Z - Z_0$. Eq. (8.6) is, of course, subject to the restriction that, by definition, $Z < 1$, which implies that, should Z approach 1, the yield as defined in Chapter 7 must obviously become small (or negative). Since both p and Z (or $Y - Y_P$) are at most only a few per cent, this does not lead to any difficulty in practice.

Equation (8.6) refers to abundances in the gas and young stars. For the average abundance of a stellar population (with $Z_0 = 0$) we have instead

$$<z> \quad = \quad \frac{1}{s}\int_0^s z(s')ds' = 1 + \frac{\mu\ln\mu}{1-\mu} \tag{8.7}$$

$$\rightarrow \quad 1, \quad \text{for } \mu < 0.1, \text{ say}, \tag{8.8}$$

i.e. the average abundance (in stellar surface layers) approaches the yield from below when the gas fraction becomes small. This result holds also in models with inflow of unprocessed material, but it should be noted that this is a **mass-weighted** average. Observations of integrated spectra from stellar populations will rather give a **luminosity-weighted** average which will be different, usually less, because luminosities of red giants are higher at low metallicities.

8.3 Some consequences of the instantaneous Simple model

8.3.1 Estimation of yields

For a stellar population,

$$<Z> \quad \leq \quad p\,; \tag{8.9}$$

$$<Z> \quad \rightarrow \quad p \text{ as } \mu \rightarrow 0. \tag{8.10}$$

E.g., in the solar neighbourhood, $<[O/H]> \simeq -0.2$, implying a yield slightly below solar abundance. This **effective yield** is in fairly good accordance with the **true yield** of $(2/3)Z_\odot(O)$ derived in Problem 3 of Chapter 7 assuming small mass-loss rates (Table 7.2) and an IMF similar to our piece-wise one, or Scalo's with $b' = 1$. With large mass-loss rates, one would need an IMF with a higher proportion of massive stars, to

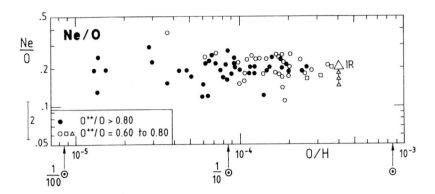

Fig. 8.1. Ne/O ratio *vs* O/H from (mainly optical) observations of H II regions. Filled and open symbols represent differing degrees of oxygen ionization. Triangles represent the Orion nebula, the large triangle being based on infra-red data. Typical error bars are ± 0.2 dex along each axis. After Meyer (1989) and Rubin *et al.* (1991). Courtesy J.-P. Meyer.

reach a similar true yield for oxygen. These ambiguities indicate that true yields are still subject to large uncertainties.

8.3.2   Abundance ratios of 'primary' elements

For any two primary elements,

$$\frac{Z_i}{Z_j} = \frac{p_i}{p_j} = \text{const.} \tag{8.11}$$

An example is dY/dZ or $(Y - Y_P)/(O/H)$, which comes out as a by-product of the determination of primordial helium abundance Y_P by extrapolation of the relation between He/H and O/H in low-abundance extragalactic H II regions to O/H $= 0$ (Lequeux *et al.* 1979; Pagel *et al.* 1992; cf. Fig. 4.9). The value $dY/dZ \simeq 4$ that is actually found is about twice the ratio predicted in the usual stellar models. The discrepancy (if real) could be due to an upper mass limit significantly under $100 M_\odot$ for stars undergoing supernova explosions (Maeder 1992, 1993), to preferential loss of supernova synthesis products in a selective galactic wind or to remaining uncertainties in the theory of single and binary star evolution. Figs. 8.1 and 8.2 show results for Ne/O and S/O derived from observations of Galactic and extragalactic H II regions, which are consistent with constant (and solar) yield ratios, while Figs. 8.3 and 8.4 show data for C/O indicating that this ratio increases with metallicity. Possible reasons are (i) effects of increasing mass loss from big stars, as predicted by Maeder (1992); or (ii) an increasing contribution from intermediate-mass stars at higher metallicities, e.g. if those systems have had more time to evolve.

Another important deviation from constancy in the abundance ratio of elements supposed to be primary is displayed by the ratio Fe/O in stars, which increases

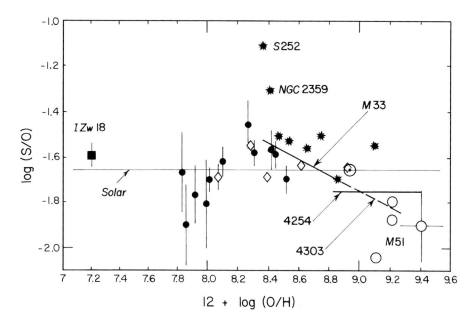

Fig. 8.2. S/O ratio *vs* O/H in Galactic and extragalactic H II regions, after a compilation by Pagel (1992a).

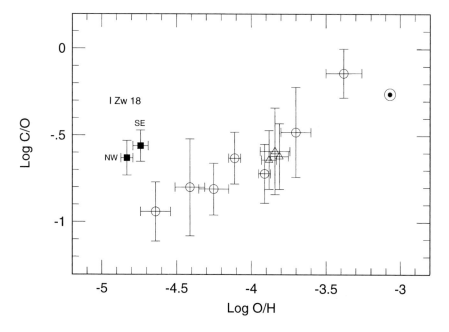

Fig. 8.3. C/O ratio *vs* O/H in Galactic and extragalactic H II regions, based on International Ultraviolet Explorer (IUE) and Hubble Space Telescope (HST) observations, after Garnett *et al.* (1995), *Ap. J.*, **443**, 64. Courtesy Don Garnett.

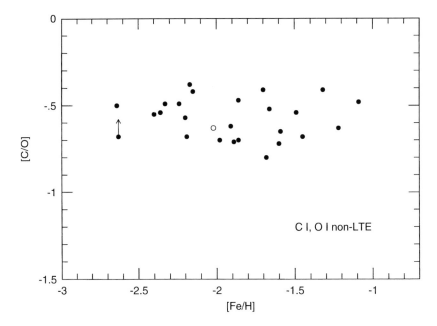

Fig. 8.4. [C/O] *vs* [Fe/H] in low-metallicity stars, after Tomkin *et al.* (1992), *Astr. J.*, **104**, 1568. The average [C/O] of −0.5 is equivalent to $\log(C/O) \simeq -0.8$ for log (O/H) between −5 and −4, in fair agreement with the H II region results shown in Fig. 8.3. Courtesy Jocelyn Tomkin.

systematically with [Fe/H] (Fig. 8.5). This is usually attributed to the existence of a substantial contribution to the production of iron found in the younger, more metal-rich stars (like the Sun) by SN Ia, which take times of the order of 1 Gyr to complete their evolution and therefore cannot be treated in the instaneous recycling approximation. As there is much more information available about stellar iron abundances than about their oxygen abundances, this introduces some complication in the interpretation of the data.

A schematic indication of how the O/Fe effect might arise from different production time-scales is given in Fig. 8.6. If star formation has taken place rapidly relative to the solar neighbourhood denoted by SN, as may apply to the Galactic bulge, the O/Fe ratio remains high up to large metallicities. Conversely, if it has been relatively slow, or has happened in bursts separated by long intervals, as may apply in the Magellanic Clouds (SMC, LMC), the O/Fe ratio becomes solar (or less) at subsolar metallicities (cf. Gilmore & Wyse 1991). The plotted data points are not very certain, however, and it may also be necessary to invoke a selective galactic wind in order to explain high Fe/O ratios in the LMC (Pilyugin 1996). The evolution of dwarf galaxies will be discussed further in Chapter 11.

Other element:iron ratios, measured in stars of the Galactic disk in the solar neighbourhood, are shown in Fig. 8.7. With the exception of C and N, all of the observed lighter elements up to Ti show a tendency to have higher abundances relative

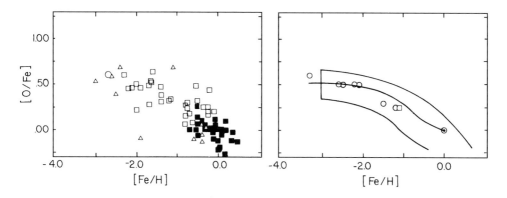

Fig. 8.5. Variation of [O/Fe] with [Fe/H] in stars, adapted from Bessell *et al.* (1991). The panel on the left shows data from the forbidden [O I] line λ 6300 (rederived by Bessell *et al.* from work of other observers) mainly seen in giant stars, while the right panel shows abundances derived by them from OH molecular bands in both giants and dwarfs and the suggested overall trend. Typical error bars are about ±0.15 dex in each coordinate. Courtesy Mike Bessell.

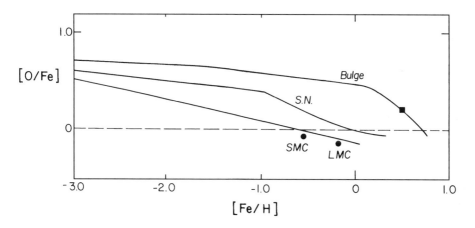

Fig. 8.6. Predicted [O/Fe] *vs* [Fe/H] relations, after Matteucci (1991).

to iron as one goes to low metallicities, except that for Na and Al this trend may be reversed for still lower metallicities in the halo.

8.3.3 'Primary' and 'secondary' elements

For a secondary element such as nitrogen resulting from CNO burning of previously existing C (and maybe O) and ejected in winds from massive or intermediate-mass stars, the yield is (roughly) proportional to the abundance of the primary progenitor,

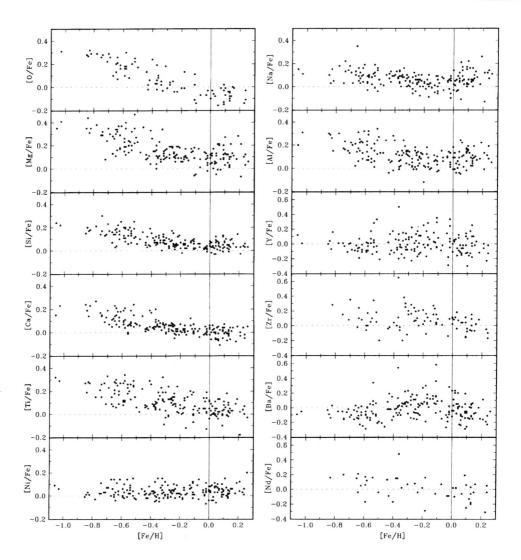

Fig. 8.7. Element:iron ratios measured in Galactic disk stars, after Edvardsson *et al.* (1993), *Astr. Astrophys.*, **275**, 101. Courtesy Johannes Andersen.

i.e.

$$\frac{Z_{sec}}{Z_{prim}} \propto Z_{prim}. \qquad (8.12)$$

Fig. 8.8 shows that there is indeed a tendency for N/O to increase with O/H, but not as fast as implied by Eq. (8.12). The overall trend is actually remarkably like that of carbon, and there has to be a primary component that dominates at low metallicities unless the effects of secondary production are offset by some complication such as preferential loss of supernova ejecta from these dwarf galaxies. There is substantial

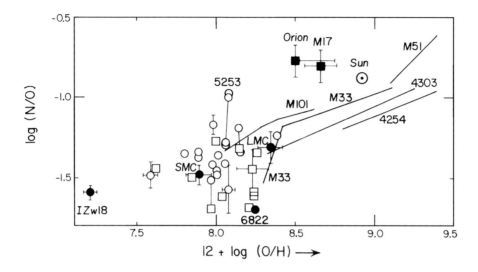

Fig. 8.8. N/O ratio in Galactic and extragalactic H II regions, after Pagel (1992a).

scatter, some of which is probably real. In NGC 5253, variations are seen across a single H II region, the highest values occurring near the central ionizing cluster which contains Wolf–Rayet stars; the effect may be due to local pollution by winds from these (or from their red supergiant or luminous blue variable progenitors). Other possible sources of scatter are differences in the ages of prominent stellar populations, either in a continuous or in a bursting star-formation mode, leading to differences in the contribution of intermediate-mass stars.

To summarise the observed behaviour of element:element ratios, one group (O, Ne, S) varies approximately in lockstep and potentially can be described by the instantaneous Simple model. There is no proof that the actual yields are constant, but their ratios may place constraints on the upper IMF slope and its variability. A second group (C, N, Fe) increases by factors between 3 and about 5 from the lowest metallicity systems observed to solar metallicity, relative to the first group. This is usually attributed to time-delay effects (plus effects of secondary production in the case of nitrogen), but their yields relative to those of the first group could also vary, e.g. as a consequence of stellar winds becoming more significant at higher metallicities. Metals up to Ti show intermediate behaviour, which is probably due to varying contributions from massive and Type Ia supernovae, while heavier metals tend to track iron more or less, except at extremely low metallicities where neutron-capture elements show special effects.

8.3.4 Relation with gas fraction

According to Eq. (8.6), the 'primary' abundances should increase as the logarithm of the gas fraction, the proportionality coefficient giving another estimate of the yield.

E.g., from the data in Table 7.7, assuming solar oxygen abundance in the local ISM, the effective yield is between 0.5 and $0.7Z_\odot$. Other gas-rich systems in which one may try to test this relationship (which obviously does not apply, or at least no longer applies, to gas-poor systems) are the dwarf irregular galaxies with O-abundance determinations from H II regions within them, but there are difficulties in estimating both the mass of gas (which may include undetected molecules as well as the observed H I) and the total stellar mass (which is difficult to separate from effects of dark matter on dynamical determinations, while 'photometric' determinations have uncertainties of their own). Furthermore, the H I gas is often much more extended than the stellar population, so that it is not clear how much of it shares the reasonably homogeneous chemical composition displayed in any one system by the H II regions themselves. Thus the comparison is best (but still not very well) carried out in irregular galaxies rather than blue compact galaxies (BCGs).

With these caveats, Fig. 8.9 shows an attempt at such a comparison, with some estimates of uncertainties. A rough relationship of the predicted form is indeed apparent, but the resulting effective yield is low, $Z_\odot/6$ or about 4 times less than the estimates for the solar neighbourhood. This is usually attributed to a preferential loss of hot metal-enriched stellar ejecta in galactic winds escaping from the shallow potential wells of these systems, although it has also been suggested that the true yield may increase with metallicity, e.g. if the low-metallicity IMF has a larger proportion of low-mass stars.

A somewhat more impressive relationship exists between abundances and luminosity of the host galaxy, whether gas-rich or otherwise (Fig. 8.10), although low surface-brightness and blue compact galaxies lie systematically to the left of the plotted line. Also, the fit between O-abundances in the ISM of irregulars and the mean iron abundance of stellar populations in dwarf ellipticals (dE) and spheroidals (dSph) is somewhat ill-defined; the plot simply assumes $[O/H]_{ISM} = <[Fe/H]>_{stars}$. The latter might be expected to be somewhat lower in systems sharing a common GCE history, but if (as the data suggest) the dEs and dSphs resulted from the stripping of gas away from dwarf irregulars due, e.g., to tidal interaction with the Milky Way or M 31, there would be a fading of the dEs and dSphs that would push them back towards the line. Studies of elliptical galaxies in general suggest that the primary factor fixing the mean metallicity of their stellar populations is the velocity dispersion which in turn may be closely related to the escape velocity (cf. Chapter 11).

8.3.5   Radial abundance gradients in spiral disks

The Simple model can account to some extent for the existence of radial abundance gradients in the ISM of spiral galaxies (cf. Figs. 3.33, 3.34, 3.38 and 3.39), if it is assumed (a) that evolution takes place in isolated concentric zones analogous to the solar cylinder, and (b) that the gas fraction decreases inwards towards the Galactic centre (Searle & Sargent 1972), implying faster time-scales for conversion of gas into

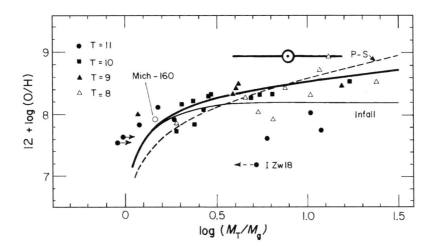

Fig. 8.9. Logarithm of oxygen abundance in irregular and blue compact galaxies plotted against the logarithm of the gas fraction, adapted from Axon *et al.* (1988). Different symbols show galaxies of different morphological types (*T* increases towards 'later' types), and the position of the Sun is shown with its uncertainties. The heavy curve shows expectation from the Simple model with a yield of $0.2Z_\odot$ and the broken curve P – S shows a model in which the yield increases linearly with *Z* (Peimbert & Serrano 1982). After B.E.J. Pagel, 'Abundances in Galaxies', in H. Oberhummer (ed.), *Nuclei in the Cosmos*, p. 93, Fig. 4, ©Springer-Verlag Berlin Heidelberg 1991.

stars in the inner regions. A much smaller gradient would apply to the stellar population according to Eqs. (8.7), (8.8). The stellar surface density falls off exponentially with Galactocentric distance with a scale length $\alpha_*^{-1} \simeq 4$ kpc, while the distribution of H I + H_2 shown in Fig. 8.11 can be quite well approximated between 4 and 17 kpc (assuming the KBH rotation curve) as an exponential with a scale length $\alpha_g^{-1} \simeq 12$ kpc. Eq. (8.6) then leads to an abundance distribution law in the ISM

$$z(R) = -\ln \mu(R_\odot) - (\alpha_* - \alpha_g)(R - R_\odot) + \ln \left(\frac{1 + g(R)/s(R)}{1 + g(R_\odot)/s(R_\odot)} \right) \tag{8.13}$$

$$\simeq 1.8 - 0.17(R - R_\odot) + 0.0013 R^2. \tag{8.14}$$

The oxygen abundance then decreases more or less linearly from about $2Z_\odot$ near the centre to $Z_\odot$ in the solar neighbourhood, reaching about $0.3Z_\odot$ at 20 kpc. This corresponds to logarithmic gradients of -0.03 dex kpc^{-1} between 5 and 10 kpc and a slightly steeper one of -0.04 dex kpc^{-1} between 10 and 15 kpc. The observed gradient is believed to be about -0.07 dex kpc^{-1} and more uniform in logarithmic terms, possibly even becoming flatter with increasing Galactocentric distance (cf. Fig. 3.33). Given all the uncertainties, this simple effect is probably at least a significant contributor to the gradient and it may be the main cause in the outer parts of the Galaxy. Other influences on the gradient, which may become dominant in the more inner regions, are dissipative processes such as inward radial flows of gas driven by viscosity or by

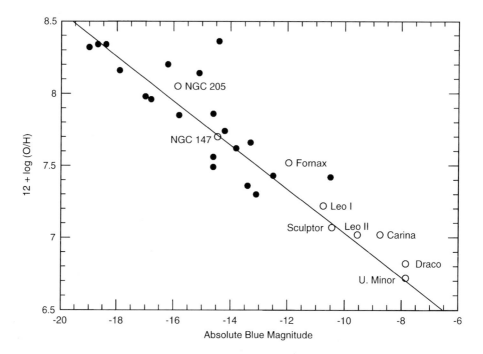

Fig. 8.10. Abundance against absolute magnitude for dwarf irregular, dwarf elliptical and dwarf spheroidal galaxies, after Skillman, Kennicutt & Hodge (1989), *Ap. J.*, **347**, 875. The vertical scale shows logarithmic oxygen abundances ($\odot = 8.9$) measured in H II regions of irregular galaxies (filled circles), while open circles show <[Fe/H]> + 8.9 for dwarf ellipticals and dwarf spheroidals deduced from colours of stars on the giant branch in the HR diagram. Courtesy Evan Skillman.

a mismatch of angular momentum of inflowing material. A generic relation between dissipation and abundance gradients in the stellar population is discussed in Appendix 5, but its effect on the abundance gradient in the gas depends on details of the velocity field. Other factors influencing abundance gradients include differential effects of inflow of unprocessed material at different radii (neglecting any dynamical effects) and radial mixing caused by perturbations of axial symmetry, e.g. by a central bar.

8.3.6   Age–metallicity relation in different stellar populations

The instantaneous Simple model predicts a monotonic increase in the abundance of any robust element with time, which can be quantified if a star-formation law is assumed. If, for example, the SFR is assumed proportional to the mass of gas as

$$ds/dt = \omega\, g \tag{8.15}$$

with ω constant, the abundance of a primary element increases linearly with time

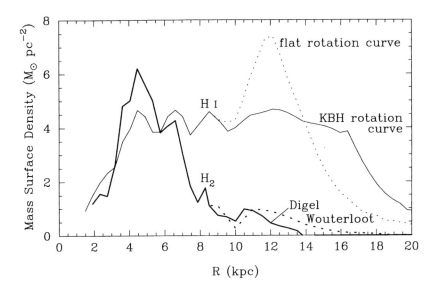

Fig. 8.11. Surface densities of atomic and molecular hydrogen in the Galaxy as a function of Galactocentric distance; the Sun is at 8.5 kpc. Beyond that distance, the deduced surface density depends on the assumed law of Galactic rotation; KBH refers to Kulkarni, Blitz & Heiles (1982). Assuming their rotation curve, the total gas surface density falls by about a factor of 2 between 4.5 and 13 kpc, corresponding to an exponential fall-off with a scale length α_g^{-1} of about 12 kpc. After T.M. Dame, in S.S. Holt & F. Verter (eds.), *Back to the Galaxy*, Amer. Inst. Phys. Publ. (1993), p. 267. Courtesy T.M. Dame.

according to

$$z = \omega t. \tag{8.16}$$

More generally, ω may be treated as variable, thus not appealing to any particular star formation law, in which case we still have

$$z(t) = \int_0^t \omega(t')dt' \equiv u, \tag{8.17}$$

say, where u is some single-valued non-decreasing function of time. An increase with time is found in a qualitative sense in the low metallicities observed in most stars of the Galactic halo population and in high red-shift absorption-line systems on the line of sight to quasars, including the damped Lyman-α systems which are believed to represent an early form of disk galaxies. However, the scatter at a given red-shift is large, which is not surprising given that star formation probably began at different times in different places. Fig. 8.12 shows the relationship between metallicity and Galactic rotation velocity for high proper motion stars, which probably give a fair sample of the halo population while undersampling the disk in a kinematically biased way. The disk stars, with numerically small relative velocities V in the direction of Galactic rotation, overlap in age and metallicity with the halo stars, which have an

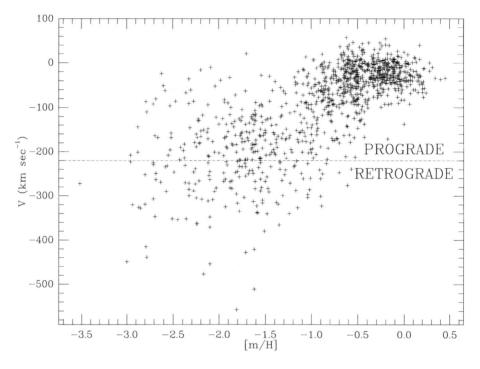

Fig. 8.12. Plot of velocity V along the direction of Galactic rotation, relative to the local standard of rest, against metallicity [m/H] (essentially [Fe/H]) for high proper motion stars, after Carney, Laird, Latham & Aguilar (1996), *Astr. J.*, **112**, 668. A relatively sparsely populated gap runs from upper left at $(-1.6, 0)$ towards lower right at $(-0.5, -200)$. Courtesy John Laird.

average $<V> \simeq -220$ km s^{-1}, but there is a diagonal gap in the diagram which enables most individual stars to be assigned to one class or the other. The two populations have very different abundance (as well as dynamical) properties, as will be described below, which reflect their distinct evolutionary histories. The halo stars are old, with ages somewhere between 12 and 18 Gyr (this range is a combination of systematic uncertainties with probably some real variation), while the ages of disk stars range from zero to values typical of halo stars.

It is therefore among the disk stars of the solar neighbourhood that one looks for quantitative evidence for an age–metallicity relation; relevant data are shown in Fig. 8.13.

An overall relation of the form of Eq. (8.16) gives as good a fit as any to the average of the data, but the scatter is substantially larger than the errors and this sample is in any case biased against old, metal-rich stars. The best that one can say is that a statistical correlation exists, with scatter presumably resulting from local inhomogeneities and/or the mixing of stellar populations from different parts of the Galaxy that have evolved at different rates. A contribution from the latter effect is suggested by a tendency noted by Edvardsson *et al.* (1993) for [O/Fe] and [α/Fe] to

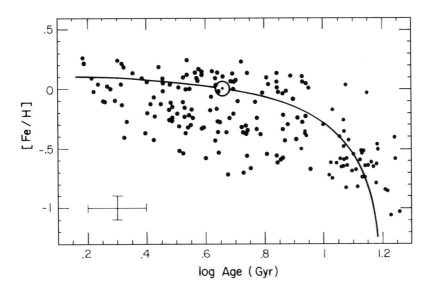

Fig. 8.13. Metallicity *vs* age for a sample of nearby F and G-type stars, adapted from Edvardsson *et al.* (1993). The curve is based on Eq. (8.16), normalised to the Sun and with an assumed age for the disk of 17 Gyr. Nissen & Schuster (1991) give the corresponding relation for high-velocity stars.

be larger (for given [Fe/H]) at smaller mean galactocentric distances (cf. the 'bulge' point in Fig. 8.6), implying faster enrichment rates there if most of the iron comes from SN Ia progenitors with significant lifetimes.

Fig. 8.7 shows a marked increase in the scatter of $[\alpha/\text{Fe}]$ as one goes down through $[\text{Fe}/\text{H}] = -0.4$, at which point there is also a marked change in kinematic properties, notably the dispersion of velocities W perpendicular to the Galactic plane (Fig. 8.14). The stars on the left of this divide are usually referred to as belonging to the 'thick disk', or 'intermediate Population II', and have ages ranging from about 8 Gyr up to that of the halo stars. There are doubts as to whether the sudden increase in velocity dispersion can be explained by conventional dynamical heating mechanisms from gravitational action by molecular clouds or spiral structure. It could be related to cooling processes in the early formation of first the halo, then the thick disk and then (after enhancement of cooling efficiency by growing metallicity) the thin disk (Burkert, Truran & Hensler 1992), but another possibility is that the disk was dynamically heated at the appropriate stage by capture of one or more satellite galaxies which themselves may or may not have been part of the primeval halo (Freeman 1990). Residual gas in the disk would then have cooled and collapsed leading to the thin disk. It is not clear whether there is any discontinuity in chemical history between the thick and thin disks, comparable to the differences between the disk and the halo.

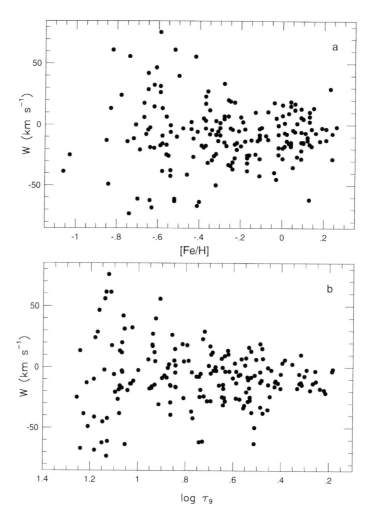

Fig. 8.14. *W*-velocities *vs* age and metallicity for the nearby F and early G star sample, after Edvardsson *et al.* (1993), *A. & A.*, **275**, 101. Courtesy Bengt Edvardsson.

8.3.7   Stellar abundance distribution functions and the 'G-dwarf problem'

Unlike the age-abundance relation, the distribution function of stellar abundances of primary elements is independent of past rates of star formation as long as instantaneous recycling holds, and this makes it a potentially powerful clue to the evolutionary histories of stellar populations.

This can be seen by rewriting Eq. (8.6) as

$$\frac{g}{M} = 1 - \frac{s}{M} = e^{-z} \tag{8.18}$$

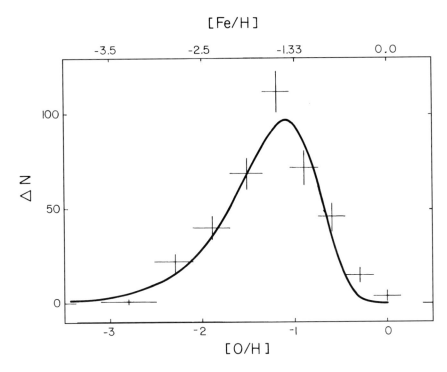

Fig. 8.15. Distribution function of oxygen abundances among halo field stars taking [Fe/H] from Fig. 8.12 and assuming an Fe–O relation similar to that in Fig. 8.5, after B.E.J. Pagel in B. Barbuy & A. Renzini (eds.), IAU Symp. 149: *The Stellar Populations of Galaxies*, p. 133, Fig. 2. ©1992 International Astronomical Union. With kind permission from Kluwer Academic Publishers.

or

$$\frac{s(z)}{M} = 1 - e^{-z}, \tag{8.19}$$

whence

$$\frac{ds}{d\log z} \propto z e^{-z}; \quad z \leq \ln\left(\frac{1}{\mu}\right). \tag{8.20}$$

This equation (with a characteristic peak value at $z = 1$ or $Z = p$) is expected to apply primarily to oxygen and α-elements in stars that have lived long enough since the origin of the Galaxy to be still observable today, but may also apply approximately to iron in cases where $O,\alpha/Fe \simeq$ const.

Fig. 8.15 shows an estimate of the distribution function of oxygen abundances among field stars of the Galactic halo and Fig. 8.16 shows the iron abundance distribution function for globular clusters with a portion of the one for halo field stars superposed.

The distribution function for field stars is reasonably well fitted by the Simple model equation (8.20), but with a very low effective yield $p \simeq 10^{-1.1} Z_\odot$ for oxygen (cf. earlier

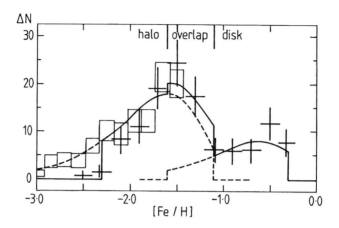

Fig. 8.16. Metallicity distribution function of globular clusters (crosses indicating error bars and bin widths) and halo field stars (boxes), after B.E.J. Pagel, 'Abundances in Galaxies', in H. Oberhummer (ed.), *Nuclei in the Cosmos*, p. 97, Fig. 7, ©Springer-Verlag Berlin Heidelberg 1991.

comments on dwarf galaxies). This was first noted (actually for globular clusters) by Hartwick (1976), who pointed out that it could be readily explained by continuous loss of gas from the halo in the form of a homogeneous wind with a mass-loss rate from the system proportional to the rate of star formation. In this case,

$$\frac{E}{\alpha\psi} = \text{const.} = \eta, \tag{8.21}$$

say. In this case we have from Eq. (7.35), with $F = 0$,

$$g\frac{dz}{ds} = 1, \ i.e. \ \frac{dz}{ds} = \frac{1}{g} \tag{8.22}$$

and then from Eq. (7.32)

$$g\frac{dz}{dg} = -\frac{1}{(1+\eta)}, \tag{8.23}$$

whence

$$\frac{g}{M_0} = e^{-(1+\eta)z}, \tag{8.24}$$

where M_0 is the initial mass of the system, so from Eq. (8.22)

$$\frac{ds}{d\log z} \propto z\frac{ds}{dz} \propto z\,e^{-(1+\eta)z}, \tag{8.25}$$

similar to the Simple model but with an effective yield $p/(1+\eta)$. (Note that a *terminal* wind, in which all the gas is expelled at a particular time when the metallicity is still low but the effective yield is undiminished, would lead to a completely different distribution function rising monotonically up to a cutoff (Yoshii & Arimoto 1987)). If there were

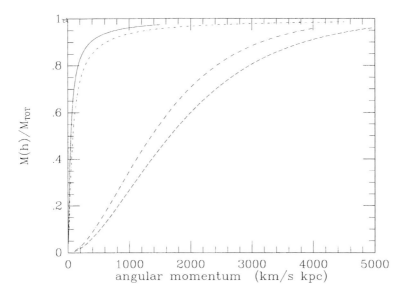

Fig. 8.17. Cumulative distribution function of specific angular momentum in the thin disk (broken line), thick disk (dots and long dashes), halo (dots and short dashes) and bulge (solid line), after Wyse & Gilmore (1992), *Astr. J.*, **104**, 144. Courtesy Rosemary Wyse.

a metal-enhanced wind rather than a homogeneous one, the relative amount of mass required to be lost would be smaller.

The distribution function for globular clusters is somewhat more complicated, as there appear to be two (probably overlapping) distributions corresponding to the halo and the thick disk, respectively. These have been tentatively fitted in Fig. 8.16 with a Simple model truncated at [Fe/H] $= -1.1$ for the halo and a model for the thick disk clusters with an initial abundance [Fe/H] $= -1.6$ (the mean metallicity of the halo) and truncated at [Fe/H] $= -0.35$. The disk-like character of the more metal-rich clusters is supported by their spatial distribution (Zinn 1985). Furthermore, there is a marginally significant shortage of globular clusters in the lowest metallicity bins which, if real, could reflect some difference in the chemical evolution of cluster and field stars. However, disregarding the disk clusters, the overall distributions of cluster and field star abundances are quite similar.

What happened to the gas expelled from the halo? A traditional answer based on monolithic models of a Galaxy collapsing through successive stages of halo, thick disk and thin disk would be that the expelled gas formed the raw material of the disk. This hypothesis faces some severe difficulties. For one thing, the remaining mass in the halo should then be about 10 per cent of the mass of the disk, whereas it is probably a factor of 5 or so less than this (Carney, Latham & Laird 1990). Another difficulty is the specific angular momentum, as is qualitatively obvious from Fig. 8.12 and is illustrated quantitatively in Fig. 8.17.

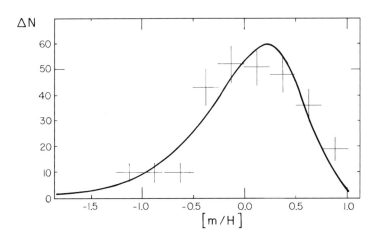

Fig. 8.18. Metallicity distribution function for giants in the Galactic bulge, after Geisler & Friel (1992), fitted with a Simple model having an 'm' yield of $2Z_\odot$. There is some doubt about the abundance scale at high metallicities, however (McWilliam & Rich 1994). After B.E.J. Pagel, 'Abundances in Galaxies', in H. Oberhummer (ed.), *Nuclei in the Cosmos*, p. 97, Fig. 8. ©Springer-Verlag Berlin Heidelberg 1991.

The bulge metallicity distribution also shows a rough compatibility with a Simple model (Fig. 8.18), this time apparently with a high effective yield, about $2Z_\odot$, although there could be some doubt about this as the metallicity scale is uncertain, particularly at the high end (McWilliam & Rich 1994); the x-axis is probably best interpreted as representing the magnesium abundance with $[Mg/Fe] > 0$. An initial enrichment corresponding to $[Fe/H] = -1.6$ (as could be expected from inflow of halo gas) is not excluded by the data.

The data discussed so far have not been in violent conflict with the ideas underlying the Simple model. However, as was pointed out long ago by van den Bergh (1962) and Schmidt (1963), the abundance distribution function for long-lived stars (G dwarfs) in the solar neighbourhood is in sharp contradiction to these ideas because there are relatively too few low-metallicity stars compared to Eq. (8.20). This 'G-dwarf problem' is illustrated in Fig. 8.19, which shows two estimates of the oxygen abundance distribution function in the solar cylinder, based on ultra-violet excess statistics for nearby stars, together with an iron–oxygen relation and allowance for the dependence of the scale height of the spatial distribution of stars as a function of metallicity. These corrections lead to the substantial uncertainties indicated by the error boxes in the figure, and the data have subsequently been improved, but the contradiction with the prediction of the Simple model is evident, the more so after its relative success with the halo and the bulge. Errors in the use of the instantaneous recycling approximation need to be considered, depending on the assumed history of star formation and mass ejection. After long times, e.g. 15 Gyr, low-mass stars eject relatively metal-poor material and modify the distribution in a similar way to what happens in some of

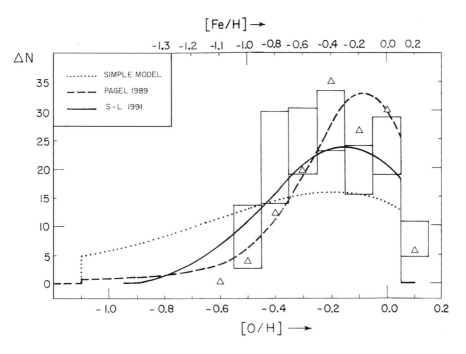

Fig. 8.19. Distribution function of oxygen abundances of 132 G-dwarfs in the solar cylinder, binned in intervals of 0.1 in [O/H]. Triangles show the data points after Pagel (1989ab), based on a reanalysis of those discussed by Pagel & Patchett (1975), and boxes show lower and upper limits based on a new discussion of the dependence of the scale height on age and metallicity by Sommer-Larsen (1991a). The dotted curve shows predictions of an instantaneous Simple model with an initial enrichment [O/H] = −1.1 from the halo. The other model curves are discussed below. After B.E.J. Pagel, 'Abundances in Galaxies', in H. Oberhummer (ed.), *Nuclei in the Cosmos*, p. 98, Fig. 9. ©Springer-Verlag Berlin Heidelberg 1991.

the inflow models discussed below. However, it is still rather doubtful whether any Simple model can explain the abundance distribution function as well as satisfying other constraints.

The 'G-dwarf problem' has inspired numerous attempts to devise more sophisticated GCE models, some of which may be more realistic than the Simple one.

8.4   Suggested answers to the G-dwarf problem

A great variety of solutions to the G-dwarf problem have been proposed, including the following:

(1) The problem is a pseudo-problem due to poor selection of stars, either because the G dwarfs are not long-lived enough or because the older ones have migrated away from the solar neighbourhood into the inner Galaxy. The first point is

believed to be an additional source of corrections to the adopted distribution law, but not to be so strong as to remove the 'problem' (Bazan & Mathews 1990); there is, after all, no obvious shortage of low-metallicity G dwarfs in the halo. Their masses are probably somewhat smaller than those of more metal-rich G dwarfs, but given the small slope of the IMF in this region, that can have only a minor influence.

The second point, which has been raised by Grenon (1989, 1990), is somewhat more interesting. Grenon's conclusions are based on a survey of parallax and proper motion stars which is kinematically biased against the 'silent majority' of ordinary stars of the (thin) disk, and the problem is to balance the number of stars that have moved out of the solar cylinder due to kinematic heating and related effects against the number that have moved in for the same reason. However, only insignificant radial abundance gradients are found for old G and K giant populations in the disk (Neese & Yoss 1988; Lewis & Freeman 1989); thus no major effect is expected, although the problem does deserve to be looked at more closely.

(2) The solution originally proposed by Schmidt (1963) was that yields were higher in the past, due to an IMF with more high-mass stars. (A similar result would follow from Maeder's oxygen yields for strong mass loss, even for fixed IMF, if all stars up to 120 $M_\odot$ undergo SN explosions, cf. Carigi 1994; if, however, only those below 25 or 30 $M_\odot$ did so, the result would be the opposite.) The problem is that the drop in numbers below [Fe/H]$= -1$ is very sudden, so the required behaviour of the yield looks rather contrived.

(3) Talbot & Arnett (1973b, 1975) suggested the hypothesis of 'metal-enhanced star formation' in which there would be large fluctuations in the metallicity of the ISM and stars would be preferentially formed in high-metallicity regions. This idea also requires contrived-looking parameters and is now only of historical interest; chemo-dynamical models (Burkert & Hensler 1989) suggest precisely the opposite effect, i.e. that the hot ISM is more enriched (from supernova ejecta) than the cold molecular clouds from which stars form.

(4) If for some reason there is a finite initial abundance Z_0 of the order of [O/H]$=$ -0.6 or [Fe/H] $\simeq -1$, due to an initial production spike or 'prompt initial enrichment' (Truran & Cameron 1971), an excellent fit to the distribution function is immediately obtained at the expense of excluding the few stars that actually do lie in the low-metallicity tail. Such initial enrichment could come from prior star formation activity in the halo (Ostriker & Thuan 1975) or preferably the bulge. Köppen & Arimoto (1990) have put forward a model in which the bulge evolves with a large true yield (1.7$Z_\odot$) and ends star formation by expelling 1/10 of its mass as gas with an abundance 2.5$Z_\odot$ in a terminal wind which is captured by the proto-disk. Assuming the latter to have a mass equal to that of the bulge, this results in an initial disk abundance of 0.25$Z_\odot$ after mixing. This picture is obviously consistent with the metallicity distribution functions of both the bulge (if the right-most data point in Fig. 8.18 is disregarded, as

it probably should be) and the disk, apart from the low-metallicity tail. The possible relevance of this picture regarding both the high yields in the bulge and its large mass:light ratio (similar to elliptical galaxies and attributed to a high proportion of white dwarfs resulting from the relatively flat IMF that causes the high yield) is contingent on confirmation of the high metallicities of bulge stars shown in Fig. 8.18.

(5) Starting with the work of Larson (1972) in connection with dynamical models of galaxy formation, it has been a popular idea to account for the G-dwarf metallicity distribution on the basis of inflow models, i.e. formation of the Galactic disk by gradual accretion of unprocessed or partially processed material which only starts to form stars after it has fallen into the disk. These models have many attractive features and will be described in some more detail in the next section.

8.5 Inflow models

Inflow models are attractive because they can shed light on dynamical processes during galaxy formation. The inflowing gas may be regarded as primordial or as having a modest amount of enrichment, e.g. if it comes from the gaseous halo as exemplified by the high-velocity HI clouds, from outer parts of the galactic disk or from the accretion of small satellite galaxies. The disk itself may be regarded as being formed entirely from such inflowing gas or as having a small (compared to its ultimate mass) initial mass and/or metallicity.

8.5.1 Extreme inflow model (Larson 1972)

In this case, which may be regarded as the opposite extreme to the Simple model, it is supposed that, through feedback processes, accretion occurs at just such a rate as to compensate the loss of gas in star formation, i.e.

$$g = \text{const.} \tag{8.26}$$

Then from Eq. (7.35) with $Z_F = 0$, $F = \alpha\psi$,

$$\frac{dZ}{ds} + \frac{Z}{g} = \frac{p}{g} \tag{8.27}$$

whence

$$Z = Z_0 e^{-s/g} + p(1 - e^{-s/g}) \tag{8.28}$$

$$\rightarrow p, \text{ as } s/g \rightarrow \infty. \tag{8.29}$$

Thus the mass of stars and that of the whole system steadily increase while z soon approaches 1 and the stellar metallicity distribution is very narrow (see Fig. 8.20). The accretion rate is constant in time if the star formation rate is any fixed function of the

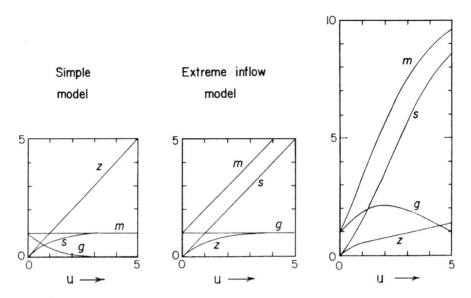

Fig. 8.20. Schematic behaviour of gas mass, total mass and metallicity in the Simple model (left), the extreme inflow model of Larson (1972) (middle) and a model with time-decaying inflow (right). The abscissa is $u \equiv \int_0^t \omega(t')dt'$ where ω is the (constant or otherwise) transition probability per unit time for gas to change into stars. The initial mass has been taken as unity in each case.

mass of gas. Other models in which the accretion rate is constant, but less than in the extreme model, have been quite often considered in the older literature (e.g. Twarog 1980), but are less popular now because they are not well motivated from a dynamical point of view, there is an upper limit to the present inflow rate into the whole Galaxy of about $1\,M_\odot/\mathrm{yr}$ from X-ray data (Cox & Smith 1976) and they do not provide a very good fit to the observed metallicity distribution function.

8.5.2   Analytical models with declining inflow rates

Larson (1974a, 1976) constructed dynamical models of the formation of the Galactic disk in which the disk develops outward from the central parts by inflow of gas at a rate which declines in the course of time as gas runs out, the mass of the system (and that of stars) approaches an asymptotic limit and the mass of gas reaches a maximum before declining at later stages more or less as in the Simple model. He noted that this picture can give a good representation of the G-dwarf metallicity distribution, and Lynden-Bell (1975) developed analytical versions of these and other models, among which his 'Best Accretion Model' is especially elegant. This model postulates a non-linear relation between gas mass and star mass which is the simplest quadratic relation having the desirable properties of leading to a maximum in the gas mass and a smooth decay at late times when star formation approaches completion:

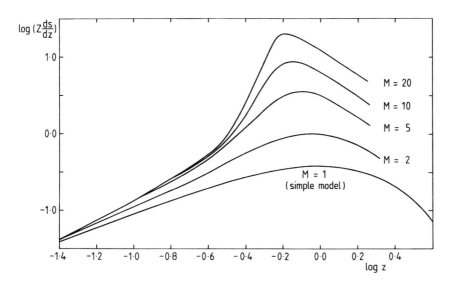

Fig. 8.21. Theoretical abundance distribution functions according to Lynden-Bell's 'Best Accretion Model' for different values of M, after Pagel (1989b).

$$g(s) = (1 - s/M)(1 + s - s/M) \qquad (8.30)$$

where all quantities are in units of the initial mass and $M(\geq 1)$ is the final mass of the system. The case $M = 1$ reduces to the Simple model whereas, for larger values of M, $g(s)$ has a parabolic dependence with a maximum when $M \geq 2$. Eq. (7.34) with $Z_F = E = 0$ can now be written

$$\frac{d}{ds}(Zg) = p - Z = p - \frac{Zg}{g} \qquad (8.31)$$

with the solution (taking $Z_0 = 0$)

$$z(s) = \frac{1}{g(s)} e^{-\int_0^s \frac{ds'}{g(s')}} \int_0^s e^{\int_0^{s'} \frac{ds''}{g(s'')}} \, ds' \qquad (8.32)$$

$$= \left(\frac{M}{1 + s - s/M}\right)^2 \left[\ln \frac{1}{1 - s/M} - \frac{s}{M}\left(1 - \frac{1}{M}\right)\right] \qquad (8.33)$$

and

$$\frac{ds}{d\ln z} = \frac{z\left[1 + s(1 - 1/M)\right]}{(1 - s/M)^{-1} - 2z\left(1 - 1/M\right)}. \qquad (8.34)$$

This generates a series of abundance distribution functions with M as a parameter (Fig. 8.21). As M increases, the distribution becomes more like a Gaussian (i.e. a parabola on this logarithmic plot) on the low-metallicity side of the peak (although there is still always a low-metallicity tail), and the peak itself shifts to lower metallicities in units of the yield.

Other analytical inflow models less sophisticatedly refer to a specific star formation law, usually linear

$$\alpha\psi \equiv \frac{ds}{dt} = \omega\, g(t)\,;\ \omega = \text{const.} \tag{8.35}$$

This, however, can be generalised (except for cases where actual time is significant, as in the age–metallicity relation or cosmochronology) by letting ω vary arbitrarily with time and defining a time-like variable u by Eq. (8.17). Clayton (1985ab) has developed a very convenient series of models in which the inflow rate is parameterised by

$$F(t) = \frac{k}{t + t_0}\, g(t),\ \text{or}\ F(u) = \frac{k}{u + u_0}\, g(u), \tag{8.36}$$

where k is a small integer and t_0 or u_0 arbitrary. With Eq. (8.35), the mass of gas (in units of the initial mass) is then

$$g(t) = \left(1 + \frac{t}{t_0}\right)^k e^{-\omega t},\ \text{or}\ g(u) = \left(1 + \frac{u}{u_0}\right)^k e^{-u} \tag{8.37}$$

giving the desirable maximum and (with $Z_F = Z_0 = 0$) one has

$$z(u) = \frac{u_0}{k+1}\left[1 + \frac{u}{u_0} - \left(1 + \frac{u}{u_0}\right)^{-k}\right] \tag{8.38}$$

and

$$\frac{ds}{d\ln z} = \frac{z(k+1)(1 + u/u_0)^k\, e^{-u}}{1 + k(1 + u/u_0)^{-(k+1)}} \tag{8.39}$$

with k and u_0 as parameters. The dashed curve in Fig. 8.19 shows a model of this kind with $k = 4$, after Pagel (1989a).

A still simpler inflow model, based on a dynamical model due to Sommer-Larsen (1991b), assumes inflow according to the law

$$F(u) = A\omega\, e^{-u}\,;\ A = \text{const.} \tag{8.40}$$

Then from Eq. (7.31)

$$g(u) = (g_0 + Au)\, e^{-u} \tag{8.41}$$

where g_0 is the initial mass of the system assumed to be pure gas, while

$$M(u) = g_0 + A(1 - e^{-u}), \tag{8.42}$$

$$s(u) = g_0(1 - e^{-u}) + A\left[1 - (1 + u)e^{-u}\right] \tag{8.43}$$

and from Eq. (7.35), taking u as the independent variable instead of s,

$$z(u) = \frac{z_F}{1 + g_0/(Au)} + \frac{z_0}{1 + Au/g_0} + \frac{u(1 + \frac{1}{2}Au/g_0)}{1 + Au/g_0} \tag{8.44}$$

$$\rightarrow\ z_F + u/2 \ \text{as}\ Au/g_0 \rightarrow \infty. \tag{8.45}$$

The full-drawn curve in Fig. 8.19 shows a model of this kind with $g_0 = Z_0 = 0$ and $Z_F = 0.1 Z_\odot$ for oxygen.

When $g_0 = Z_0 = Z_F = 0$, there are some very simple relations:

$$M(u) = A(1 - e^{-u}) \tag{8.46}$$

$$g(u) = Au e^{-u} \tag{8.47}$$

$$s(u) = A[1 - (1 + u)e^{-u}] \tag{8.48}$$

$$\mu(u) = u/(e^u - 1) \tag{8.49}$$

$$z(u) = u/2 \tag{8.50}$$

$$<z> = 1 - \frac{\frac{1}{2}u^2}{e^u - (1 + u)} \tag{8.51}$$

$$\frac{ds}{d\ln z} = 4Az^2 e^{-2z} \; ; \; z \le u/2. \tag{8.52}$$

The last relation is essentially the square of the corresponding one for the Simple model, Eq. (8.20); this does not go far enough to fit the G-dwarf metallicity distribution, but the above relations are useful in giving a quick estimate of the kind of departures from Simple model predictions that may be caused by inflow.

8.6 Analytical models for the Galactic halo and disk

8.6.1 Did the Galaxy form by collapse?

The many models that have been put forward to describe the chemical evolution of the Galaxy (e.g. Hartwick 1976; Tinsley & Larson 1979; Matteucci & François 1989; Pagel 1989a) have been largely inspired by the classic paper of Eggen, Lynden-Bell & Sandage (1962) who used correlations between the ultra-violet excess (i.e. metal-deficiency) of stars and their orbital eccentricity and motions perpendicular to the Galactic plane (cf. Fig. 3.32) to infer that the Galaxy had collapsed on a 'rapid' timescale from a protogalactic cloud, eventually forming the disk by a dissipative process. Since then, complications have appeared in this simple picture, notably an age spread among globular clusters, indications of a possible duality in the halo (Hartwick 1987; Norris 1994) and the suggestion that the latter is composed of merged satellite galaxies or 'fragments', as opposed to being a well-preserved relic of the initial collapse (Searle & Zinn 1978). The original ELS picture has been stoutly defended (with some modifications) by Sandage (1990), who sums up his point of view as 'SZ = ELS + noise!' Eventually some synthesis of the two viewpoints is likely to emerge, but one significant modification of the original ELS picture is occasioned by the identification of the Thick Disk (Gilmore & Reid 1983) — broadly overlapping what had previously been known as 'Intermediate Population II' (Strömgren 1987 and references therein) — which consists of metal-deficient stars with only moderate eccentricities (cf. Fig. 8.12) and was missed by ELS owing to observational selection effects. Subsequent studies based on more extensive data for high-velocity stars (Sandage & Fouts 1987;

Carney, Latham & Laird 1990; Ryan & Norris 1991; Nissen & Schuster 1991) and on kinematically unbiased objective-prism surveys (Norris, Bessell & Pickles 1985; Morrison, Flynn & Freeman 1990; Beers & Sommer-Larsen 1995) have revealed an ever-increasing degree of overlap in metallicity between thick-disk and halo stars. At the same time, the angular momentum distribution of halo stars (Fig. 8.17) suggests that the halo is quite disconnected from the evolution of at least the local part of the disk, so that the sort of connections studied by ELS should rather be sought among stars of the thick disk; within 1 kpc of the Galactic plane, these account for a significant proportion (about 30 per cent) of the stars with $[Fe/H] \leq -1.5$ as well as essentially all stars with $-1.0 \leq [Fe/H] \leq -0.5$ (Beers & Sommer-Larsen 1995).

8.6.2   Developments in analytical modelling

Pagel (1989a) put forward an analytical model that was intended to give a comprehensive description of the relevant data for the solar neighbourhood as they appeared at the time: the abundance distribution functions in the halo and disk, the relation between abundances of oxygen and magnesium, taken to be instantaneously produced, and elements involving a time delay, mainly iron and barium, and the age-metallicity relation for disk stars. Barium was taken to be produced at early times just by the r-process (Truran 1981; Lambert 1989), which accounts for about one tenth of its abundance in the Solar System (see Table 6.3), and its s-process contribution was taken to be 'primary' with a dimensionless time delay $\omega\Delta = 0.06$, to be compared with 0.5 for the delayed component of iron. Halo evolution was modelled following Hartwick (1976), assuming $1 + \eta = 10$ in Eqs. (8.21) to (8.25) (so as to fit an abundance distribution function like that in Fig. 8.15) and the linear star formation law Eq. (8.15). With the delayed production approximation this leads to enrichment of the ISM according to

$$Z(t) = p\,e^{(1+\eta)\omega\Delta}\,\omega\,(t - \Delta); \quad t > \Delta \tag{8.53}$$

where p is the true yield. Halo evolution was assumed to end at $u \equiv \omega t = 0.5$, i.e. just before the onset of delayed iron, and to lead to deposition of gas with a mean $[O/H] = \log(p/10Z_\odot) = -1.1$ which formed the proto-disk. Disk evolution was then followed according to Clayton's formalism with Eq. (8.38) modified to allow for an initial abundance and delayed production as follows:

$$Z(t) = (1 + t/t_0)^{-k}\left[Z_0 + \frac{\omega t_0}{k+1}p\,e^{\omega\Delta}\left\{\left(1 + \frac{t - \Delta}{t_0}\right)^{k+1} - 1\right\}\right]; \quad t > \Delta. \tag{8.54}$$

k was chosen to be 4 and ωt_0 to be 2 (corresponding to a final to inital mass ratio of 10.5), so as to give the disk abundance distribution function shown by the broken-line curve in Fig. 8.19.

Since then, there have been several developments suggesting that the above assumptions need to be modified. The r-process itself appears to involve a small time delay (Mathews & Cowan 1990) and the short time delay for the s-process gives too high

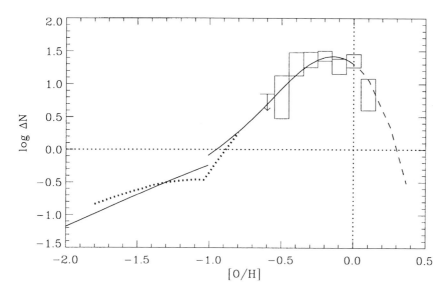

Fig. 8.22. Oxygen abundance distribution function for disk stars. Data deduced from observation are shown by the boxes (same as in Fig. 8.19, after Sommer-Larsen 1991a) and by the dotted line and curve at lower left (after Beers & Sommer-Larsen 1995). The solid curve shows the distribution given by the model of Pagel & Tautvaišienė (1995), *MNRAS*, **276**, 505.

abundances relative to iron at moderately low metallicities (Pagel 1991). The G-dwarf abundance distribution function, based on the old data of Pagel & Patchett (1975), was somewhat too narrow because of insufficient allowance for the larger scale height of low-metallicity stars resulting from poor statistics, and so more recently I have used the distribution shown by the boxes in Fig. 8.19. There are now better data on the relation between oxygen, α-elements and iron (Ryan, Norris & Bessell 1991; Bessell, Sutherland & Ruan 1991; Nissen & Edvardsson 1992; Edvardsson *et al.* 1993); cf. Figs. 8.5, 8.24. Finally, there no longer seem to be any strong grounds for believing that gas expelled from the halo provided the initial configuration of the disk. It seems rather that the thick disk began with zero metallicity and that the disk and halo have evolved independently.

Pagel & Tautvaišienė (1995) have developed a simple analytical disk model which can reproduce the abundance distribution functions found by Sommer-Larson (1991a) for the solar neighbourhood and by Beers & Sommer-Larsen (1995) for the metal-weak thick disk and also the new data on certain element-to-element ratios. The thick-disk data indicate that we should now begin the game with a Simple model that goes on for a time t_1 such that $u_1 \equiv \omega t_1 = 0.14$ (assuming a true yield $p = 0.7Z_\odot$ for oxygen), whereafter we switch on inflow according to Clayton's formalism with $k = 3$, $u_0 = 1.3$ corresponding to an ultimate mass multiplication factor of 7. This reproduces the data in Fig. 8.22 fairly well and leads to an age-metallicity relation similar to the one in Fig. 8.13; our age-metallicity relation is reproduced in slightly different form in Fig.

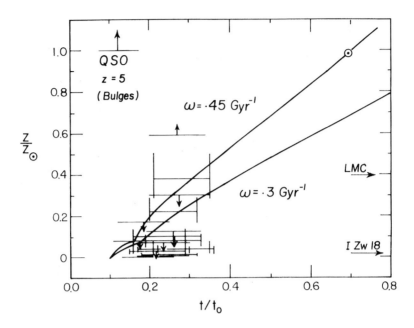

Fig. 8.23. Age-metallicity relation by Pagel & Tautvaišienė (1995), where $\omega = 0.3$ Gyr^{-1} applies to the solar neighbourhood and $\omega = 0.45$ Gyr^{-1} to the inner disk of the Galaxy, compared to zinc abundances measured by Pettini *et al.* (1994) in damped Lyman-α absorption-line systems at red-shifts 1.9 to 3.0. t/t_0 is the age of the universe in units of its present age and the horizontal error bars represent ages deduced from the Einstein–de Sitter model (to the left) and from an open Friedman model (to the right). After Pagel (1997).

8.23, where it is shown in relation to abundances measured at high red-shifts. In this comparison, one should note that the starting time of the model relative to the Big Bang is rather arbitrary.

Table 8.1 shows some of the parameters in the model that determine the element:element ratios. Each element is assumed to have an instantaneous component, due to production by massive stars, giving a yield p_1, and a delayed component (treated by the delayed-production approximation) due to the contribution of SN Ia (for Si, Ca, Ti and Fe) or of low-mass SN II (for the r-process element Eu), giving a yield p_2 with a time-delay Δ. These are all independent of the mean galactocentric distance R_m of the stars, but it is assumed that ω increases from 0.3 Gyr^{-1} in the solar neighbourhood to 0.45 Gyr^{-1} in the inner Galaxy, so that the dimensionless time-delay $\omega\Delta$ (as well as the metallicity at a given time) is larger in the latter case. Table 8.2 gives an indication of the time-dependence of various quantities of interest, according to the model. The quantities given are the dimensionless time parameter u, the gas mass g, the total mass M, the gas fraction $\mu \equiv g/M$, the abundance z_1 of a promptly produced element in units of its yield, the corresponding abundance distribution function and abundances of representative elements. Delayed Eu comes in already at $u = 0.008$ (corresponding

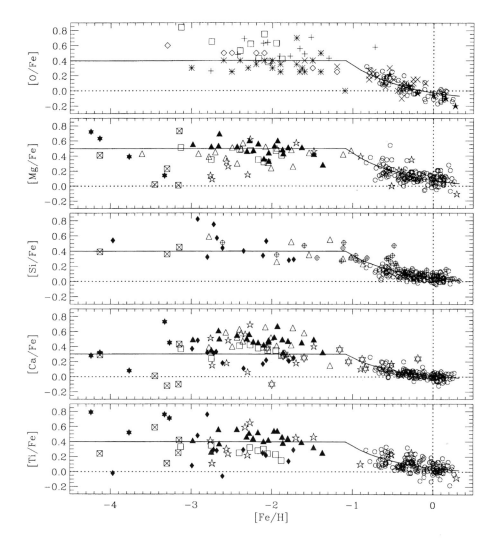

Fig. 8.24. Fits of the model to element-to-iron ratios measured by Edvardsson *et al.* (1993) for disk stars and by various authors for Population II stars which include a significant component from the metal-weak thick disk. The model curve is for $\omega\Delta = 0.4$. After Pagel & Tautvaišienė (1995), *MNRAS*, **276**, 505.

to a time delay of, very roughly, 25 Myr, which fits a $\sim 10M_\odot$ star) and at $u = 0.14$ the abundance distribution function changes owing to the onset of inflow. At $u = 0.4$ ($t \simeq 1.3$ Gyr), SN Ia start to contribute a major portion of Fe and minor portions of Si–Ti. $u = 1.60$ is the phase of maximum star-formation rate (2.3 times the present rate), $u = 3.20$ is the time of formation of the Solar System and $u = 4.50$ at the present time. $u = 4.10$ at the time of formation of stars now undergoing SN Ia outbursts. Fig. 8.24 shows the fits of the model to oxygen and α-element to iron ratios derived

Table 8.1. *Yields and time delays assumed in the
analytical model for the Galactic disk*

El.	$p_1/Z_\odot$	$p_2/Z_\odot$	$\omega\Delta$ $R_m \geq 7$ kpc	$\omega\Delta$ $R_m < 7$ kpc
O	0.70	0.00		
Mg	0.88	0.00		
Si	0.70	0.12	0.4	0.6
Ca	0.56	0.18	0.4	0.6
Ti	0.70	0.12	0.4	0.6
Fe	0.28	0.42	0.4	0.6
Eu	0.08	0.66	0.008	

Table 8.2. *Model of chemical evolution of the disk in the solar neighbourhood*

u	g	M	μ	z_1	$z_1\,ds/dz_1$	[O/H]	[Ca/H]	[Fe/H]	[Eu/H]
0.001	1.00	1.00	1.00	0.001	0.001	−3.15	−3.25	−3.55	−4.10
0.008	0.99	1.00	0.99	0.008	0.008	−2.25	−2.35	−2.65	−3.19
0.14	0.87	1.00	0.87	0.14	0.12	−1.01	−1.11	−1.41	−1.00
0.14					0.172				
0.40	1.10	1.49	0.74	0.29	0.66	−0.69	−0.79	−1.09	−0.67
1.60	1.65	3.78	0.44	0.70	4.1	−0.31	−0.31	−0.37	−0.28
3.20	1.24	5.79	0.22	1.12	5.5	−0.11	−0.08	−0.10	−0.08
4.10	0.87	6.38	0.14	1.35	4.7	−0.02	0.01	0.00	0.00
4.50	0.75	6.55	0.11	1.45	4.2	0.00	0.05	0.05	0.03

from observation and Fig. 8.25 shows the corresponding fits for the r-process elements Eu and Th, the latter involving a small correction for radioactive decay. The overall fits in the two figures are quite good, bearing in mind that a few halo stars have composition anomalies such as are displayed by the remarkable iron-poor r-process rich CS 22892–052 ([Fe/H] = −3.1; [m/Fe] ≃ 1.5 for all r-process elements; Sneden *et al.* 1996). The behaviour of elements like Ba, which have an s-process contribution at least in the case of the more metal-rich stars, seems to require two components: one with a short time delay as assumed by Pagel (1989a) and another one with a time-delay slightly exceeding that for SN Ia (Edvardsson *et al.* 1993; Pagel & Tautvaišienė 1997).

Notes to Chapter 8

Properties of the **1-zone model** have been systematically studied by Talbot & Arnett (1971), who contrasted the instantaneous recycling approximation with the 'initial burst' approximation which is an opposite extreme. In the latter case, most of the gas is quickly used up in forming the first generation of stars and later the little gas

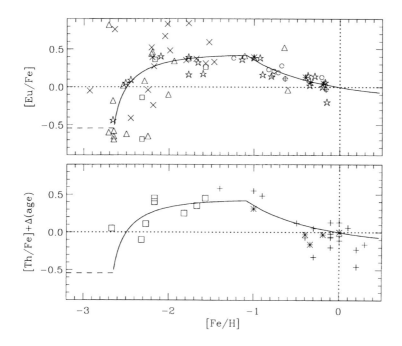

Fig. 8.25. Europium and thorium to iron ratios plotted against metallicity [Fe/H]. Curves represent the model predictions as in Table 8.2 and the symbols represent observational results by different authors. After Pagel & Tautvaišienė (1995), *MNRAS*, **276**, 505.

that remains is heavily diluted by their ejecta so that metallicity actually decreases. This model does not seem to apply to the Galactic disk, where there is no evidence for an excess of old stars (Soderblom *et al.* 1991) and it appears from Fig. 8.13 that the presence of some old, metal-rich stars is more naturally explained by infiltration from the inner part of the Galaxy where enrichment was quicker (and possibly more efficient). However, repeated star-formation bursts may influence the evolution of dwarf galaxies (see Chapter 11).

The predictions of **ratios of primary elements** formed by massive stars are not specific to the Simple model; that model just provides a convenient framework in which to discuss them. The case for other element ratios may be more complicated; there is an illuminating discussion by Wheeler, Sneden & Truran (1989) who comprehensively review the older literature. Regarding the data in Figs. 8.1 and 8.2, all stellar nucleosynthesis models lead one to expect a nearly constant O/Ne ratio, but the O/S ratio could conceivably vary because sulphur is more affected by explosive (as opposed to hydrostatic) nucleosynthesis with a yield per star that is estimated not to increase greatly with the star's initial mass (Thielemann, Nomoto & Hashimoto 1993; Tsujimoto *et al.* 1995). Fig. 8.2 provides only marginal evidence for any variation in practice; the two high points, in particular, are probably spurious.

The α/**Fe effect** was found in extreme subdwarfs by Aller & Greenstein (1960) and in disk stars by Wallerstein (1962). The O/Fe effect was noted in Arcturus by Gasson & Pagel (1966) and in a sample of several giants by Conti *et al.* (1967). The [O/Fe] data have been subject to many discrepancies and controversies in recent years; in particular the change in slope near [Fe/H] = −1 (cf. Fig. 8.24) has been interpreted as a transition from the halo to the disk, but it is not clear that factors other than the simple time delay are involved. A recent investigation indicates that the high O/Fe ratio characteristic of stars with [Fe/H] < −1 actually continues towards higher metallicities in the thick disk, leading to a bifurcation where the curve defined mainly by thin-disk stars in Fig. 8.24 starts to go down (Gratton *et al.* 1997). The most reliable oxygen abundance data are probably those derived from the forbidden line λ 6300, which unfortunately becomes too weak to measure in the most metal-deficient stars, particularly in dwarfs, and so one then has to rely on the stronger permitted lines, which are highly sensitive to temperature and non-LTE effects, or on OH bands which also involve some uncertainties. Part of the scatter in Fig. 8.24 is due to these problems, and it is still not clear at the time of writing whether O/Fe is flat for [Fe/H] < −1 as assumed in the analytical model, or whether it actually continues to increase towards lower metallicities.

The suggestion that **Type I** (or as one would rather say now, Ia) **supernovae** with a significant evolution time contribute substantially to the iron found in the Solar System and in disk stars goes back to Chevalier (1976) and was treated in some detail by Tinsley (1979). A modern discussion is given by Matteucci & Tornambé (1987). An alternative explanation, that the O/Fe and α/Fe yield ratios are a decreasing function of metallicity, has been suggested by Edmunds *et al.* (1991); but the order-of-magnitude suitability of SN Ia ages, frequency and composition, and indications of a high Mg/Fe ratio in Galactic bulge stars (McWilliam & Rich 1994), make the time-delay explanation look more promising.

s-process elements have often been naively regarded as 'secondary', but it is obvious from the behaviour of Ba/Fe in Fig. 8.7 that this is not so. A plot of Ba/O against O/H might suggest some dependence on chemical composition in that sense, but a more important factor is probably a time delay similar to, but somewhat exceeding, the one that applies to iron (Edvardsson *et al.* 1993; Pagel & Tautvaišienė 1997). This corresponds quite nicely with the model by Käppeler *et al.* (1990) illustrated in Fig. 6.5, but an additional s-process source seems to be needed (over and above the r-process) to account for heavy-metal abundances in the more metal-rich halo stars. The whole 'primary/secondary' concept is much too simplistic to be applied in this case; cf. Clayton (1988) and Mathews, Bazan & Cowan (1992).

A Galactic **O-abundance gradient** of 0.07 dex kpc^{-1} was derived from observations of H II regions by Shaver *et al.* (1983), and this is in good agreement with deductions from observations of H II regions in external spirals (e.g. Pagel & Edmunds 1981; Edmunds & Pagel 1984; Belley & Roy 1991; Chapter 3) and of Galactic planetary nebulae (Maciel 1992), but observations of old and young stars lead to more ambiguous results, e.g. very little gradient is found among B-type stars (Rolleston *et al.* 1993) which ought to match the surrounding ISM and indeed do so in Orion (Gies & Lambert 1992). A small or even zero gradient has been found for samples of distant old red giants (Neese & Yoss 1988; Lewis & Freeman 1989), but these investigations are subject to some selection bias owing to the range in colours. Fig. 3.34 shows evidence for a gradient among F and early G stars of the same order as the one found for H II regions.

Many authors have investigated the possibility of gas flows along the Galactic plane and their effects on abundance gradients, starting with some simple considerations by Tinsley (1980) who found that a slow inward flow (compared to the star formation time scale) could enhance a (negative) gradient, whereas a rapid one could smooth it out. Mayor & Vigroux (1981) showed that inflow of gas from a low angular-momentum halo would lead to inward flows of gas that could lead to a steepening of the abundance gradient towards the Galactic centre; the same effect has been discussed by Pitts & Tayler (1989). Lacey & Fall (1985) examined the consequences of various analytical velocity laws, as did Edmunds (1990), Götz & Köppen (1992) and Edmunds & Greenhow (1995), who have derived various theorems and clarified the conditions under which inward flows can either increase or decrease the gradient. Sommer-Larsen & Yoshii (1990), Clarke (1989, 1991) and others have made numerical models of GCE with gas inflow driven by viscous effects.

Evidence for a **spread in ages** of 2 or more Gyr in the halo comes from colour-magnitude diagrams of globular clusters of equal metallicity (Chaboyer, Demarque & Sarajedini 1996) and from photometric determinations of surface temperatures and gravities of field stars (Nissen & Schuster 1991). (An age–metallicity relation, if any, for halo stars cannot be established beyond doubt because of uncertainties about abundance-dependent systematic errors in the age determination.) The clump of stars in the upper right corner of Fig. 8.12 represents mostly the 'thick disk' (Gilmore & Reid 1983; Gilmore & Wyse 1985) or 'Intermediate Population II' (Strömgren 1987) whose properties stand out quite clearly in Fig. 8.14. The scale height of the thick disk population is of the order of 1 kpc, compared to 300 pc for the old thin disk, so that thick-disk stars are relatively under-represented (compared to their column density) in a sample taken from the local volume. The selection effects on their relative frequency in Fig. 8.13 are difficult to judge, because larger volumes have been sampled for the rarer low-metallicity stars.

The main causes of **scatter in the age–metallicity relation** have not been completely clarified. Wielen, Fuchs & Dettbarn (1996) suggest that it can all be accounted for by a combination of a constant abundance gradient of -0.09 dex kpc^{-1} combined with stellar orbital diffusion in the course of time, but another factor that is probably important is a spasmodic variation in the inflow rate, which can cause a temporary decrease in metallicity (Edvardsson *et al.* 1993). A third factor could be a measure of self-enrichment in star-forming regions (Pilyugin & Edmunds 1996), for which there is some evidence in the Orion Association for the elements oxygen and silicon (Cunha & Lambert 1992, 1994), but no variation has been found for iron (a result consistent with the SN Ia hypothesis) and the AMR scatter for iron is at least as great as for oxygen and α-elements.

Fig. 8.15 differs from the more composite-looking **halo abundance distribution func-**

tion given by Laird *et al.* (1988) for essentially the same data set because their *z* mainly represents iron. The iron–oxygen conversion tends to 'iron out' the bulge that appears on the high-metallicity side of their diagram, and this will also affect Fig. 8.16 if [Fe/H] is replaced by [O/H]. In view of the possible compositeness of the halo, the fit to Hartwick's modified Simple model should not necessarily be taken literally. On the other hand, no better model of comparable simplicity has emerged, and one may perhaps think of the distribution as a mixture of modified Simple (or other) models (cf. Norris 1994). Malinie, Hartmann & Mathews (1991) have put forward a chemical evolution model with a density gradient in a collapsing halo which can reproduce the metallicity distribution function with a yield near solar abundance and no continuous wind, but this model predicts strong abundance gradients and the question of whether these could have been smoothed out by subsequent violent relaxation or other processes is a difficult one.

Many discussions of the **'G-dwarf problem'** have been based on an influential paper by Pagel & Patchett (1975) which with hindsight could have been much better. In particular, I now regret having used mainly cumulative distribution functions in that paper, as opposed to the differential distribution shown in Fig. 8.19. The cumulative distribution (in which one plots $Z(s)$ against s) does have the advantage of being independent of binning, but the errors are harder to quantify and the influences of yield and cutoff are less apparent. Also, the data used were of rather poor quality and the corrections for differing scale heights were insufficient owing to poor statistics. The whole analysis has been done again *ab initio* by Gilmore, Wyse & Jones (1995) and Rocha-Pinto & Maciel (1996) resulting in a somewhat narrower distribution than found by Sommer-Larsen (1991a); the new function resembles more the one shown by the broken-line curve in Fig. 8.19 taken from Pagel (1989a). Other ways of presenting the (differential) distribution function have been explored by Pagel (1989b).

Ideas on **'prompt initial enrichment'** have undergone some evolution in step with those on the nature of the halo. Dynamical evidence for a massive halo motivated Truran & Cameron (1971) to postulate a proto-galactic population of massive stars which enriched the proto-disk and collapsed into black holes or 'collapsars', while Ostriker & Thuan (1975) envisaged a large population of normal stars in the halo which enriched the proto-disk with supernova ejecta. Later the idea was accepted that the massive dynamical halo (as opposed to the sparse stellar halo) consists of dark matter, e.g. brown dwarfs or 'Machos', for which there is now evidence from micro-lensing effects (Alcock *et al.* 1993; Aubourg *et al.* 1993), and maybe also non-baryonic matter, which has not been involved at all in the chemical evolution of the Galaxy, but 'prompt initial enrichment' has returned as a consequence of a possible high yield in the bulge. I still find the idea somewhat *ad hoc*.

Problems

1. Assuming instantaneous recycling, write an equation to replace Eq. (7.33) for the case of deuterium, which has no stellar production and is destroyed in recycled gas.

Show that, in the Simple Model (no inflow and no galactic wind), the evolution of deuterium abundance in the interstellar medium is given by

$$X_D/X_{D0} = \mu^{(\alpha^{-1}-1)} = e^{-z(\alpha^{-1}-1)}, \tag{8.55}$$

where μ is the gas fraction, $\alpha \equiv 1 - R$ is the lock-up fraction and z the abundance of a primary element in units of its yield. Supposing $\mu = 0.2$, contrast the effects of assuming $\alpha = 0.8$ or 0.5.

2. Consider the model of Hartwick (1976) in which gas is lost to the system in a homogeneous wind at a rate η times the net star formation rate with $\eta = $ const. Assuming instantaneous recycling, derive the following relations:

$$m = 1 - \eta s; \quad g = 1 - (1 + \eta)s; \quad s(\infty) = 1/(1 + \eta), \tag{8.56}$$

where m is the mass of the system in units of its initial mass and s and g the masses of stars and gas in the same units;

$$z(t) = u \equiv \int_0^t \omega(t')dt' = \frac{1}{1+\eta} \ln \frac{1}{1-(1+\eta)s} \tag{8.57}$$

where $\omega(t)$ is the transition probability for gas to change into stars;

$$\frac{ds}{d\ln z} = z\, e^{-(1+\eta)z}, \tag{8.58}$$

so that the abundance distribution function has a peak at $(1 + \eta)^{-1}$ times the yield, and that the average abundance in the stars approaches $(1 + \eta)^{-1}$ times the yield asymptotically from below.

Now consider the destruction of deuterium in this model and show that the second expression in Eq. (8.55) still applies, although the first one does not.

3. Neglecting the variation in the denominator of Eq. (8.39), show that the abundance distribution function in Clayton's inflow model peaks in the neighbourhood of $z = 1$.

4. Using the relation

$$\frac{N_{Ia}}{N_{II}} = \frac{p_2}{p_1} \frac{m_{II}}{m_{Ia}} \frac{\dot{s}(u_{now} - \omega\Delta)}{\dot{s}(u_{now})}, \tag{8.59}$$

with the yields p_1 and p_2 for iron given in Table 8.1 and assuming that a SN II ejects $m_{II} = 0.09 M_\odot$ and a SN Ia ejects $m_{Ia} = 0.74 M_\odot$ of iron, find the relative rates of SN Ia and SN II.

5. Show that, in Twarog's model with inflow of unprocessed material at $f \times$ the net star formation rate ($f = $ const. < 1), in instantaneous recycling, the abundance grows as

$$Z = \frac{p}{f}(1 - g^{f/(1-f)}), \tag{8.60}$$

where g is the mass of gas in units of its initial mass.

9 Origin and evolution of light elements

9.1 Introduction

The light elements (D to B, apart from ^{4}He) have such fragile nuclei (see Table 9.1) that they tend to be destroyed, rather than created, in thermonuclear burning, although certain special processes can lead to stellar production of ^{3}He, ^{7}Li and ^{11}B.

B^{2}FH accordingly postulated for their creation an 'x'-process involving spallation by fast particles at high energy, but low temperature and density. They considered stellar flares and supernova shells as possible sites, while also envisaging the possibility of ^{7}Li creation in H-free helium zones in stars. Since then, it has been accepted that all D, some ^{3}He and some ^{7}Li come from the Big Bang (see Chapter 4), where rapid expansion and cooling allow traces of these elements to be preserved; and that a significant clue to the origin of Li, Be and B comes from their relative overabundance (by factors of 10^4 to 10^5) in Galactic cosmic rays (Fig. 9.1).

9.2 Sketch of cosmic-ray physics

Cosmic rays reaching the ground are secondary particles resulting from the impact of primary cosmic rays coming mainly from the Galaxy. The latter are mostly protons and α-particles with a sprinkling of heavier nuclei, coming in with a broad distribution of energies. The most energetic, with energies up to 10^{20} eV or so, are quite rare but are detected occasionally in the form of extensive air showers.

The origin and cause of acceleration of cosmic rays is somewhat uncertain, but the power-law energy spectrum suggests some kind of electromagnetic process analogous in some ways to laboratory accelerators. The most promising mechanisms, due originally to Fermi (1949, 1954), are based on repeated reflections from moving magnetized gas clouds. These could well take place in interstellar shock fronts due to supernova ejecta if the particles are injected with sufficient energy to make the process efficient in the face of energy losses by ionization; such particles, with energies up to a few GeV, are in fact ejected by the Sun and presumably other stars during phases of high solar activity associated with flares and the ejecta from supernovae have a kinetic energy of the order of an MeV per nucleon. The initial spectrum of cosmic rays is expected to

Table 9.1. *Destruction of light nuclei in stellar interiors*

^{2}D	destroyed by	$(p,\gamma)\,^3$He	for $T >$	0.5×10^6 K
^{6}Li	" "	$(p,\alpha)\,^3$He	for $T >$	2×10^6 K
^{7}Li	" "	$(p,\alpha)\,^4$He	for $T >$	2.5×10^6 K
^{9}Be	" "	$(p,\alpha)\,^6$Li;\ $(p,$D$)\,^8$Be$\rightarrow 2\,^4$He	for $T >$	3.5×10^6 K
^{10}B	" "	$(p,\alpha)\,^7$Be$(EC)\,^7$Li	for $T >$	5.3×10^6 K
^{11}B	" "	$(p,\alpha)\,^8$Be$\rightarrow 2\,^4$He	for $T >$	5×10^6 K
^{3}He	" "	$(^3$He,$\alpha)\,^4$He $+ 2\,^1$H	for $T >$	$\sim 10^7$ K

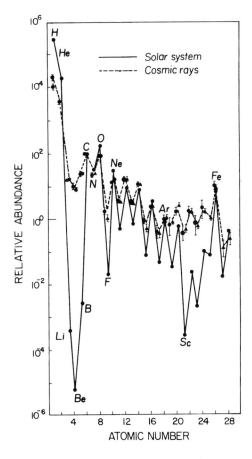

Fig. 9.1. Abundances in primary cosmic rays reaching the top of the Earth's atmosphere, compared to solar-system abundances. (Both normalized to C = 100). After C.E. Rolfs & W.S. Rodney, *Cauldrons in the Cosmos*, University of Chicago Press, 1988, Fig. 10.4, p. 509. ©1988 by the University of Chicago. Courtesy Claus Rolfs.

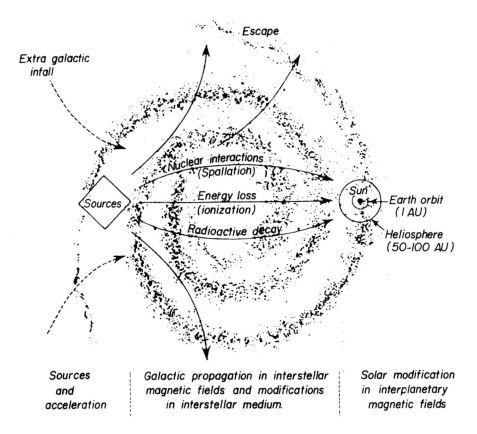

Fig. 9.2. Schematic view of the life history of a cosmic ray from acceleration in the source through propagation in the Galaxy to observation above the Earth's atmosphere. Adapted from Rolfs & Rodney (1988).

be a power law in the magnetic rigidity or momentum p of particles, e.g.

$$N(E)\, dE \propto p^{-2} dp \equiv \frac{E + E_0}{[E(E + 2E_0)]^{3/2}}\, dE, \tag{9.1}$$

where E is the kinetic energy per nucleon and E_0 the rest-mass energy 0.93 GeV. The initial spectrum thus has a negative energy dependence, steepening from $E^{-3/2}$ at low energies ($\ll 1$ GeV) to E^{-2} at high energies ($\gg 1$ GeV). The cosmic rays then diffuse through the Galaxy scattering off irregularities in the magnetic field. The Larmor radius is

$$R_{\mathrm{Larmor}} = \frac{pc}{eB} = 10^{-7} E_{\mathrm{GeV}}\ \mathrm{pc} \tag{9.2}$$

(for a Galactic magnetic field $B = 10^{-5}$ gauss) so that the cosmic rays are closely confined by the field up to energies of 10^{18} eV or so. During their propagation through

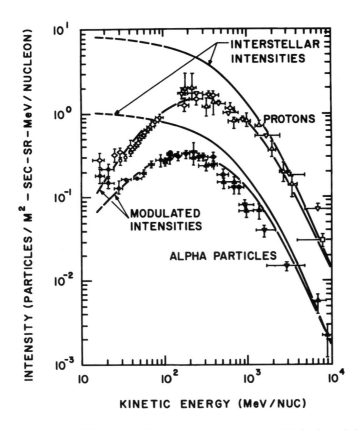

Fig. 9.3. Proton and α-particle spectra of primary cosmic rays and their demodulated versions. After M.A. Goldstein, L.A. Fisk & R. Ramaty, *Phys. Rev. Lett.*, **25**, 832, 1970. Courtesy Reuven Ramaty.

the Galaxy, the cosmic rays suffer a number of effects (see Fig. 9.2):

(1) Energy loss by ionization, especially at low energies; this tends to flatten the spectrum.

(2) Fragmentation by high-energy nuclear (spallation) reactions with nuclei in the interstellar medium. This converts abundant nuclei into their lighter neighbours, accounting for the abundances in Fig. 9.1. The source abundances, derived by correcting for this propagation effect, are more like those of the Solar System[1] with some modifications such as a relative enhancement of elements with a low first ionization potential (FIP effect) and a few isotopic anomalies.

(3) Escape from the Galaxy. This is often represented by the 'leaky box' model, of which a simple form represents the escape probability as an exponential

---

[1] Hubert Reeves once summed up the similarity of cosmic-ray source and solar-system abundances in the form of a graffito seen at times in Paris: 'CRS = SS'!

function of the mass column density (or 'grammage') passed through before escaping:

$$P(M)\,dM = e^{-M/\Lambda}\,dM/\Lambda, \tag{9.3}$$

where from the amount of fragmentation observed one finds

$$\Lambda \simeq 8 \text{ gm cm}^{-2}. \tag{9.4}$$

From the radioactive decay of ^{10}Be, one also has an idea of the time scale for escape, which seems to be of the order of 10^7 yrs and perhaps an order of magnitude longer than the time to pass through Λ at an average density of one atom (2×10^{-24} gm) cm^{-3} in the Galactic plane.

(4) Finally, some reach the Solar System, where they suffer solar modulation: they are decelerated by the solar wind, especially at lower energies (< 0.5 GeV per nucleon) and during high levels of solar activity. The observed spectrum from satellites is demodulated on the basis of theoretical models down to about 100 MeV per nucleon and more or less guessed below that (Fig. 9.3). The interstellar proton flux is deduced to be given by

$$\phi_p(E)\,dE \simeq 12.5\,(E + E_0)^{-2.6}dE \text{ cm}^{-2}\text{s}^{-1} \tag{9.5}$$

(E in GeV), giving a flat spectrum for $E \ll E_0 \simeq 1$ GeV. Heavy nuclei have similar spectra in energy per nucleon.

9.3   Light element production

The light elements present in cosmic rays are partly thermalized, i.e. brought down to low velocities by ionization losses, and thus make a minor contribution to their abundance in the ISM (perhaps about 20 per cent). The main source is usually thought to come from reactions between cosmic-ray protons and α-particles and stationary nuclei of He, C, N and O in the ISM.

The reactions can be divided into $\alpha - \alpha$ fusion reactions

$$\alpha + \alpha \quad \rightarrow \quad {}^{6}\text{Li} + p + n - 36\,\text{MeV} \tag{9.6}$$
$${}^{7}\text{Li} + p - 44\,\text{MeV} \tag{9.7}$$
$${}^{7}\text{Be}\,(EC){}^{7}\text{Li} + n - 44\,\text{MeV} \tag{9.8}$$

and spallation reactions

$$p, \alpha + \text{C}, \text{N}, \text{O} \rightarrow^{6,7} \text{Li}, {}^{9}\text{Be}, {}^{10,11}\text{B} - \text{about } 30\,\text{MeV}. \tag{9.9}$$

About 80 per cent of the cosmic ray flux is at energies above 150 MeV/nucleon where the cross-sections are more or less constant (Fig. 9.4).

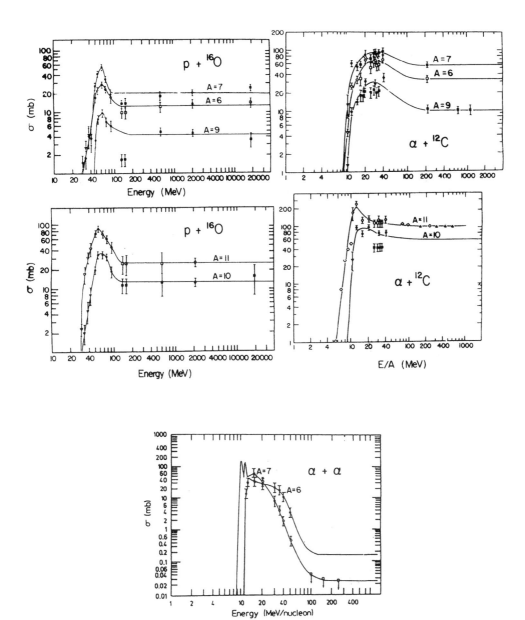

Fig. 9.4. Reaction cross-sections, as a function of energy per nucleon, for the production of light elements for some typical cases. Adapted from S.M. Read & V.E. Viola, *Atomic Data Nuclear Data Tables*, **31**, 359, 1984. Courtesy Vic Viola.

A simple calculation neglecting GCE refinements is based on the assumption of a constant production rate:

$$\frac{dN_L}{dt} = N_{CNO} \int \bar{\sigma}_{CNO}(E)\,\phi(E)\,dE \simeq N_{CNO} <\bar{\sigma}> \phi \tag{9.10}$$

for each incident particle type, i.e. p, α. Assuming the rhs to be constant, the abundance by number of atoms relative to hydrogen is

$$\frac{L}{H} = \frac{CNO}{H} <\bar{\sigma}> \phi_p\, T = 4 \times 10^{-12} \left(\frac{T}{10\,\mathrm{Gyr}}\right) <\bar{\sigma}>_{mb}, \tag{9.11}$$

where CNO/H is assumed to be solar, T is the age of the Galaxy and only protons are considered. This treatment thus neglects the effects of lock-up and destruction of light elements in stars and any variations with time in the cosmic ray flux and in the abundances of CNO in the ISM. Results of this simple calculation are shown in Table 9.2, where it can be seen that, while meteoritic abundances of the three nuclear species ^{6}Li, ^{9}Be, ^{10}B are reproduced within a factor of 2 or so, ^{7}Li is underproduced by an order of magnitude (which is probably also the case for cosmological production; see Fig. 4.10). This suggests a major component from thermonuclear synthesis in stars, e.g. AGB stars — see Section 5.10 — and/or supernovae. Also, ^{11}B is underproduced by a smaller factor ~ 3 which comes either also from stellar production, e.g. in supernovae, or from a low-energy component of cosmic rays over and above the flat spectrum that was assumed (see Fig. 9.4).

9.4 Galactic chemical evolution of light elements

9.4.1 Introduction

Special features of the light elements are

(1) Initial abundance Z_0 and/or abundance in inflowing material Z_F from the Big Bang.
(2) Creation in both stars and cosmic rays.
(3) Destruction by astration.

Using the instantaneous recycling approximation (Section 7.4.2), Eq. (7.33) now has to be replaced by

$$\frac{d}{dS}(gZ) = q + \beta R Z - Z + \frac{N_{CNO}\sigma\phi A m_H + Z_F F - Z_E E}{\psi}, \tag{9.12}$$

where $0 \leq \beta \leq 1$ allows for astration (for complete destruction, $\beta = 0$) and we have replaced $<\bar{\sigma}>$ by σ for brevity. A is the atomic mass number of the relevant light element, and $\alpha - \alpha$ fusion reactions have been neglected. We need also to bring in time, which we do using the variable u (Eq. 8.17), noting that

$$\frac{\psi}{\omega} \equiv \frac{dS}{du} \equiv \frac{g}{\alpha} \equiv \frac{g}{1 - R}, \tag{9.13}$$

Table 9.2. *Results of simple calculations of light element abundances*

	$\overline{\sigma}_{CNO}$ (mb)	$10^{11} \frac{L}{H}$	$10^{11} \frac{L}{H}$	$10^{11} \frac{L}{H}$	
	$E > 150$ MeV	Eq. (9.11)	WMV 85[a]	meteoritic	
^{6}Li	13	5	7	17	$\checkmark$
^{7}Li	20	8	10	210	$\times\times$
^{9}Be	4	1.6	1.5	2.6	$\checkmark$
^{10}B	16	6	7	15	$\checkmark$
^{11}B	34	14	18	62	$\times$
^{6}Li/^{9}Be/^{10}B		3/1/4	5/1/5	7/1/6	$\checkmark$

[a] Numerical calculation by Walker, Mathews & Viola (1985).

whence

$$\frac{d}{du}(gZ) = g\left(p + \frac{A}{A_{CNO}}\sigma Z_{CNO}\frac{\phi}{\omega} - \frac{Z}{\alpha'}\right) + \frac{Z_F F - Z_E E}{\omega}, \tag{9.14}$$

where p is the true yield from stellar production and

$$\alpha' \equiv \frac{\alpha}{1 - \beta(1 - \alpha)} \tag{9.15}$$

$$\rightarrow \alpha \quad \text{for complete destruction} \tag{9.16}$$

$$\rightarrow 1 \quad \text{for no destruction.} \tag{9.17}$$

We now consider two special cases:

(1) D, ^{7}Li: $Z_0, Z_F > 0$, cosmic ray production neglected.
(2) Cosmic ray nuclei, $Z_0 = Z_F = 0$, stellar production neglected.

9.4.2   D, ^{7}Li in the Simple or homogeneous outflow model

The assumptions of these models are:

$$F = 0, \tag{9.18}$$

$$E = \eta\frac{ds}{dt} = \eta\omega g, \tag{9.19}$$

$$Z_E = Z, \tag{9.20}$$

$$\alpha' = \alpha \text{ (complete destruction)}, \tag{9.21}$$

$$\frac{dg}{du} = -(1+\eta)g. \tag{9.22}$$

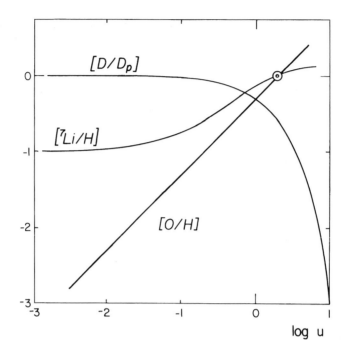

Fig. 9.5. Schematic of decline of D and growth of ^{7}Li and oxygen abundances in the Simple or homogeneous outflow model, assuming instantaneous recycling with $\alpha = 0.6$ and yields $p(O) = 0.5Z_\odot(O); \; p(^7\text{Li}) = 0.88Z_\odot(^7\text{Li}); \; Z_0(^7\text{Li}) = 0.1Z_\odot(^7\text{Li})$.

From Eq. (9.14),

$$g \frac{dZ}{du} - Z(1+\eta)g = g\left(p - \frac{Z}{\alpha}\right) - Z\eta g. \tag{9.23}$$

g, η conveniently cancel out and the differential equation simplifies to

$$\frac{dZ}{du} + Z\left(\frac{1}{\alpha} - 1\right) = p, \tag{9.24}$$

with the solution

$$Z = Z_0 e^{-v} + \frac{p}{\alpha^{-1} - 1}(1 - e^{-v}), \tag{9.25}$$

where

$$v \equiv \left(\frac{1}{\alpha} - 1\right)u \simeq \frac{2}{3}u \text{ if } \alpha \simeq 0.6. \tag{9.26}$$

Results for the Simple model are shown in Fig. 9.5. Oxygen abundance simply increases in proportion to u, whereas that of ^{7}Li starts from a finite base and tends to a limiting value $p_{\text{Li}}/(\alpha^{-1} - 1)$ owing to its eventual destruction by astration. Deuterium ($p = 0$ in Eq. 9.25) suffers little destruction as long as the oxygen abundance (or metallicity) is less than about 0.1 of the true yield (assumed to be comparable to solar abundance), e.g. in high red-shift absorption-line clouds, but declines as e^{-v} or $\mu^{(1/\alpha-1)}$ in the Simple model (Ostriker & Tinsley 1975) reaching a cut-down factor at solar metallicity of nearly 4 in the case shown in Fig. 9.5. This factor is very sensitive to α, however, which is an uncertain quantity owing to uncertainty in the initial mass function (cf. Section 7.3.3), but it is unlikely in any case that $\alpha < 0.5$.

9.4.3   D, ^{7}Li in the simple inflow model

Using the simplest inflow model of Section 8.5.2, our assumptions are:

$$g_0 = E = 0; \tag{9.27}$$

$$F = A\omega e^{-u}; \tag{9.28}$$

$$g = Aue^{-u} \text{ (Eq. 8.47)}. \tag{9.29}$$

Then Eq. (9.14) becomes

$$\frac{d}{du}(Zg) = pAue^{-u} - \frac{Zg}{\alpha} + Z_\mathrm{F}Ae^{-u}, \tag{9.30}$$

with the solution

$$Z = Z_\mathrm{F}\frac{1-e^{-v}}{v} + \frac{p}{\alpha^{-1}-1}\left(1 - \frac{1-e^{-v}}{v}\right). \tag{9.31}$$

E.g. at the time of formation of the Solar System, $\mu \simeq 0.16$, $u = 3$, $v = 2$, assuming $Z_\mathrm{F}(\mathrm{D})$ is the primordial value,

$$\frac{Z(\mathrm{D})}{Z_\mathrm{F}(\mathrm{D})} \simeq 0.4, \tag{9.32}$$

i.e. the cut-down factor for deuterium is considerably reduced compared to the no-inflow models.

With the yield $p_\mathrm{Li} = 1.13Z_\odot(^7\mathrm{Li})$ appropriate to this model,

$$Z(^7\mathrm{Li}) = 0.1Z_\odot(^7\mathrm{Li})\frac{1-e^{-v}}{v} + 1.75Z_\odot(^7\mathrm{Li})\left(1 - \frac{1-e^{-v}}{v}\right), \tag{9.33}$$

leading to a somewhat higher asymptotic limit than in the no-inflow case because a higher yield has had to be assumed.

9.4.4   Pure cosmic ray spallation products: ^{6}Li, Be, B

In this case, Eq. (9.14) becomes

$$\frac{d}{du}(gZ) + \frac{gZ}{\alpha'} = \frac{A}{A_\mathrm{CNO}}gZ_\mathrm{CNO}\,\sigma\,\frac{\phi}{\omega} + \frac{Z_\mathrm{F}F - Z_\mathrm{E}E}{\omega}. \tag{9.34}$$

In making models, it is usually assumed that ϕ is proportional to the supernova rate and therefore to the star formation rate ψ (e.g. Reeves & Meyer 1978), i.e.

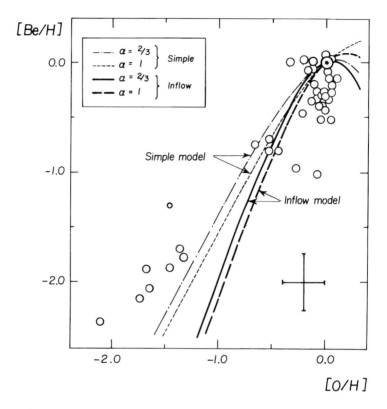

Fig. 9.6. Beryllium abundance as a function of oxygen abundance, according to models (curves) and observations (open circles) by Gilmore *et al.* (1992). After B.E.J. Pagel in D. Lynden-Bell (ed.), *Cosmical Magnetism*, p. 113, Fig. 1, ©1994 Kluwer Academic Publishers. With kind permission from Kluwer Academic Publishers.

$$\frac{\phi/\omega}{(\phi/\omega)_1} = g/g_1, \tag{9.35}$$

where the suffix $_1$ refers to the present epoch. Thus in the Simple model we have

$$E = F = 0, \quad g = e^{-u}, \tag{9.36}$$

$$Z_{\text{CNO}} = p_{\text{CNO}}u, \tag{9.37}$$

$$\frac{dZ}{du} + (\alpha'^{-1} - 1)Z = \frac{A}{A_{\text{CNO}}} \sigma\, p_{\text{CNO}} \frac{\phi_1}{\omega_1} u\, e^{-(u-u_1)}, \tag{9.38}$$

$$= k\left(\frac{\sigma\phi}{\omega}\right)_1 u\, e^{u_1-u}, \tag{9.39}$$

$$Z = k\left(\frac{\sigma\phi}{\omega}\right)_1 e^{u_1} \frac{e^{-u}[e^{v'} - (1+v')]}{(2-1/\alpha')^2}, \tag{9.40}$$

where

$$k \equiv \frac{A}{A_{CNO}} p_{CNO} \qquad (9.41)$$

and

$$v' \equiv (2 - 1/\alpha')u \simeq u/2 \quad \text{if } \alpha' \simeq 2/3. \qquad (9.42)$$

This leads to the limiting behaviours

$$Z \propto \frac{1}{2}u^2 \qquad \text{as } u \to 0 \text{ ('secondary dependence');} \qquad (9.43)$$

$$\propto \frac{e^{-(u-v')}}{(2-1/\alpha')^2} \quad \text{as } u \to \infty. \qquad (9.44)$$

The predictions of this model (normalized to meteoritic abundance for solar metallicity) are illustrated in Fig. 9.6 and compared with observational data for beryllium in stars, based on ground-based measurements of the near-UV Be II doublet λ 3130. Assuming that surface Be can suffer some destruction in metal-rich disk stars like the Sun (where the photospheric abundance is about half of that in meteorites), there is fair agreement down to about 0.1 of solar abundance, but the 'secondary' trend predicted at still lower metallicities is too steep.

Considering now the simple inflow model, we again have Eqs. (9.27) to (9.29) together with

$$Z_{CNO} = \frac{1}{2} p_{CNO} u \quad \text{(Eq. 8.50)} \qquad (9.45)$$

and Eq. (9.14) becomes (assuming $Z_F = 0$)

$$\frac{d}{du}(gZ) = \frac{A}{A_{CNO}} \frac{\sigma\phi}{\omega} gZ_{CNO} - g\frac{Z}{\alpha'} \qquad (9.46)$$

$$= \frac{k}{2} \left(\frac{\sigma\phi}{\omega}\right)_1 \frac{e^{u_1}}{u_1} u^3 e^{-2u} - \frac{gZ}{\alpha'} \qquad (9.47)$$

with the solution

$$Z = \frac{k}{2} \left(\frac{\sigma\phi}{\omega}\right)_1 \frac{e^{u_1}}{u_1} \frac{e^{-u}[6(e^{v'} - 1) - 6v' - 3v'^2 - v'^3]}{u(2 - 1/\alpha')^4}, \qquad (9.48)$$

where v' is defined by Eq. (9.42). The limiting behaviours in this case are

$$Z \propto \frac{u^3}{24} \qquad \text{for } u \to 0; \qquad (9.49)$$

$$\propto \frac{6}{(2-1/\alpha')^4} \frac{e^{-(u-v')}}{u} \quad \text{for } u \to \infty. \qquad (9.50)$$

This solution is also shown in Fig. 9.6, where it can be seen that its fit at low metallicities is still worse than for the (otherwise less satisfactory) Simple model. This is basically because of the assumption that the cosmic ray flux is proportional to the total star formation rate. The latter is low at early times in this model because the mass

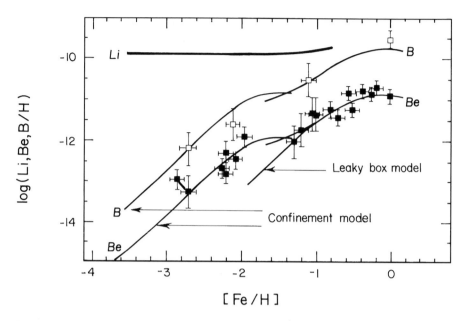

Fig. 9.7. Evolution of Be and B abundances according to the model based on confinement of cosmic rays in the early Galaxy (solid curves). Some observational data points are shown with error bars. Adapted from Prantzos, Cassé & Vangioni-Flam (1993).

of gas at such times is small and the assumption may well not apply, especially in this case. The discovery of beryllium (and boron from the UV B I line λ 2497 measured using the Hubble Space Telescope) in these metal-deficient stars was nevertheless a surprising one, which was made in part because of predictions (now no longer taken very seriously) that there might be cosmological production of these elements in an inhomogeneous Big Bang (cf. Chapter 4). As it is, the nearly constant ratio of Be and B to primary elements like iron and oxygen raises a problem to which no definitive solution has yet been found, although there have been various ideas.

9.4.5   Halo models with homogeneous outflow

Ryan *et al.* (1992) noted that in models for the Galactic halo assuming homogeneous outflow (Hartwick 1976), the limiting behaviour of Eq. (9.44) sets in at a low metallicity, thus counteracting the steep u^2 dependence of the Simple model. In this case, Eq. (9.36) is replaced by

$$E = \eta\,\omega g; \quad g = e^{-(1+\eta)u}, \tag{9.51}$$

where $\eta \simeq 9$ is the ratio of mass lost in a homogeneous wind to mass locked up in stars (cf. Section 8.3.7). Eq. (9.39) then becomes (bearing in mind that $Z_{CNO} = p_{CNO}\,u$

as in the Simple model and using the initial value $(\sigma\phi/\omega)_0$ in place of the present-day value of that parameter used before)

$$\frac{dZ}{du} + Z(1/\alpha' - 1) = k \left(\frac{\sigma\phi}{\omega}\right)_0 u\, e^{-(1+\eta)u} \tag{9.52}$$

with the solution

$$Z = k \left(\frac{\sigma\phi}{\omega}\right)_0 \frac{e^{-(1/\alpha'-1)u}}{(\eta + 2 - 1/\alpha')^2} [1 - e^{-w}(1 + w)], \tag{9.53}$$

where

$$w \equiv (\eta + 2 - 1/\alpha')u \simeq (1 + \eta)u. \tag{9.54}$$

Thus the limiting behaviour is

$$Z \propto u^2 \quad \text{for } u \to 0; \tag{9.55}$$
$$\propto (\eta + 2 - 1/\alpha')^{-2} \simeq (1 + \eta)^{-2} = \text{const. for } u \simeq 0.1, \tag{9.56}$$

whereafter ordinary disk evolution takes over. The trend is shown by the lower parts of the solid curves in Fig. 9.7. However, there is an obvious difficulty in that, to get an adequate yield, $(\sigma\phi/\omega)_0$ has to be much larger than $(\sigma\phi/\omega)_1$ applying at the present time. Specifically, if we compare Eq. (9.40) with Eq. (9.53) for $u = 0.02$ ([O/H] $= -2$) and $u_1 = 2$, we find that, to match the production of light elements from the Simple Model, we need to have

$$\left(\frac{\sigma\phi}{\omega}\right)_0 = 8 \left(\frac{\sigma\phi}{\omega}\right)_1 \tag{9.57}$$

and the Simple model itself fails by about an order of magnitude (see Fig. 9.6). Thus a cosmic-ray boosting factor of nearly two orders of magnitude is needed. Apart from having to think of a mechanism for this, it has been pointed out by Steigman & Walker (1992) that such a boost would result in unacceptably high primary production of lithium isotopes by $\alpha - \alpha$ fusion if the cosmic-ray spectrum is assumed to be unchanged. An interesting solution to both dilemmas was proposed by Prantzos *et al.* (1993), who suggested that, because of the greater extent of the Galaxy at right angles to the plane in early phases of collapse, the grammage parameter Λ could have been up to 2 orders of magnitude greater than at present (cosmic-ray confinement model), leading to both a higher flux and a flatter spectrum, which latter serves to avoid the Li overproduction (cf. Fig. 9.4). However, this model still leads to secondary-like production at the lowest metallicities (which does not fit more recent data; see Fig. 9.8) and the amount by which Λ can ever be enhanced is well below 2 orders of magnitude because of nuclear fragmentation processes (Malaney & Butler 1993). The required boost factor can be reduced by invoking a 'chemo-dynamical' model of Galactic evolution (Burkert & Hensler 1989) in which there is a hot medium consisting of metal-rich stellar ejecta and a cool medium of lower metallicity from which stars form, but this still offers no escape from quadratic behaviour (Pagel 1994a).

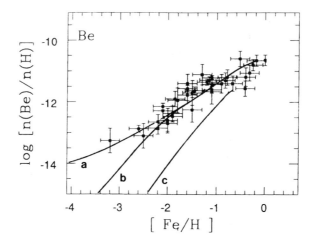

Fig. 9.8. Trend of beryllium abundance with metallicity (crosses) compared to predictions from 3 models: (a) Orion-type energetic C,N,O nuclei combined with Galactic cosmic rays; (b) GCR confinement model as in Fig. 9.7; (c) Simple model with GCR as in Fig. 9.6. After Vangioni-Flam & Cassé (1996, in S.S. Holt & G.G. Sonneborn (eds.), *Cosmic Abundances*, ASP Conf. Series, Vol. **99**, p. 366). Courtesy of the Astronomical Society of the Pacific Conference Series and Elisabeth Vangioni-Flam.

9.4.6   Other possibilities

The above models are all rather unsatisfactory, because they involve somewhat arbitrary assumptions about the time-dependence of the cosmic ray flux and spectrum and because they predict a secondary-like behaviour for Be and B abundances, whereas the overall trend indicated by the data is more like a primary one. Furthermore, there is a long-standing discrepancy between the predicted $^{11}B/^{10}B$ isotope ratio and the value in the Solar System (see Table 9.2). In the case of ^{11}B, there is a possible primary mechanism for stellar production in supernovae by neutrino spallation processes (Woosley *et al.* 1990; Woosley & Weaver 1995), but the constancy of the B/Be ratio in metal-poor stars (see Fig. 9.7), combined with the absence of any known similar process for Be, indicates that this does not solve the problem unless a primary process can be found for Be as well. The isotopic ratio in stars other than the Sun is not known, but the overall B/Be ratio in metal-poor or other stars is consistent (within errors) with either the solar-system ratio of 30 or that predicted by conventional cosmic-ray models, about 15 to 20. Anomalous isotope ratios for $^{7}Li/^{6}Li$ are also found in some nearby interstellar diffuse clouds, with a ratio apparently as small as 2 in one case (Lemoine *et al.* 1994), compared to the solar-system ratio of 12; the 'anomaly' here is that the low ratio in such clouds is consistent with cosmic-ray spallation whereas that in the Solar System is not.

   A possible explanation for the $^{11}B/^{10}B$ ratio in the Solar System was put forward long ago by Meneguzzi & Reeves (1975), who suggested the existence of a significant low-energy component of cosmic rays (see Fig. 9.4 for the behaviour of the production ratio near threshold) which, owing to the magnetic field and ionization losses, would be confined to the neighbourhood of star-forming regions. This idea more recently received striking confirmation from observations of the Orion nebula with the Comptel telescope on the Compton γ-ray Observatory satellite, which revealed line transitions of ^{12}C and ^{16}O at energies of 4 to 6 MeV (Bloemen *et al.* 1994). It appears that these γ-ray lines are most probably excited by the impact of ^{12}C and ^{16}O nuclei in supernova ejecta on protons and α-particles in the interstellar medium and that, with certain assumptions, such particles could make a still larger contribution to light-element formation than do the Galactic cosmic rays. If so, then that could explain a number of otherwise mysterious facts: the boron isotope ratio and the B/Be ratio in the Solar System, the lithium isotope ratio in some diffuse clouds (if this process has been unusually dominant) and the primary behaviour of Be and B abundances in low-metallicity stars, since the spallated material is essentially the supernova ejecta (Vangioni-Flam *et al.* 1996). A stellar source for the bulk of the ^{7}Li in the Solar System and other Population I stars is still needed, however, and the ^{11}B/^{10}B isotopic ratio is quite likely enhanced by the supernova neutrino process.

Notes to Chapter 9

The idea of cosmic-ray spallation as the major contributor to light-element production was first put forward in some detail by Reeves, Fowler & Hoyle (1970) and detailed physical models of cosmic-ray acceleration and propagation have been discussed by Meneguzzi, Audouze & Reeves (1971) and Meneguzzi & Reeves (1975). Calculations of both cosmic-ray and stellar flare type production, using updated cross-sections, have been made by Walker, Mathews & Viola (1985), whose results for the cosmic-ray case are presented alongside my more simple-minded calculation in Table 9.2.

Beryllium was first demonstrated to be present in low-metallicity stars by Gilmore, Edvardsson & Nissen (1991) and soon confirmed by others, e.g. Ryan *et al.* (1992) who gave the argument for a flattening in the Be/Fe ratio at [Fe/H] ~ -1. Extensive data are now available from their work and from that of Gilmore *et al.* (1992) and Boesgaard & King (1993). Boron was discovered in a few metal-deficient stars by Duncan, Lambert & Lemke (1992), but their abundance estimates have had to be corrected for line blending and non-LTE effects (Edvardsson *et al.* 1994; Kiselman & Carlsson 1994).

The discussion of the GCE of light elements formed by cosmic-ray spallation is based on Pagel (1994a) and has a somewhat different outlook from many other treatments in the literature, e.g. Fields, Olive & Schramm (1995), which should be consulted to get another viewpoint. A numerical treatment of the Simple model by Vangioni-Flam *et al.* (1990) gave similar results to the ones derived more simply here, as did the discussion of inflow models by Prantzos (1994b).

A number of authors have considered ways in which Be and B might be expected to show primary behaviour in the early Galaxy. Duncan, Lambert & Lemke (1992) suggested a possible dominance of the 'decelerated' component, an idea which is somewhat related to recent considerations of the fast C and O nuclei in Orion, while Feltzing & Gustafsson (1994) and Tayler (1995) have considered other ways in which spallation might take place preferentially in a metal-enriched supernova environment. Silk & Schramm (1992) have considered the contribution of a hypothetical early enhanced cosmic-ray flux to the γ-ray background. The most detailed discussions that I am aware of of the implications of the discovery of γ-ray lines in Orion, which probably outdate much of the previous work, are those by Vangioni-Flam *et al.* (1996) and Ramaty *et al.* (1996).

A good overview of the abundance and evolution of light elements is available in P. Crane (ed.), *The Light Element Abundances*, Springer-Verlag 1995.

Problems

1. Calculate the energy in MeV of a proton in supernova ejecta travelling at 10^4 km s^{-1}.

2. Verify that the momentum and energy expressions in Eq. (9.1) are equivalent.

3. Calculate the total number density and (kinetic) energy density (in eV cm^{-3}) of cosmic-ray protons from Eq. (9.5) and deduce their average energy.

4. Calculate the solar-system Be/H ratio predicted by the Simple model (Eq. 9.40) using the cross-section from Table 10.2, the present-day cosmic ray flux from Eq. (9.5) and assuming $u = u_1 = 2$, $\alpha' = 2/3$, $\omega^{-1} = 3$ Gyr.

5. Calculate the solar-system Be/H ratio predicted by the simple inflow model, Eq. (9.48), with the same assumptions as in Problem 4, except that now $u = 3$.

10 Radioactive cosmochronology

10.1 Introduction

Early in the 20th century, Rutherford and his colleagues developed the use of the fixed lifetimes of radioactive nuclei as a chronometer to measure ages of terrestrial and meteoritic rocks, and in 1929 Rutherford extended their use to make arguments about the age of the elements since their mean epoch of creation, related to the age of the Galaxy and the universe. In the 1950s and afterwards, Fowler and Hoyle and others have refined these arguments on the basis of improved understanding of nucleosynthesis; complications raised by questions related to Galactic chemical evolution will form a major topic of this chapter. In recent years, the discovery of dead short-lived radioactivities (e.g. from the isotopic anomalies mentioned in Chapter 3) has led to further inferences of time scales related to the formation of the Solar System. Some of the relevant species are listed in Table 10.1.

10.2 Age-dating of rocks

The basic idea in radioactive age-dating of rocks (from the Earth, Moon and meteorites) is to find the ratio of daughter to parent in an isolated system. Thus the age inferred is usually the 'solidification age' which is the time since the last occasion when chemical fractionation was halted by solidification. (K-Ar dating gives a 'gas-retention age' which can be slightly shorter.)

In general, the daughter nucleus will have an initial abundance at time zero, which can be assumed to bear a constant ratio to another isotope of the same element (one that is not enhanced by the radioactive decay). This is because isotope ratios are usually not affected by chemical fractionation processes, whereas the initial ratio of parent to daughter can vary widely. By comparison of the three species, parent ^{A}P, daughter ^{A}D (assuming that we are dealing with β-decay) and isotope of daughter $^{A'}$D in different samples drawn from one or several meteorites, it is possible to generate an *isochrone* by plotting ^{A}D$/^{A'}$D against ^{A}P$/^{A'}$D (see Fig. 10.1), since after time t

$$
\frac{^A\mathrm{P}(t)}{^{A'}\mathrm{D}} = \frac{^A\mathrm{P}(0)}{^{A'}\mathrm{D}} e^{-\lambda t}, \tag{10.1}
$$

Table 10.1. *Some radioactive species*

Parent	Daughter	Mean life λ^{-1}		
^{187}Re	^{187}Os	72	Gyr	
^{87}Rb	^{87}Sr	68	,,	
^{232}Th	^{208}Pb	20.3	,,	living
^{238}U	^{206}Pb	6.45	,,	
^{40}K	^{40}Ar (^{40}Ca)	1.7	,,	
^{235}U	^{207}Pb	1.02	,,	
^{244}Pu	^{232}Th	120	Myr	
^{129}I	^{129}Xe	23	,,	dead,
^{247}Cm	^{235}U	22.5	,,	but live
^{107}Pd	^{107}Ag	9.4	,,	in ISM
^{26}Al	^{26}Mg	1.03	,,	

and

$$\frac{^{A}D(t)}{^{A'}D(t)} = \frac{^{A}D(0)}{^{A'}D} + \frac{^{A}P(0)}{^{A'}D}(1 - e^{-\lambda t}) \tag{10.2}$$

$$= \frac{^{A}D(0)}{^{A'}D} + \frac{^{A}P(t)}{^{A'}D}(e^{\lambda t} - 1). \tag{10.3}$$

The first term in Eq. (10.3) is a constant, being the natural isotope ratio of, e.g., ^{87}Sr/^{86}Sr, while the second has a slope ($e^{\lambda t} - 1$), representing the growth of the daughter nucleus ^{87}Sr due to decay of ^{87}Rb in the course of time. Thus the solidification age of meteorites is found to be 4.56 Gyr, which is identified as the age of the Solar System (and of the Sun). Subtle differences of the intercept relative to carbonaceous chondrites (Allende has the smallest initial ^{87}Sr/^{86}Sr ratio known) reveal brief melting episodes in certain meteorites prior to the last solidification.

A few meteorites have significantly younger ages; these are believed to come from the Moon and in some cases from Mars, rather than from asteroids.

10.3　Galactic cosmochronology

10.3.1　Some historical landmarks

Rutherford (1929) noted that the abundance ratio ^{235}U/^{238}U is 0.007 now and was 0.3 at the birth of the Solar System, and he pointed out that extrapolation still further back to the (mean) epoch of nucleosynthesis would give the production ratio; the assumption at that time was that elements had been synthesized in the Sun and that the planets were formed out of material extracted from the Sun by a passing star. Reasoning that the production ratio was likely to be a reasonably small number (< 10,

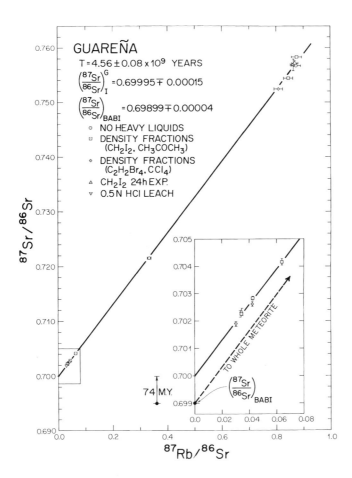

Fig. 10.1. Rb-Sr isochrone measured from separated components of the stony meteorite Guareña. The initial $^{87}Sr/^{86}Sr$ ratio is slightly higher than that inferred in basaltic achondrites (BABI) because of a period of metamorphism. After G.J. Wasserburg, D.A. Papanastassiou & H.G. Sanz, *Earth Planet. Sci. Lett.*, **7**, 33 (1969), with permission. Courtesy G.J. Wasserburg.

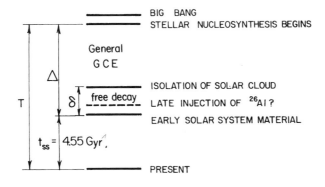

Fig. 10.2. Time-scales related to Galactic cosmochronology.

say), he deduced that the mean nucleosynthesis epoch could not have been more than 4.3 Gyr before the formation of the Solar System.

A theory for the production ratio became available with the work of B^2FH and Cameron (1957), who established that the actinides (U, Th etc.) result from the r-process, probably in supernovae. A first estimate of production ratios results from the simple consideration that an actinide nucleus (Z, A) can be formed either directly via β-decay or indirectly by α-decay of a higher nucleus $(Z + 2n, A + 4n)$, where n is limited by the occurrence of fission along the $A + 4n$ isobaric chain at a lower Z. Thus $^{238}_{92}U$ can be formed either directly or by α-decay of $^{242}_{94}Pu$ and $^{246}_{96}Cm$, but the chain becomes weaker at $^{250}_{98}Cf$, owing to partial fission of $^{250}_{96}Cm$, and stops altogether short of $^{254}_{100}Fm$ owing to complete fission of $^{254}_{98}Cf$. Thus ^{238}U has effectively about 4 progenitors (including itself). Similar considerations result in about 6 progenitors for ^{235}U and ^{232}Th and 3 for ^{244}Pu[1] so that from this factor alone one might expect production ratios relative to ^{238}U of about 1.5 for ^{235}U and ^{232}Th, and 0.75 for ^{244}Pu. The theoretical production ratios have been steadily refined over the years by Hoyle, Fowler, Clayton, Schramm, Cameron, Thielemann, Truran, Cowan and others (see Chapter 6 and Table 10.2). The same authors have then proceeded to apply these ratios to the problem of estimating the age of the Galaxy from the abundance ratios of $^{235,238}U$ and ^{232}Th in the Solar System. Clayton (1964) suggested an additional 'aeonglass' in the form of the ratio $^{187}Re/^{187}Os$. ^{187}Re has a very long lifetime (see Table 10.1), which has some advantages in reducing sensitivity to GCE details, but associated uncertainties (including enhancement of β-decay rates at stellar temperatures and the s-process contribution to ^{187}Os) have prevented any significant results from this so far (Yokoi, Takahashi & Arnould 1983). Butcher (1987) invented a completely new technique of judging stellar ages from the abundance of thorium, which has the advantage of being independent of either production ratios or solar-system abundances, but involves some technical difficulties of its own. Schramm & Wasserburg (1970) and especially Tinsley (1977) and Clayton (1988) noted the sensitivity of Galactic age estimates based on radioactivity to the details of the GCE model adopted ('the aeonglasses leak'), but in principle (if perhaps not yet in practice) it may still be possible to obtain interesting constraints on both the GCE model and the age of the Galaxy and/or of individual stars by combining solar-system and stellar data.

10.3.2 Galactic cosmochronology: theory

We assume that actinides are formed along with other elements in some sort of GCE history during a period Δ up to the formation of the Solar System (Fig. 10.2). Thus

$$T = \Delta + t_{SS} = \Delta + 4.56 \text{ Gyr} \tag{10.4}$$

---

[1] An illustration of the relevant tracks in the (N, Z) plane is given by Thielemann, Metzinger & Klapdor (1983).

is the age of the Galaxy since the beginning of nucleosynthesis. Possible complications are

(1) The Solar System could have been enriched by a late nucleosynthesis 'spike' due to a nearby supernova or AGB star. Supernova effects are not now usually considered significant, at least as a contributor to the r-process, because of the small abundance of ^{129}I (see Section 10.4.2), but there is a problem with ^{26}Al, and a late injection from an AGB star has been suggested by Wasserburg *et al.* (1994); this would not affect the actinides since they come from the r-process.

(2) There is generally assumed to be a period δ of free decay while the parent cloud of the Solar System was isolated from the general ISM for some time before the system was formed and/or during solidification of metorite parent bodies over periods dependent on individual mineralogies. Since the dead activities indicate $\delta \leq 10^8$ yr (see Section 10.4), the long-lived actinides will again be unaffected.

(3) Actinides can undergo destruction by photo-fission in He zones (Malaney, Mathews & Dearborn 1989) which may lead to a modest degree of destruction by astration; we shall neglect this effect, however.

In the theoretical development, we assume instantaneous recycling with constant yields and generalize Eq. (7.34) to the case of a decaying element with $Z_F = 0$ using now time as the independent variable, bearing in mind that $E \equiv \eta \alpha \psi$ and neglecting free decay:

$$\frac{dN_i}{dt} \equiv \frac{d}{dt}\left(\frac{Z_i g}{A_i m_H}\right) = p_i \frac{ds}{dt} - (1+\eta)\frac{ds}{dt}\frac{N_i}{g} - \lambda_i N_i, \tag{10.5}$$

where N_i is the number of atoms in the ISM of a representative region, e.g. a cylinder at right angles to the Galactic plane through the Sun, and p_i is the yield expressed in number of atoms rather than mass. The solution of Eq. (10.5) is

$$N_i(t) = e^{-\lambda_i t - v(t)}\left[N_{i0} + p_i \int_0^t \underline{\alpha\psi(t')e^{v(t')}e^{\lambda_i t'}}\,dt'\right], \tag{10.6}$$

where

$$v(t) \equiv \int_0^t [(1+\eta)\alpha\psi(t')/g(t')]\,dt'. \tag{10.7}$$

The underlined factor in Eq. (10.6) is sometimes called the 'effective production function'.

To proceed further, we assume for simplicity (as has mostly been done in the literature) a linear model, i.e.

$$\alpha\psi \equiv \frac{ds}{dt} = \omega g \quad \text{with} \quad \omega = \text{const.}, \tag{10.8}$$

Table 10.2. *Some estimates of K-ratios in the Solar System*

	Meyer & Schramm 1986	Fowler 1987	Cowan *et al.* 1987,91
p_{235}/p_{238}	$1.5^{+.4}_{-.3}$	$1.34 \pm .19$	1.16
N_{235}/N_{238} [a]	0.310	0.33	0.32
p_{232}/p_{238}	$1.6 \pm .2$	$1.71 \pm .07$	$1.65 \pm .05$
N_{232}/N_{238} [a]	$3.5 \pm .2$	2.30	2.32
p_{244}/p_{238}	0.12 to 1		
N_{244}/N_{238} [a]	$0.005 \pm .001$		
$K_{235,238}$	$0.21 \pm .05$	$0.246 \pm .035$	0.273
$K_{232,238}$	$2.2 \pm .3$	$1.35 \pm .10$	$1.40 \pm .05$
$K_{244,238}$	0.004 to 0.05		

[a] At formation of Solar System

and (without significant loss of generality) we take $\eta = 0$. Then $v = \omega t$ and Eq. (10.6) simplifies to

$$N_i(t) = e^{-(\omega+\lambda_i)t} \left[N_{i0} + p_i \int_0^t \alpha\psi(t') e^{(\omega+\lambda_i)t'} dt' \right].$$

(10.9)

For solar-system actinides, given a set of production ratios p_i/p_j calculated from the theory of the r-process, one can use meteoritic abundance measurements to derive the 'observed' quantities

$$K_{ij}(\Delta) \equiv \frac{N_i(\Delta)/p_i}{N_j(\Delta)/p_j}$$

(10.10)

at the time of formation of the Solar System.

10.3.3 Some data and models

Table 10.2 gives some of the data on K-ratios used by different authors, including some rather uncertain data for ^{244}Pu. There are significant differences among the K-ratios adopted, and age estimates based on them are also dependent on the GCE model used, as will be illustrated with two such models.

The first one is that of Fowler (1987), which is a Simple closed model with an initial production spike (cf. Section 8.4, item (iv), in relation to the 'G-dwarf problem'). In

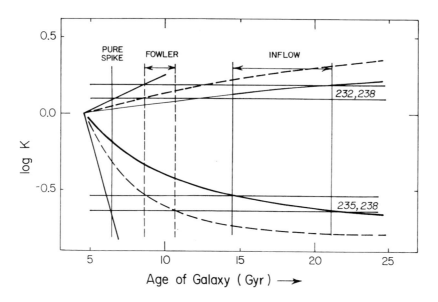

Fig. 10.3. 'Observed' K-ratios (estimated error limits shown by pairs of horizontal lines) and their theoretical variation with the age of the Galaxy according to three models: a pure initial spike (continuous straight lines); Fowler's (1987) model (broken-line curves); and the simple inflow model (continuous curves). After Pagel (1993).

this model the net star formation rate is

$$\alpha\psi = \omega g = Ae^{-\omega t} \tag{10.11}$$

and the abundances are

$$N_i(\Delta) = e^{-\omega\Delta}\left[N_{i0}e^{-\lambda_i\Delta} + A\frac{p_i}{\lambda_i}(1 - e^{-\lambda_i\Delta})\right] \tag{10.12}$$

for a radioactive element and

$$N_k(\Delta) = e^{-\omega\Delta}\left[N_{k0} + Ap_k\Delta\right] \tag{10.13}$$

for a stable element. If a fraction S of the stable element in the Solar System comes from the initial spike and the stable element is produced *pari passu* with the radioactive one, then

$$\frac{N_{i0}}{Ap_i} = \frac{N_{k0}}{Ap_k} = \frac{S\Delta}{1-S}, \tag{10.14}$$

so that, from Eq. (10.12),

$$\frac{e^{\omega\Delta}}{\Delta}\frac{N_i(\Delta)}{Ap_i} = \frac{S}{1-S}e^{-\lambda_i\Delta} + \frac{1-e^{-\lambda_i\Delta}}{\lambda_i\Delta}. \tag{10.15}$$

Fowler uses $K_{235,238}$ and $K_{232,238}$ to solve for the two unknowns S and Δ (note that the

K-ratios are independent of ω) finding

$$S = 0.17 \pm .02; \tag{10.16}$$

$$T = \Delta + 4.6 = 10.0 \pm 1.5 \text{ Gyr} \tag{10.17}$$

(cf. Fig. 10.3).

For these parameters, $K_{244,238} = 0.03$. It is to be noted that a pure initial spike ($S = 1$) would just fit the U-Th data in Fig. 10.3 with T=6.4 Gyr (the rock-bottom limit to the age of the universe), but at the expense of making $K_{244,238} = 4 \times 10^{-7}$, which is far too small unless one can appeal to some special factor (like a nearby supernova) that introduced Pu shortly before the formation of the Solar System. On the other hand, with S just slightly less than 1, this difficulty could be overcome, giving an age of still only 7 Gyr, say. However, there are other reasons for believing nucleosynthesis to have been a continuous process in our Galaxy.

As an alternative to Fowler's model, we now consider the (probably more realistic) simple inflow model of Section 8.5.2. In this model we assume

$$g_o = z_0 = z_F = 0 \tag{10.18}$$

and take the final mass of the system to be 1. We then have from Eq. (10.9) with $N_{i0} = 0$, $\alpha\psi = \omega g = \omega u e^{-u} = \omega^2 t e^{-\omega t}$,

$$N_i(\Delta) = e^{-(\omega+\lambda_i)\Delta} p_i \omega^2 \int_0^\Delta t \, e^{\lambda_i t} dt \tag{10.19}$$

$$= e^{-(\omega+\lambda_i)\Delta} p_i \omega^2 \frac{1}{\lambda_i^2} \left[\lambda_i \Delta e^{\lambda_i \Delta} - (e^{\lambda_i \Delta} - 1) \right], \tag{10.20}$$

whence

$$\frac{e^{\omega\Delta}}{(\omega\Delta)^2} \frac{N_i(\Delta)}{p_i} = \frac{1}{\lambda_i \Delta} - \frac{1 - e^{-\lambda_i \Delta}}{(\lambda_i \Delta)^2}. \tag{10.21}$$

This formula fits the same U, Th data for $13 \leq T \leq 20$ Gyr, and it can be seen from Fig. 10.3 that the upper limit could be infinity for slightly different models, basically because the beginning of nucleosynthesis is something that is now assumed to happen gradually. The inflow model gives $K_{244,238} = 0.04$, much the same as in Fowler's model. Thus *solar-system actinide data do not provide any upper limit to the age of the Galaxy, nor do they strongly constrain GCE models.*

10.3.4 Stellar thorium chronology

Butcher (1987) invented a new original approach to Galactic cosmochronology based on spectroscopic observations of thorium in stars (cf. Fig. 10.4). The idea is to judge the ages of stars by the amount by which thorium has decayed compared to some expected standard: older stars may be expected to have relatively less thorium than younger ones, and this can act as an independent check on the age of the star deduced conventionally from its position in the HR diagram. This could be of importance in

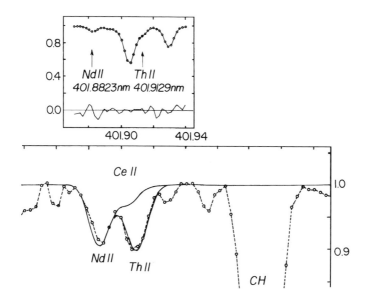

Fig. 10.4. Above: Spectrum of the solar-type G-dwarf star HR 509 showing features of Th II and Nd II near λ 4019 Å, after Butcher (1987). Th II is blended with a strong feature due to Fe and Ni, as well as weaker features. The tracing around the zero level shows 10 × the difference between the observed spectrum (dots) and the fitted synthetic spectrum (continuous curve). Reproduced with permission from *Nature*, **328**, 127, ©1987, Macmillan Magazines Ltd. Below: Spectrum of the same region in the very metal-poor giant star CS 22892–052 ([Fe/H] $\simeq$ −3) with a large relative excess of r-process elements ([r/Fe] = 1.7), adapted from Sneden *et al.* (1996).

cosmology, since conventional ages of globular clusters and some field stars (12 to 20 Gyr) are in some cases too large to be reconciled with observational estimates of the Hubble constant, at least in Einstein–de Sitter cosmology.

The advantages of this method are

(1) Compared to the onset of nucleosynthesis in the Galaxy, which may be rather poorly defined, in this case one is dating a very well-defined event, namely the birth of the particular star. Consequently there is less sensitivity to GCE models, although there is still some, as discussed below.

(2) No theoretical production ratios are needed. The Sun has a well known age, so that its thorium abundance can be used as a standard.

There are also problems with the method:

(1) The abundance determination is technically difficult in solar-type stars because of blending with other spectral lines (see Fig. 10.4).

(2) There are difficulties in the choice of a stable element for comparison, so as to cancel out effects of differing metallicity. Butcher chose neodymium, because

Table 10.3. *Data for stellar thorium chronology*

Star	Age (Gyr)	[Fe/H] [a]	[Th/Nd] [b]	[Th/Nd] [a]	[Th/Nd] [c]	[Th/Eu] [d]	direct [e]
Sun	4.6	0.0	.00	.00	.00	.00	.00
HR 4523	8 ± 3	−0.3	−.14	−.08	−.26	−.20	−.27
HR 509 (τ Cet)	9 ± 4	−0.5	.08	.19	−.05	−.06	−.23
HR 98 (β Hyi)	10 ± 2	−0.3	−.06	−.03	−.16	−.13	−.16
HR 3018	19 ± 3	−1.0	.12	.22	−.22	−.12	−.24
4 halo stars	15 ± 3	≥ −2.4					−.15
3 halo stars	15 ± 3	≤ −2.4					≤ .4
CS 22892−052	15 ± 3	−3.0	.16				−.27 ± .10 :

[a] Morell *et al.* (1992), except for the last three entries
[b] Lawler *et al.* (1990), except for the last entry
[c] Th/Nd (Lawler) × Nd/Eu (da Silva)
[d] Th/Nd (Morell) × Nd/Eu (da Silva)
[e] da Silva *et al.* (1990); François, Spite & Spite (1993); Sneden *et al.* (1996).

there is a Nd II line very nearby in the spectrum to the Th II line which has similar excitation potential making the ratio quite insensitive to stellar temperature and gravity. This choice is flawed, however, by differential GCE effects on the two elements, since about half of the Nd in metal-rich stars comes from the s-process (cf. Table 6.3) and Th exclusively from the r-process.

Butcher argued that differential GCE effects between Th and Nd are negligible (because the crude abundance measurements available at the time had not been able to detect any changes in s/r/Fe ratios at least in Galactic disk stars); from his failure to detect any change in Th/Nd correlated with the conventional ages of the stars, he deduced an age of 10 Gyr or less for the Galaxy, in agreement with Fowler. This led to a number of papers and counter-papers discussing the above points, and also to some new measurements which mostly reaffirmed Butcher's original results. Pagel (1989b) proposed that Th should be compared to another pure r-process element, specifically Eu, and this has been done in a few investigations which give mild support to larger ages, particularly in the case of the very metal-deficient r-process enhanced star CS 22892−052 (see Fig. 10.4) where the Th/Eu ratio is approximately half solar; at the same time, numerous stable r-process species in that star have abundance ratios in excellent agreement with those of the r-process contributions to the abundances of the corresponding elements in the Solar System (Sneden *et al.* 1996).

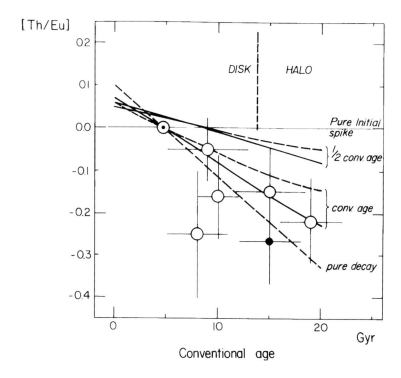

Fig. 10.5. Logarithmic differential Th/Eu ratios plotted against stellar age derived from HR diagrams, assuming a straight mean of the entries in the last three columns of Table 10.3 and an error of ±0.15 dex in most cases (open circles). The filled circle represents CS 22892–052, which is a kind of 'Rosetta stone' for the r-process. Continuous curves show the prediction from the simple inflow model and broken curves that from Fowler's model, for the two assumptions about the correctness or otherwise of conventional ages. Adapted from Pagel (1993).

10.3.5 Galactic cosmochronology and stellar thorium abundances: theory

Two extreme models of the evolution of stellar thorium abundances can be considered:

(1) The pure initial spike model for the r-process considered in Section 10.3.3. In this case, all the thorium was created at an initial instant and decayed at a constant rate up to the present, whether in stellar atmospheres or elsewhere. Hence no variation in Th abundance with stellar age is to be expected ('pure initial spike' in Fig. 10.5).

(2) A model in which each star has a mix of r-process elements derived from a nearby supernova at the time of formation. In this case, the Th would be formed in a constant ratio to stable r-process nuclides and then undergo free decay with a half-life of 13 Gyr through the age of the star; this leads to the greatest effect that can ever be expected ('pure decay' in Fig. 10.5), but it is clearly unrealistic, at least for the Solar System, since it implies that all the K-ratios should $= 1$.

To be (a bit) more realistic, we consider again the two models of Section 10.3.3. The abundance in a stellar atmosphere is the initial abundance given by Eq. (10.12) with Δ replaced by the formation time t of the star (with age $T - t$), modified by a factor $e^{-\lambda_i(T-t)}$ due to subsequent free decay. Thus in Fowler's model we have from Eqs. (10.13) to (10.15)

$$\text{for Th} \quad e^{\omega t} \frac{N_i(t, T)}{Ap_i} = \frac{S\Delta}{1 - S} e^{-\lambda_i T} + \frac{e^{-\lambda_i(T-t)} - e^{-\lambda_i T}}{\lambda_i} ; \tag{10.22}$$

$$\text{for Eu} \quad e^{\omega t} \frac{N_k(t, T)}{Ap_k} = \frac{S\Delta}{1 - S} + t, \tag{10.23}$$

where $\Delta = T - 4.6$ Gyr.

In the simple inflow model we have from Eq. (10.21)

$$\text{for Th} \quad \frac{e^{\omega t}}{(\omega t)^2} \frac{N_i(t, T)}{p_i} = \frac{e^{-\lambda_i(T-t)}}{(\lambda_i t)^2} \left[\lambda_i t - (1 - e^{-\lambda_i t}) \right] ; \tag{10.24}$$

$$\text{for Eu} \quad \frac{e^{\omega t}}{(\omega t)^2} \frac{N_k(t)}{p_k} = \frac{1}{2}. \tag{10.25}$$

For the halo stars, we can take $S = t = 0$ and their age is just T.

We can now put a well-defined question: Do the Th/Eu ratios measured in stars confirm the ages deduced conventionally from the HR diagram, or not? Table 10.3 gives some observational data and these are compared in Fig. 10.5 with the theoretical predictions from the two models for two alternative hypotheses: conventional ages correct, or conventional ages a factor of 2 too large (except for the Sun!), which is essentially what was originally claimed in the case of the oldest stars.

Two points emerge from the comparison:

(1) The difference between ages deduced from Fowler's model and the simple inflow model is in the opposite sense to that in Fig. 10.3. Fowler's model requires greater ages (for a given value of [Th/Eu]) because of the initial spike, which dilutes the effect of Galactic evolution.

(2) The data are not very good, since the errors in even good spectroscopic stellar abundance determinations are nearly as large as the effect sought. However, so far as they go, they tend to favour the conventional stellar ages.

10.4   Short-lived radioactivities

10.4.1 Introduction

As was mentioned in Chapter 3, a number of decay products of dead, short-lived radioactive nuclei are found in meteorites; some of these are listed in Table 10.1. These could come from a variety of sources including:

(1) Inheritance from the interstellar medium (where they have some steady-state abundance from continuing nucleosynthesis), possibly after some interval(s) of free decay.

(2) 'Cosmic chemical memory' (Clayton 1982), whereby decay products from short-lived activities formed in dust grains ejected from supernovae or AGB stars and these grains were later incorporated without substantial change in the meteorites. This explanation is clearly preferred for ultra short-lived activities, e.g. ^{22}Na$(\beta^{+}, \nu)^{22}$Ne, with a 2.6 yr half-life, which leads to Ne-E, and for a number of other isotopic anomalies found among fine grains in the matrix of carbonaceous chondrites. On the other hand, it can hardly apply to ^{129}I, because the ^{129}Xe is found associated with ^{127}I (actually ^{128}Xe produced from ^{127}I by neutron activation) in a variety of meteoritic sites comprising a substantial bulk of material in objects that have been thermally processed. Thus ^{129}I must have been alive in the early Solar System and similar arguments (plus the fission tracks) apply to ^{244}Pu. ^{107}Pd must also have been alive because it is found in the metal phase of iron meteorites, and so probably also was ^{26}Al, because its daughter ^{26}Mg is correlated with aluminium in bulky and thermally processed material. Such heating episodes would have destroyed any correlation with aluminium unless the ^{26}Mg was actually in the form of ^{26}Al at the time. However, there is also decayed ^{26}Al (often with much larger ratios of ^{26}Al/^{27}Al than the standard 5×10^{-5}) in fine oxide, graphite and SiC grains of extra-solar origin.

(3) A late-time contribution to the Solar System from a nearby supernova (Cameron 1993) or an AGB star or planetary nebula.

(4) Production in the Solar System itself by solar-flare or cosmic-ray activity, e.g. ^{14}C.

10.4.2 Short-lived activities and Galactic chemical evolution

For a sufficiently short-lived activity ($\lambda\Delta \gg 1$), Eqs. (10.15) and (10.21) both lead to

$$K_{ij} = \lambda_j/\lambda_i, \tag{10.26}$$

representing a steady state in the ISM where the decay is fast enough to balance creation and dominate other processes. Following Reeves (1991), we may define a related parameter

$$A_{ij} \equiv K_{ij}\,\lambda_i/\lambda_j = 1 \ \text{ in a steady state.} \tag{10.27}$$

For a standard of comparison, Reeves chooses ^{235}U, whose mean life of 1 Gyr is short enough to lead to a steady state and long enough to integrate over fluctuations in production rates and to survive periods of free decay. (Actually, even for ^{238}U and Th, the A-ratios relative to ^{235}U are within a factor 2 of unity.)

The results of Reeves's analysis are shown in Fig. 10.6, where it is seen that ^{244}Pu fits in quite nicely assuming $K_{244,238} \simeq 0.02$. Then comes a group of shorter-lived elements for which $A_{ik} < 1$ (where k refers to a neighbouring stable element with an assumed lifetime of 10 Gyr), and this is interpreted as an indication of a free decay interval δ

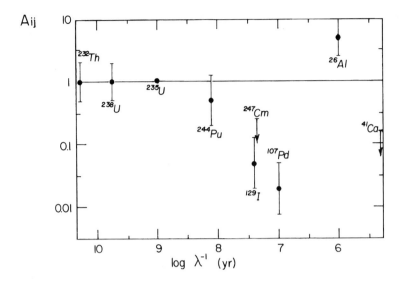

Fig. 10.6.   Log–log plot of A_{ij} against decay constant (yr^{-1}). Adapted from Reeves (1991).

between 40 and 100 Myr. Reeves identifies this as the result of a phase of absent or reduced exposure to nucleosynthesis during a passage of the parent cloud of the Solar System between spiral arms just before the Solar System was formed. The abundance of ^{129}I relative to ^{127}I, which have comparable production ratios, is 10^{-4}, and that number then places an upper limit on any contribution to r-process material in the Solar System from a hypothetical nearby supernova. However, since not all supernovae contribute to the r-process, some other kind of supernova is not ruled out.

The situation for ^{26}Al, however, is very different, since the solar-system abundance ratio ^{26}Al/^{27}Al $\simeq 5 \times 10^{-5}$ is a few times higher than that inferred for the interstellar medium from γ-ray observations, and the latter could not contribute to the Solar System in any case because of the short lifetime, so that in this case a late addition to the Solar System is indicated. Wasserburg *et al.* (1994) suggest that H-burning and s-process products from a nearby AGB star, contributing about 3×10^{-4} of the mass of the Solar nebula, could account for the presence of ^{26}Al and several other isotopic anomalies.

Notes to Chapter 10

The theory of nucleo-cosmochronology (and developments up to that time) are described by
D.N. Schramm, *Ann. Rev. Astr. Astrophys.*, **12**, 383, 1974.
    More recent developments are described in
J.J. Cowan, F.-K. Thielemann & J.W. Truran, *Ann. Rev. Astr. Astrophys.*, **29**, 447, 1991.
D.D. Clayton, in W.D. Arnett & J.W. Truran (eds.), *Nucleosynthesis: Challenges and New Developments*, University of Chicago Press, 1985, p. 65, gives a description and

simple algorithms for GCE and nucleocosmochronological calculations on the basis of his standard inflow models. Our simple inflow model shares some of the characteristics of Clayton's standard model with $k = 1$. Cf. also Clayton (1985b).

Interpretations of Butcher's Th-Nd chronology have been discussed by Clayton (1988), Mathews & Schramm (1988) and Malaney & Fowler (1989b); these are outdated by the more recent use of Th/Eu ratios, and the treatment in this chapter is based on papers by Pagel (1989b, 1993).

For more detailed discussions of isotopic anomalies and short-lived radioactivities in the Solar System, see

G.J. Wasserburg & D.A. Papanastassiou, in C.A. Barnes, D.D. Clayton & D.N. Schramm (eds.), *Essays in Nuclear Astrophysics*, Cambridge University Press, 1982, p. 77,

D.D. Clayton, *Quart. J. R. astr. Soc.*, **23**, 174, 1982,

G.J. Wasserburg, *Earth Plan. Sci. Lett.*, **86**, 129, 1987,

F.A. Podosek & T.D. Swindle, in J.F. Kerridge & M.S. Matthews (eds.), *Meteorites and the Early Solar System*, University of Arizona Press, Tucson, 1988, p. 1093,

A.G.W. Cameron, in E.H. Levy, E.H. Lunine & M. Matthews (eds.), *Protostars and Planets III*, University of Arizona Press, Tucson, 1993, p. 47,

E. Anders & E. Zinner, *Meteoritics*, **28**, 490, 1993, and

U. Ott, *Nature*, **364**, 25, 1993.

Problems

1. Supposing that Fig. 3.27 were to be interpreted as an isochrone (which it is not, in fact), what age would be deduced?

2. Show that, from the decay of ^{235}U and ^{238}U, one can derive a lead–lead isochrone given by

$$\frac{^{207}\text{Pb}(t)}{^{204}\text{Pb}} = \text{const.} + k \frac{e^{\lambda_{235}t} - 1}{e^{\lambda_{238}t} - 1} \frac{^{206}\text{Pb}(t)}{^{204}\text{Pb}}, \tag{10.28}$$

where k is the present-day abundance ratio $^{235}\text{U}/^{238}\text{U} = 0.0073$. (^{204}Pb has no radioactive progenitors.) Compare the slope of this isochrone with that of the Rb-Sr isochrone in Fig. 10.1.

3. Re-writing Eq. (10.6) in the form (neglecting N_{i0})

$$N_i(t) = e^{-\lambda_i(t - <\tau>) - v(t)} \, p_i \int \alpha \psi e^v e^{\lambda_i(t' - <\tau>)} dt' \tag{10.29}$$

and expanding $e^{\lambda_i(t' - <\tau>)}$ in the integral as a power series, verify that

$$N_i(t) = e^{-\lambda_i(t - <\tau>) - v(t)} \, p_i \int \alpha \psi e^v \left[1 + \frac{1}{2}\lambda_i^2 <(t' - <\tau>)^2> + \ldots\right] dt', \tag{10.30}$$

where $<\tau>$ is the mean time of formation of elements up to time t.

Hence show that, for sufficiently small λ's, independently of any model,

$$K_{ij}(t) = e^{-(\lambda_i - \lambda_j)(t - <\tau>)}. \tag{10.31}$$

Apply this result to solar-system ^{232}Th and ^{238}U to find the corresponding value of $\Delta - <\tau>$ and make a rough estimate of the error. What does that make the age of the Galaxy, (a) if $\alpha\psi e^{\nu} = $ const.; (b) if all nucleosynthesis had taken place in a single event at the beginning?

4. If K_{ik} is the K-ratio for a radioactive element i relative to a stable element k, and $\lambda_i\Delta \gg 1$, show that, in Fowler's model,

$$K_{ik} = \frac{(1-S)}{\lambda_i\Delta} \tag{10.32}$$

whereas in the simple inflow model

$$K_{ik} = \frac{2}{\lambda_i\Delta} \tag{10.33}$$

(ignoring any period of free decay before incorporation into metorites).

11 Chemical evolution in other sorts of galaxies

11.1 Dwarf galaxies

11.1.1 Applications of the Simple model with extensions

The Simple model achieves a modest degree of success in accounting for abundances measured in dwarf galaxies, particularly the gas-rich dwarf irregulars and blue compact/H II galaxies where several element abundances are measured from observations of H II regions and vary little across the parent galaxy. The relation with gas fraction (Fig. 8.9) is somewhat problematic and controversial, but as far as it goes it suggests a rather low effective yield of about $0.2Z_\odot$ (compared to about $0.7Z_\odot$ for the solar neighbourhood). This could arise from a bottom-heavy IMF (Peimbert & Serrano 1982; Peimbert, Colin & Sarmiento 1994)[1] or from galactic winds, which are discussed in Section 11.2.3 below. The wind models lead to certain scaling laws relating metallicity to galaxy mass or luminosity. Such scaling laws can also be derived on the assumption that the star formation bursts are driven by mergers in which the gas fraction successively decreases, followed eventually by expulsion of the remaining gas in a (homogeneous) terminal wind or by ram pressure in an intra-cluster medium (Tinsley & Larson 1979; Faber & Lin 1983 and Section 11.2.4 below). A tight relation between metallicity (as estimated from colours or line features, particularly Mg_2) and escape velocity or velocity dispersion (probably closely related) has been found in elliptical galaxies by Vigroux, Chièze & Lazareff (1981), Franx & Illingworth (1990) and Bender (1992); cf. Fig. 11.5. Removal of gas by one of the above processes could perhaps lead to the conversion of a gas-rich dwarf galaxy into a dwarf elliptical with roughly similar metallicity, in accordance with Fig. 8.10 (Skillman, Kennicutt & Hodge 1989). Metal-enhanced winds involve much smaller overall mass loss rates (relative to star formation rates) than do homogeneous winds, and consequently are less likely to have dynamical effects on galaxy evolution although the chemical effects could be quite similar in many respects.

Element-to-element ratios depend on the time and metallicity dependences of stellar

---

[1] Kumai & Tosa (1992) have suggested that deviations from a solar-like effective yield could result from the presence of varying relative amounts of dark matter which affect the total (dynamically inferred) mass without contributing to chemical evolution.

yields as well as on details of galactic evolution. The C,O relation (Figs. 8.3, 8.4) is well explained by Maeder's metallicity-dependent yields provided that effectively the full range of stellar masses is assumed to undergo supernova explosions (Prantzos, Vangioni-Flam & Chauveau 1994) and that carbon comes predominantly from massive stars, so that at low metallicities where stellar winds are weak the C/O ratio is relatively insensitive to any metal-enhanced wind. This will also be the case for Fe/O when Fe comes predominantly from SN II.

The cases of helium and nitrogen, in which a convincing degree of scatter is superposed on the overall trends with oxygen, are more complicated. The large value of $dY/dZ \simeq 3$ or 4 found by, e.g., Lequeux *et al.* (1979) and Pagel *et al.* (1992) has presented a long-standing challenge to stellar nucleosynthesis theory which Maeder (1992, 1993) proposed to solve by assuming that stars initially above about $25M_\odot$ become black holes after emitting winds but without becoming supernovae. Serious objections have been raised against this idea, related both to the total yields and to the C/O ratio (Prantzos 1994a; Peimbert, Colin & Sarmiento 1994); a more likely explanation appeals to metal-enhanced winds, since helium comes predominantly from intermediate-mass stars that would be less affected. Doubts have also been raised as to whether dY/dZ is really so large anyway (Izotov, Thuan and Lipovetsky 1997). As far as nitrogen is concerned, it is not easy to say anything definite beyond the evidence from Fig. 8.8 that it has both primary and secondary components with efficiencies that may vary from one place to another (cf. Pagel & Kazlauskas 1992; Henry & Howard 1995; the scatter in both elements is discussed further below). At the same time there is a remarkable resemblance between the overall relationships with oxygen of all the three elements C, N and Fe. This could be related to the partially 'secondary' nature of carbon production according to Maeder; at the time of writing there are no predictions as to how the iron yield may depend on stellar metallicity.

11.1.2 Effects of star formation bursts

In the instantaneous recycling approximation, changes in element-to-element ratios are attributed to primary/secondary effects. When this approximation is relaxed there can be significant consequences, either due to overall galactic time-scales relative to production time-scales for, say, nitrogen (Edmunds & Pagel 1978), or due to effects of star formation bursts localized in time and space. This would apply especially in dwarf galaxies that may have experienced only a few such bursts over a Hubble time and may have lost gas in wind-ejection episodes accompanying the bursts (Matteucci & Tosi 1985; Pantelaki & Clayton 1987; Gilmore & Wyse 1991).

Clear signs of the effects of separated star formation bursts are shown in the age-metallicity relation of the Large Magellanic Cloud (Fig. 11.1), as well as in the age distribution of star clusters in that system, whereas in the Small Cloud the age distribution is more continuous (van den Bergh 1991). A few clusters are 10 Gyr old or more, but after that relatively little happened until a major burst about 2 to 3 Gyr

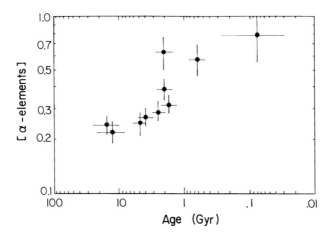

Fig. 11.1. Age-metallicity relation for α-particle elements in the Large Magellanic Cloud, deduced from ground-based and Hubble Space Telescope observations of planetary nebulae, after Dopita (1996). Published by courtesy of Editions Frontières. M.A. Dopita, *The Sun and Beyond*, Proceedings of the IInd Rencontre du Vietnam, 22-28 October 1995, edited by J. Tran Thanh Van, L. Celnikier, Hua Chon Trong & S. Vauclair, Ed. Frontières, 1996.

ago, which led to a relatively sudden increase in metallicity (as judged from oxygen and α-particle elements) up to essentially its present value of the order of 1/2 solar. During the hiatus, iron continued to be produced by type Ia supernovae, leading to an enhanced Fe/O ratio.

Fig. 11.2 shows a scenario described by Garnett (1990; based on work in Pantelaki's thesis) to explain scatter in the N/O ratio. In the top left corner of the vectors, a dwarf galaxy with mass $10^8 M_\odot$ turns 1 per cent of its mass into massive stars in a burst, quickly increasing the oxygen abundance of the whole galaxy by a factor of 6 and moving it down a 45° line to the right. Ejection of nitrogen from intermediate-mass stars then increases N/O at constant O/H after the starburst has faded. A second burst of the same size again moves the galaxy to the lower right, but by a much smaller amount, etc.

This model assumes that the galaxy behaves like a closed box and that newly synthesized elements are instantaneously mixed over the whole galaxy; these two assumptions have been relaxed in the work of Pilyugin (1993). Thus, in Pilyugin's model, the vertical tracks in Fig. 11.2 are replaced by lines sloping upwards to the left, due to gradual escape of oxygen from the initially self-enriched H II region. Similar tracks are predicted for helium with a mean trend $dY/dZ \sim 2$ in the closed-box model, which is consequently abandoned in favour of a model with metal-enhanced galactic wind expelling mass at up to 90 per cent of the star formation rate (Fig. 11.3). The scatter in (N,O) and (He,O) relations is then mainly attributed to the zig-zags in the diagrams resulting from the dilution of self-enrichment by mixing, with some contribution from differing metal-enhanced mass loss rates, inter-burst periods etc. The

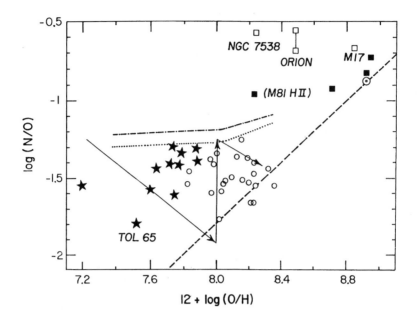

Fig. 11.2. Nitrogen *vs* oxygen abundance in extragalactic and Galactic H II regions with a schematic indication of effects of successive star-formation bursts. The dashed line represents a lower bound to the N/O ratio predicted on the basis of secondary production in massive stars, while the dotted and dash-dot lines represent mainly primary production from intermediate-mass stars accompanied by homogeneous winds, after Matteucci & Tosi (1985). Adapted from Garnett (1990).

question of self-enrichment of H II regions is quite complicated: the uniformity of composition over the Magellanic Clouds (Pagel *et al.* 1978) suggests that this is not very common, but there is some tentative evidence in its favour from studies of I Zw 18. Here the H I gas was found by Kunth *et al.* (1994) to have still lower oxygen abundance than the two major H II regions, but doubt has been cast on this result (Pettini & Lipman 1995). A natural expectation would be that supernova ejecta would appear somewhat after the major ionization phase and rapidly escape from the H II region, whereas the slower ejecta in the form of stellar winds would appear sooner and hang around for a while. Pagel *et al.* (1992) attributed all the scatter in the (He,O) diagram (which for most objects is only barely significant in relation to errors) to self-enrichment in He and N by winds from embedded massive stars that show their presence in the form of broad Wolf–Rayet features in the underlying stellar spectra. This is supported by the existence of inhomogeneous N and He abundances in the single giant H II region of NGC 5253 (cf. Walsh & Roy 1989), although there is no evidence for any such effect in other 'WR' galaxies. Thus, apart from NGC 5253, there is no evidence for any real scatter in the He,O relation; we fitted it with a line intermediate between those marked 0(A) and 0.5(A) in the right-hand panel of Fig. 11.3. The scatter in N/O remains and is in part due to temporary self-enrichment,

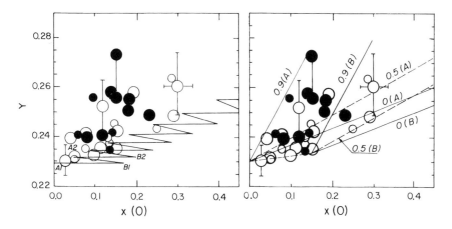

Fig. 11.3. Evolution of helium abundance relative to oxygen, adapted from Pilyugin (1993). x(O) is O/H in units of solar. *Left*, closed box model. *Right*, models with metal-enhanced winds with efficiencies (relative to star formation rate) marked. The upper and lower parallel lines in the diagram on the right are loci of corresponding points marked A, B in the one on the left, indicating abundances after and before dilution of self-enrichment, respectively. Observational data are from Pagel *et al.* (1992) and Skillman *et al.* (1994). Filled circles represent H II regions or galaxies with definite identification of a broad stellar Wolf–Rayet feature, and the two filled circles joined by a vertical line represent central and off-centre zones in NGC 5253. Larger circles represent data with a standard error of 0.012 or less in Y; error bars are shown for three cases. The linear regression adopted by Pagel *et al.* lies between the lines marked 0(A) and 0.5(A).

but galaxy-wide differences, e.g. between the Magellanic Clouds and NGC 6822 (see Fig. 8.8), are more likely due to galaxy-wide processes associated with time delays, as discussed by Edmunds & Pagel (1978) and Garnett (1990). There is no real evidence from abundance arguments that would enable us to choose between homogeneous and selective galactic winds.

Regardless of the details concerning self-enrichment and winds, the existence of isolated star formation bursts will also affect the iron–oxygen and iron–α relations, introducing scatter in Fe/O and possibly gaps in the iron abundance distribution function. When the interval between successive bursts exceeds the evolution time for SNIa (maybe about 1 Gyr), iron will build up in the ISM resulting in an enhanced Fe/O ratio in the second burst so that one can end up with [Fe/O] > 0 (Gilmore & Wyse 1991); cf. Fig. 8.6.

11.2 Chemical evolution of elliptical galaxies

11.2.1 Data sources

With the exception of nearby dwarf ellipticals and dwarf spheroidals for which some information can be obtained from magnitudes, colours and spectra of individual stars, all the information we have about abundances in E-galaxies comes from integrated colours and spectra of an entire stellar population which can be composite in both age and metallicity, while at the same time the number of independent parameters provided by observation is quite small (Faber 1973). Consequently, the interpretation of the data is not straightforward and some sort of modelling is needed. One technique is stellar population synthesis from a standard library of individual stellar types (e.g. Pickles 1985; Rose 1985) and another is evolutionary population synthesis on the basis of theoretical models of star formation and evolution. Such models in some cases assume a fixed metallicity (e.g. Guiderdoni & Rocca-Volmerange 1987; Bruzual & Charlot 1993) while in others they are computed for a grid of metallicities (Buzzoni 1990) or for a self-consistent model including chemical evolution (e.g. Yoshii & Arimoto 1987). These models also need a standard stellar library in order to tie in with observable quantities.

A vital role in the interpretation of integrated stellar properties is played by globular clusters, in which both integrated and individual stellar features can be measured and related to the (usually unique) age and metallicity deduced from the latter. One feature that seems to be particularly useful in this regard is the Mg_2 index (see Section 3.2.1.5). A combination of single-burst evolutionary synthesis models and empirical data from globular clusters gives the approximate calibration (in magnitude units)

$$Mg_2 \simeq 0.1 \left[\frac{Z}{Z_\odot} t(\text{Gyr}) \right]^{0.41} \tag{11.1}$$

(Bender, Burstein & Faber 1993), where $\log Z/Z_\odot$ is more or less equivalent to [Mg/H] (cf. Edmunds 1992).

A single feature like Mg_2 can only provide one parameter, e.g. some abundance equivalent if the entire stellar population is old. Other features such as hydrogen lines can be used to test whether this is so and whether different metals vary in constant ratios to one another or at least in some systematic mutual relation. Bica & Alloin (1986, 1987) have developed a very interesting system for building up composite galaxy spectra from a cluster base containing integrated spectra of both Galactic globular clusters (old, with differing metallicities) and Magellanic Cloud clusters (having various ages). In this way, a whole model of chemical evolution can in principle be built up by fitting a small number of cluster spectra and imposing a few theoretical constraints, such as that metallicity should not decrease with time (Bica 1988). A great advantage of this method is that the IMF and stellar evolution details are already built in to the clusters themselves. However, the number of really independent parameters that can be deduced with the currently available spectral range is still only 2 or 3 (Schmidt *et al.* 1991) and all globular cluster applications suffer from a lack of known clusters having an accurately determined metallicity $\geq$ solar; at these compositions (relevant to the nuclei of giant galaxies) risky extrapolations have to be used.

11.2.2 Systematics of elliptical galaxies and bulges

Parameters of 'dynamically hot galaxies', i.e. various classes of ellipticals and the bulges of spirals, generally lie close to a 'fundamental plane' in the 3-dimensional space of central velocity dispersion, effective surface brightness and effective radius or equivalent parameter combinations (Fig. 11.4). This is explained by a combination of three factors: the Virial Theorem, some approximation to homologous structure (e.g. de Vaucouleurs' surface brightness law) and a fixed mass:luminosity ratio at a given mass. Within this plane, however, there are distinct locations and sequences that are presumably related to various effects in galaxy formation and evolution (Bender, Burstein & Faber 1992).

Most points in Fig. 11.4 lie near one or other of two distinct sequences in the fundamental plane. One of these is a dwarf sequence in which surface brightness increases with luminosity (perhaps because of loss of gas from shallow potential wells and consequent dynamical adjustment) and there is also a nearly perpendicular 'gas-star' sequence where the opposite occurs, perhaps because of differing degrees of dissipation (depending on the gas fraction) in mergers of different sizes. Given this complexity, it is hardly surprising that there should be a large scatter around the statistical relation between the metallicities and luminosities as expressed, e.g., by the colour–magnitude relation or the Mg_2–magnitude relation (Terlevich *et al.* 1981; cf. also Fig. 3.37). Despite this, however, galaxies from all parts of the fundamental plane lie on a remarkably tight relation between Mg_2 and central velocity dispersion (Fig. 11.5).

Colour and line-strength gradients are also observed across elliptical galaxies, as is to be expected from dissipative effects (see Appendix 5). However, the detailed character varies widely from one galaxy to another and these differences presumably give clues to differing evolutionary histories, e.g. from mergers at differing stages of star formation (Gorgas, Efstathiou & Aragón Salamanca 1990). Franx & Illingworth (1990) found a universal relation between colour and local escape velocity in extranuclear as well as nuclear regions, but the interpretation of colours without line strength data is not too clear. A significant complication arises from the fact that not only metallic line strengths but also their ratios vary between elliptical galaxies, in the sense that the ratio Mg_2/Fe $\lambda5270$ increases with central velocity dispersion (Gorgas *et al.* 1990; Worthey, Faber & Gonzales 1992), i.e. the outer parts of giant ellipticals have a larger Mg/Fe ratio than the central parts of smaller ellipticals with the same Fe line strength (Fig. 11.6). This differential α-rich effect presumably reflects faster rates of star formation as well as a higher degree of enrichment in the bigger objects; cf. Fig. 8.6. The models that have been put forward to account for the higher degree of enrichment in larger galaxies can be classified as either wind or merger models.

11.2.3 Wind models

Effects of a homogeneous continuous galactic wind were discussed in Section 8.3.7 in connection with the evolution of the Galactic halo. Some kind of wind model

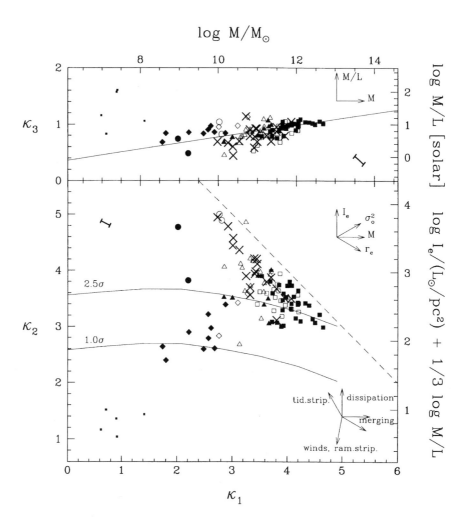

Fig. 11.4. Edge-on (above) and face-on (below) views of the Fundamental Plane in κ-space. κ_1 is essentially the total mass within the optical radius, κ_2 the effective surface brightness and κ_3 the mass:luminosity ratio, all in logarithmic units. Squares represent giant ellipticals, crosses spiral bulges, circles compact ellipticals, triangles intermediate-luminosity ellipticals, diamonds dwarf ellipticals and small dots dwarf spheroidals. Filled symbols represent galaxies with anisotropic velocity dispersion, open symbols nearly isotropic velocity systems supported by rotation. Arrows in the upper right-hand corners indicate the direction of variation of related parameters and those in the lower right corner indicate possible physical effects that could move galaxies around within the plane. The curves marked 2.5σ and 1.0σ represent hypothetical merger tracks from primordial density fluctuations with the corresponding amplitudes. Tick marks represent errors due to uncertainties in distance. After Bender, Burstein & Faber (1993), *Ap. J.*, **411**, 153. Courtesy Ralf Bender.

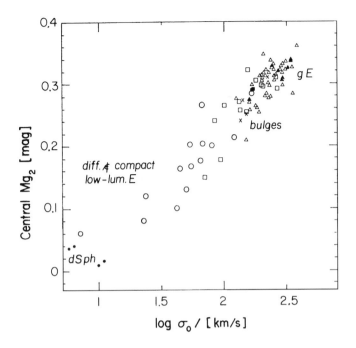

Fig. 11.5. Relation between central Mg$_2$ index and central velocity dispersion for dynamically hot galaxies. Coding as in Fig. 11.4. Adapted from Bender (1992).

for elliptical galaxies is favoured by the relation with velocity dispersion and by the presence of heavy elements in intra-cluster gas in clusters of galaxies, which was actually predicted (Larson & Dinerstein 1975) before it was discovered from X-ray observations. A number of models use the idea of a terminal wind caused by the gradual accumulation of hot gas from the interiors of supernova remnants which eventually overlap, making the whole ISM hot enough to reach escape velocity (Mathews & Baker 1971). The Salpeter IMF, Eq. (7.11), gives an idea of the energy involved, assuming that all stars above $10M_\odot$ explode as supernovae releasing 10^{51} erg. The number of supernovae per solar mass of stars formed is $\int_{10}^{100} \phi(m)dm = 5 \times 10^{-3}$ giving about 2.5×10^{15} erg/gm. This corresponds to a velocity of 700 km s^{-1}, well above the escape velocity from a large galaxy like our own, if the losses due to cooling are not too great; it is the details of these losses, and in particular the ratio of cooling time to dynamical or star-formation time, that fix whether, and if so when, the terminal wind occurs. The depth of the potential well which fixes the escape energy varies as M/R or roughly as $M^{1/2}$ (Fish 1964). (Probably only about a tenth of the supernova energy is available to drive the wind, but this is compensated by the return fraction of gas being of the same order.) The first such model to be worked out quantitatively is that of Larson (1974b), some of whose results are shown in Fig. 11.7. The condition for a wind is approached after some time (typically 2 Gyr), as the mass of gas and its volume density, which

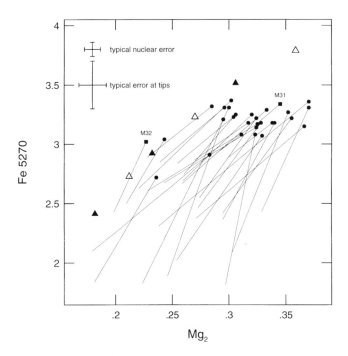

Fig. 11.6. Plot of an iron feature against Mg$_2$. Filled circles and squares represent the nuclear regions (central 5 arcsec) of elliptical galaxies, while the sloping lines show the mean trend with galactocentric distance in each one. Triangles show model predictions for ages of 9 (solid) and 18 Gyr (open), based on a single burst of star formation, which fit the features in globular clusters and assume [Mg/Fe] = 0. A young model with [Fe/H] = 0 fits the nucleus of M32 quite well, and the predicted trends with metallicity run roughly parallel to several of the observational lines, but the trend among nuclei is not fitted at all. After Worthey, Faber & Gonzalez (1992), *Ap. J.*, **398**, 69. Courtesy Guy Worthey.

affects the SNR cooling rate, diminish, and the amount of gas then expelled relative to the mass remaining in stars varies inversely as the depth of the potential well, i.e. as $M^{-2/3}$ according to the scaling law assumed by Larson. This model, which uses the instantaneous recycling approximation and a simple bulk yield for some combination of heavy elements, was based on Larson's (1974a) earlier theory of the formation of elliptical galaxies which predicted steep gradients in surface density and metallicity; these do not have time to develop when the gas fraction lost is 50 per cent or more, and the overall results are in good qualitative agreement with observational data.

A related analysis is the study of dwarf galaxies by Dekel & Silk (1986), who note the existence of two sequences of dwarf galaxies in the Fundamental Plane (Fig. 11.4) and base their model on the existence of dark-matter halos (cf. White & Rees 1978). Fig. 11.4 shows a dwarf spheroidal–dwarf elliptical sequence ascending at about 45° from the lower left corner to meet a normal-compact elliptical sequence that goes more or less at right angles. Dekel & Silk's suggestion is that there is a critical gravitational

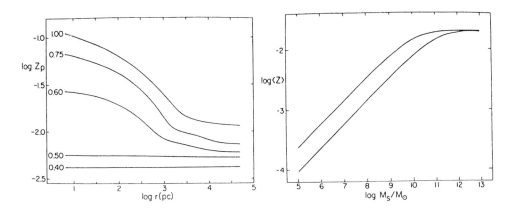

Fig. 11.7. Left panel: projected heavy-element abundance Z_p along lines of sight as a function of galactocentric distance, for differing values of the mass fraction in form of stars at the onset of a terminal galactic wind. Right panel: (Mass-weighted) mean abundance as a function of final total stellar mass, for two different assumptions as to the dependence of the amount of gas lost on the initial mass. The assumed bulk yield is 0.02 and the trend along the linear part of the curves is approximately $Z \propto M^{3/8}$. Adapted from Larson (1974b).

potential, or virial velocity, below which a gaseous proto-galaxy forming stars can lose the bulk of its gas in the first burst of star formation, whereas for deeper potential wells the situation more closely resembles Larson's analysis. Dekel & Silk assume a number of scaling laws. A mass g_i of gas settles into the potential well of a dark spherical halo with a much larger mass $M \propto g_i$, radius R and virial velocity V satisfying a scaling law

$$M \propto R^{r_m} \propto g_i. \tag{11.2}$$

M, R and V are unaltered by the mass loss. The visible matter is assumed to satisfy the scaling laws

$$L \propto R^{r_l} \propto V^v \propto Z^z, \tag{11.3}$$

where L is the luminosity and the second proportionality is a generalization of the the Faber–Jackson (1976) relation. It follows from Eqs. (11.2), (11.3) and the Virial Theorem that

$$V^2 \propto M/R \propto R^{r_m-1} \propto L^{(r_m-1)/r_l} \tag{11.4}$$

so that

$$v = 2r_l/(r_m - 1), \tag{11.5}$$

which is one relation between the 4 unknowns v, r_l, r_m, z. Supposing now that a mass s of stars has formed, this fixes both the mass:luminosity ratio

$$M/L \propto M/s \tag{11.6}$$

and the stellar heavy-element abundance (Eq. 8.7 with $s/g \simeq 1 - \mu \ll 1$)

$$Z \propto 1 - \mu = s/g_i \propto s/M \propto L/M \propto R^{r_l - r_m} \propto L^{1 - r_m/r_l} = L^{1/z}. \tag{11.7}$$

This leads to a second relation between the 4 unknowns:

$$z = r_l/(r_l - r_m). \tag{11.8}$$

A third relation comes from the condition that the number of supernovae, proportional to s or L, should have been sufficient to supply the escape energy to the bulk of the gas mass g_i, itself assumed proportional to M, i.e.

$$L \propto MV^2, \tag{11.9}$$

which leads to

$$v = 2z. \tag{11.10}$$

The equations are closed by appealing to the observational result that $r_l = 4$ for these diffuse systems, whence

$$r_m = 5/2; \quad v = 16/3; \quad z = 8/3. \tag{11.11}$$

Thus $M/L \propto L^{-3/8}$ or $M^{-3/5}$, increasing towards low luminosities (cf. the position of dwarf spheroidals in the top panel of Fig. 11.4), and $Z \propto L^{3/8}$, $M^{3/5}$ or V^2. The latter relation compares reasonably well with the lower part of Fig. 11.5, calibrated with Eq. (11.1), and the first one compares very well with Fig. 8.10. The above assumptions are expected to apply only under certain conditions, fixed by supernova rates, their efficiency in transferring energy to the ISM, the cooling rates and the dynamical time; in Dekel & Silk's calculation, by far the most important parameter turns out to be the virial velocity, other parameters like the gas density more or less cancelling out. Thus their analysis applies below a certain critical velocity that is found by them to be very close to 100 km s^{-1} (line-of-sight velocity dispersion 60 km s^{-1}). Above this velocity, a 'normal' galaxy is formed, which manages to convert most of its gas into stars.

Dekel & Silk argue that the alternative hypothesis of a self-gravitating cloud that becomes more diffuse as a result of losing most of its mass leads to a significantly flatter dependence of Z on L, $Z \propto L^{1/4}$, so that the observed relations are evidence in support of the dominance of cold dark matter halos in these systems. However, this argument relies heavily on the $L \propto R^4$ law, and some dynamical adjustment could be a natural explanation for their diffuseness, the alternative being that they have failed to undergo dissipative episodes experienced by other galaxies.

Two reservations need to be expressed about all of these analytical models:

(1) The 'observed' metallicity from integrated light is not mass-weighted, as assumed in Eq. (8.7), but luminosity-weighted. This generally will lead to the mass-weighted metallicity being underestimated, because metal-weak giants are more luminous.

(2) The (mass-weighted) average abundance according to Eq. (8.7) cannot exceed the true yield, either in a closed model or in one with inflowing unprocessed material. A mass-weighted average $Z > Z_\odot$ in giant elliptical galaxies, even if it applies to α-elements alone, requires a higher true yield than is believed to hold in the solar neighbourhood (cf. Section 8.3.1), if it is representative of the galaxy as a whole. If it is confined to the nucleus, on the other hand, then it can be accounted for by inflow of already processed material into the nuclear regions as envisaged in Larson's models (Fig. 11.7) and explained in Appendix 5. Such dissipative effects may be revealed by metallicity gradients in certain ellipticals that show a marked steepening towards the centre, suggesting effects of a merger (Bender 1992).

Detailed numerical models for the photometric and chemical evolution of elliptical galaxies have been computed by Yoshii & Arimoto (1987), Matteucci & Tornambé (1987), Brocato *et al.* (1990) and Matteucci (1992, 1994). These are terminal wind models similar in their basic assumptions to that of Larson (1974b), but with differing assumptions as to IMFs and scaling laws that specify how the rate of star formation and the depth of the potential well formed by an initial dissipational collapse depend on the mass of the galaxy. The star formation efficiency parameter ω is generally assumed to be lower in the more massive galaxies, corresponding to longer free-fall times $\propto (M/R^3)^{-1/2}$ with $R \propto M^{1/2}$, but various models differ in that Yoshii & Arimoto assumed a rather flat IMF to get the high yield, computing only one abundance parameter [Fe/H], neglecting SNIa and arriving at rather high mass:luminosity ratios because of a preponderance of white dwarfs, whereas Matteucci and colleagues assume a Salpeter IMF, include SNIa and predict abundances of several elements. Both sets of models give a fair account of mass:light ratios (Yoshii & Arimoto marginally so) and colour–luminosity relations, and Yoshii & Arimoto can also reproduce something resembling the dwarf sequence in the fundamental plane from dynamical effects of mass loss without any dark halos; Matteucci (1992) suggests that, while such halos probably exist, they could be so diffuse as to have but little effect on chemical evolution. A serious difficulty encountered by Matteucci and her colleagues (and probably also by Yoshii & Arimoto, though to a lesser degree) is that the predicted magnitude of Mg/Fe and its trend with luminosity are in flagrant disagreement with the observational effects discussed above (Matteucci 1994). Matteucci suggests, as had already been done by Worthey, Faber & Gonzalez (1992), that either star formation efficiency increases or the IMF becomes flatter with increasing galaxy mass, and considers certain other possibilities; among these it seems to me that the real solution lies in violent star formation provoked by tidal effects or mergers and not described by simple linear (or other power) laws in the gas mass. A further problem with these terminal wind models is that they predict an abundance distribution function similar to that of the Simple model, truncated at metallicities that increase with the depth of the potential well. However, the application of synthetic photometric models to observed galaxy spectra suggests the existence of a 'G-dwarf problem' (Bressan, Chiosi & Fagotto 1994); this

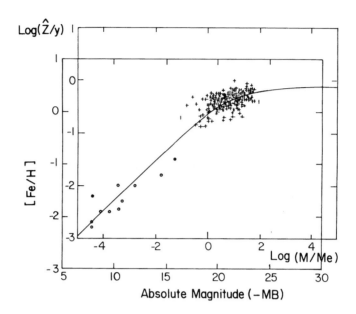

Fig. 11.8. Median metal abundance per unit true yield as a function of final mass with observations of dwarf spheroidal and elliptical galaxies superposed. The offsets of axes determine the characteristic mass M_e that constrains supernova debris and the true yield y (p in the text). *MB* is the corresponding blue absolute magnitude (neglecting any dark halos). The trend along the linear branch of the curve is for metallicity to increase approximately as $M^{1/2}$. After Lynden-Bell (1992).

was already taken care of (at least as far as the bright central regions of elliptical galaxies and bulges of spirals are concerned) in Larson's models, but the more recent numerical models cited above do not include internal gas flows despite being more refined in other ways.

Alternative wind models assume a selective outflow of enriched material, either taking place at the time of star formation bursts in dwarf galaxies (Matteucci & Tosi 1985) or taking place continuously from the hot component of the ISM which is also assumed to be metal-enriched (Vader 1987). An analytical model by Lynden-Bell (1992) supposes that, of the return fraction $1 - \alpha$ from a generation of stars, a fraction $1 - f$ escapes, leading via Eq. (7.33) to

$$\frac{d}{ds}(gZ) = pf - Z[1 - f(1 - \alpha)]/\alpha. \tag{11.12}$$

Based on the scaling law $R \propto M^{1/2}$, i.e. $M/R \propto M^{1/2}$, Lynden-Bell suggests for f the *Ansatz*

$$f = ks^{1/2}/(V^2 + ks^{1/2}), \tag{11.13}$$

where V is a characteristic virial velocity, and assumes inflow of unprocessed material

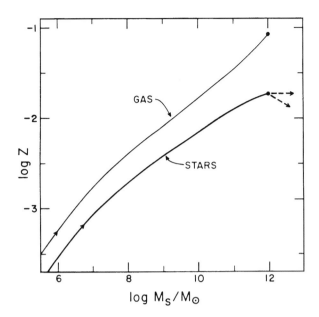

Fig. 11.9. Metallicities of stars and gas as a function of the total mass of stars in an elliptical galaxy growing by mergers, assuming a true yield of 0.02. The trend is for stellar Z to increase approximately as $M^{1/2}$ for small masses, flattening to $M^{1/4}$ for larger ones. Filled circles show the point beyond which there will be little star formation in mergers because the gas cannot cool sufficiently between collisions; arrows indicate possible outcomes of further mergers without star formation. After B.M. Tinsley & R.B. Larson (1979), *MNRAS*, **186**, 503.

following a similar law as in Lynden-Bell (1975) (cf. Section 8.5.2, but in this case the initial mass is zero), which takes care of any 'G-dwarf problem'. The analytical solution is somewhat involved and will be passed over here, but Fig. 11.8 shows the results compared with some observational data. It is notable that the true yield required is rather high, about $5Z_\odot$, because in early stages of the model most of the metals escape even from the (ultimately) largest galaxies to enrich the intergalactic medium. The true metallicity of those largest galaxies, and how representative the observed values are of the galaxy as a whole, are uncertain, however, and the results depend in any case on which element is used to define 'metallicity'. We return to the yield question in Section 11.2.5 below.

11.2.4 Merger models

Luminous star formation bursts, e.g. as observed by IRAS, are associated with mergers of gas-rich systems, and there are cogent arguments suggesting that at least a large subset of elliptical galaxies could have been formed in this way (Toomre 1977; Kormendy & Sanders 1992; Barnes 1995), perhaps following mergers of dark-matter

Table 11.1. *Properties of a sequence of mergers with star formation*

n	$\log M_n/M_\odot$	$\log s_n/M_\odot$	$\log z_g$	$\log z_s$	$\log R$ (pc)	Min. $\log t_{dyn}$ (yr)	Max. $\log t_{cool}$ (yr)
0	7.08	$-\infty$	$-\infty$	$-\infty$	–	–	–
2	7.68	6.33	-1.35	-1.65	(3.2)	(8.2)	–
6	8.89	8.20	-0.63	-0.95	(3.6)	(8.2)	–
10	10.09	9.80	-0.14	-0.50	4.0	8.1	4.7
12	10.69	10.54	0.09	-0.31	4.2	8.2	5.7
16	11.898	11.888	0.58	-0.04	4.8	8.4	8.1
17	12.199	12.197	0.73	-0.01	4.9	8.5	9.1

Data from Tinsley & Larson (1979).

halos (White & Rees 1978). The collision of large gas clouds leads to star formation, energy dissipation and possibly eventual expulsion of gas by a supernova-driven galactic wind, transforming gas-rich metal-poor systems with low phase-space density into gas-poor metal-rich systems of high surface brightness (cf. the 'merging' and 'dissipation' arrows in Fig. 11.4).

Tinsley & Larson (1979) developed a simple chemical evolution model along these lines. They assume that ordinary matter is initially clumped into gas clouds of equal mass M_0, so that a galaxy is built up by a hierarchy of binary mergers leading to an increase in mass by a factor of 2 at each stage. A crucial assumption is that the efficiency of star formation, i.e. the fraction of gas that is converted into stars at each stage, is an increasing function of the total mass, varying as $(M/M_e)^p$, say. This leads to the recurrence relations for the nth merger

$$s_n - 2s_{n-1} = 2g_{n-1}(M_n/M_e)^p = 2^{np+1}g_{n-1}(M_0/M_e)^p, \tag{11.14}$$

and

$$g_n = 2g_{n-1}[1 - 2^{np}(M_0/M_e)^p]. \tag{11.15}$$

The initial values are $g_0 = M_0$, $s_0 = 0$, $\mu_0 = 1$. Using the Simple model equations (8.6), (8.7) and assuming numerical values $p = 1/3$, $M_0 = 1.2 \times 10^7 M_\odot$, $M_e = 3.2 \times 10^{12} M_\odot$, Tinsley & Larson obtain numerical values some of which are shown in Table 11.1. The abundance as a function of final mass is shown in Fig. 11.9. Tinsley & Larson note that collisions between galaxies tend to shock-heat the gas to approximately the virial temperature, which increases with total mass according to the Faber–Jackson relation, so that eventually the gas may become too hot to be able to cool enough to form stars before the next collision. The table accordingly gives some estimates of cooling times and dynamical times, based on an empirical mass–radius relation, and it turns out that there is a critical mass, estimated to be in the region of $10^{12} M_\odot$, above which the

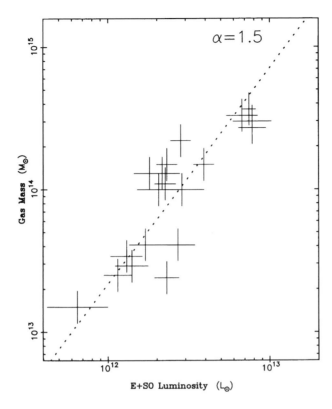

Fig. 11.10. Gas mass in clusters of galaxies plotted against the total luminosity of elliptical and lenticular (S0) galaxies in the cluster. The 1σ error in the slope α is ± 0.25. After Arnaud *et al.* (1992), *A. & A.*, **254**, 49. Courtesy Monique Arnaud.

cooling rate becomes too small to allow further star formation. Subsequent mergers with equal-sized objects leave the metallicity unaltered (horizontal arrow in Fig. 11.9), while mergers with smaller objects will lead to a reduction (downward-sloping arrow). This could perhaps be the process going on in the central-dominant (cD) galaxies in large clusters that are swallowing their neighbours. Once mergers have ceased at any stage, or possibly before, the galaxy can evolve to the point where its residual gas is driven out in a terminal wind, as discussed in the previous section, or it can become the central bulge of a spiral by accreting diffuse material that settles into a plane before itself forming stars.

11.2.5 Metal supply to the intra-cluster medium

An additional constraint on the chemical evolution of elliptical galaxies comes from the supply of metals to the intra-cluster gas. This has been discussed by Arnaud *et al.* (1992), who find a good correlation between the total mass g of gas in the cluster and the total luminosity L_V of E and S0 galaxies within the cluster (see Fig. 11.10) in the range

$$30 \leq \frac{g}{L_V} \leq 40 \tag{11.16}$$

which implies

$$3 \left(\frac{10}{M/L} \right) \leq \frac{g}{s} \leq 4 \left(\frac{10}{M/L} \right) \tag{11.17}$$

where M/L is the visual mass:luminosity ratio (in solar units) of the total stellar population of mass s. Such a large ratio of gaseous to stellar mass (given a return fraction probably no greater than 0.5) implies that at least 2/3 of the intra-cluster gas is primordial. The total iron mass in the gas, $g\,Z_{Fe}$, is found to be proportional to L_V, i.e. to s.

Thus the typical value [Fe/H] ~ -0.4 found in the intra-cluster gas, while not looking particularly remarkable in itself, becomes so when the large mass of gas is taken into account, since we now have

$$p_{Fe} \simeq \frac{(g+s)Z_{Fe}}{s} \geq 1.6\,Z_{Fe}^\odot \tag{11.18}$$

for $(M/L) \leq 10$, requiring a high efficiency of iron production.[1] One suggestion has been that the bulk of the iron comes from SNIa which have exploded over the years and given rise to iron-rich winds (Ciotti *et al.* 1991); this would lead one to expect a high Fe/α ratio in the gas and a low one in the stars (Renzini *et al.* 1993). The latter effect is in good agreement with the Fe/Mg observations of the stellar populations, but the former in disagreement with data concerning the gas, where the ratio of oxygen and α-elements to iron is typically about twice solar (Mushotzky *et al.* 1996; Arnaud 1996) and thus bears a closer resemblance to the output from SN II. An alternative hypothesis (Arnaud *et al.* 1992) supposes a bimodal model of star formation similar to that of Larson (1986), which assumes a top-heavy IMF in the early star-bursting stages of the formation of E/S0 galaxies; this predicts large α/Fe ratios for both stars and gas and is therefore in better agreement with existing estimates based on X-ray observations, although there could be some contribution from SNIa as well (Mihara & Takahara 1994; Reisenegger 1996). Assuming only massive stars to be relevant, for which the iron yield in the solar neighbourhood is only about $0.3Z_{Fe}^\odot$ (Table 8.1), it is evident from Eq. (11.18) that one indeed needs a yield in elliptical galaxies enhanced by a factor of 5 or so, as envisaged by Yoshii & Arimoto (1987), Lynden-Bell (1992) and others; if this is associated with a flat IMF slope, this could help to explain the Mg/Fe trends.

---

[1] As pointed out by Mihara & Takahara (1994), the ratio g/s is proportional to the assumed distance of the cluster, since the distance affects the mass:light ratio deduced from observations. Arnaud *et al.* assumed a Hubble constant $h = 0.5$, which is probably sufficiently good (within 30 per cent) given other uncertainties.

Problem

1. Assuming a continuous and homogeneous galactic wind with a mass flux $\eta \alpha \psi$ ($\eta = $ const.), combined with a selective (metal-enhanced) wind $\eta^* \times$ the mass of Type II supernova ejecta ($\eta^* = $ const. < 1), show that Eq. (8.6) for the Simple model is replaced by

$$\frac{Z}{\ln 1/g} = \frac{p(1 - \eta^*)}{1 + \eta + \eta^* R^*/\alpha} \simeq \frac{p(1 - \eta^*)}{1 + \eta}, \tag{11.19}$$

where R^* is the contribution of stars above $10 M_\odot$ (winds and SN ejecta) to the return fraction. (An upper limit to R^* can be found from Eq. 7.11); g is the mass of gas remaining in units of the initial mass of the system.

12 Cosmic chemical evolution and diffuse background radiation

12.1 Introduction

Observations of distant objects, notably high red-shift absorption line systems on the line of sight to quasars, give some information on chemical evolution at epochs not too far from when the first stars and most galaxies were presumably formed. Other information comes from two related effects:

- The overall abundance of helium and heavy elements in the universe today. This reflects the total effect of fuel consumption and nucleosynthesis by all the stars that ever existed. Roughly speaking, one may consider this in two ways: (i) nucleosynthesis of 'Z' elements by massive stars in starbursts; these stars emit copious ultra-violet radiation which is subsequently red-shifted and may also be reprocessed into the far infra-red by dust. Alternatively, (ii), one may consider all the products of nuclear burning in stars including the non-cosmological helium ΔY and C, O etc. locked up in stellar interiors and remnants, mainly white dwarfs.

- Diffuse background radiation from the night sky (other than the cosmological microwave background). This is another global manifestation of past stellar activity which is directly related to the resulting abundances. It is quite difficult to measure, owing to the existence of strong foregrounds from Galactic starlight, zodiacal light and far infra-red emission from dust clouds in the Galaxy, known as Galactic cirrus. Consequently there are still only upper limits at most wave-lengths. Lower limits are provided by galaxy counts at the faintest attainable magnitudes. An overview of current limits and some models based on various assumptions is shown in Fig. 12.1, where it appears that typical values of νI_ν or λI_λ (the energy in a relative frequency band of 1 in the natural logarithm) are of the order of 10^{-12} wt cm^{-2} sterad^{-1} or less. This is about 1 per cent of the peak of the cosmic microwave background, which is shown by the parabola-like curve at the left end of the diagram. Background radiation is attenuated by red-shift, especially at the shorter wave-lengths, so that most of it comes from relatively nearby objects unless there has been extreme luminosity evolution, whereas the products of fuel consumption and nucleosynthesis remain whatever the red-shift at which they were formed. There is, however, an exception to the former

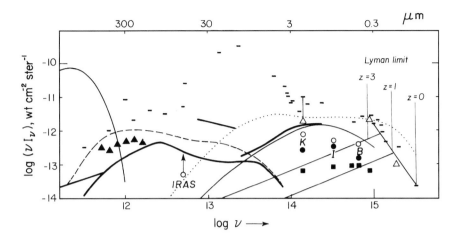

Fig. 12.1. Background radiation from the night sky above the Earth's atmosphere. Short horizontal dashes represent upper limits from various sources (chiefly the COBE satellite, launched in 1989) taken from the literature by Pagel (1994b). Three sloping lines are, in order of increasing frequency, from Mather *et al.* (1994) based on the accurate black-body spectrum of the microwave background shown by the parabola-like curve; from Dwek & Slavin (1994) based on the non-detection of attenuation of high-energy γ-rays from the AGN Markarian 231; and a tentative locus of ultra-violet upper limits based on the work of Martin & Bowyer (1989). The two parallel lines sloping upward from $\log \nu = 13.2$ and 13.9 result from nucleosynthesis arguments applied to starburst galaxies by Songaila, Cowie & Lilly (1991). Triangles represent tentative detections (the filled triangles come from Puget *et al.* 1996) and the open circles show lower limits derived from galaxy counts. The various curves are based on models discussed in the text.

statement in the case of sub-mm radiation from dust, because in this case the black-body peak is actually shifted into the observable region (Franceschini *et al.* 1991; Blain & Longair 1993). Conversely, the onset of hydrogen absorption shortward of the Lyman limit ensures that there will be virtually no contribution to blue light from red-shifts much greater than 3.

12.2 Luminosity evolution and the diffuse background

The pioneering investigation in this field was that by Partridge & Peebles (1967; hereinafter PP), who computed a number of evolutionary synthesis models for a 'typical' galaxy, assumed similar to our own. These were based on the present-day luminosity density of the universe (cf. Eq. 4.59) and the relation

$$I(\nu_0, t_0) = \frac{c}{4\pi} l(\nu_0, t_0) \int_0^{t_0} \frac{L(\nu(t), t)}{L(\nu_0, t_0)} \, dt, \tag{12.1}$$

where $l(\nu_0, t_0)$ is the luminosity density, per unit frequency interval, at frequency ν_0 at the present time t_0, and the ratio under the integral sign represents a combination of luminosity evolution and red-shift. Replacing $L_{B\odot}$ in Eq. (4.59) with $\nu L_{\nu\odot}(4400 \text{ Å}) = 2 \times 10^{33}$ erg s$^{-1} \simeq L_\odot/2$, we have

$$\nu_0 \, l(\nu_0, t_0) \simeq 1.4 \times 10^{-32} h \ \text{erg s}^{-1} \ \text{cm}^{-3}. \tag{12.2}$$

To get a very rough idea of the resulting background intensity, one can simply assume the factor under the integral sign in Eq. (12.1) to have been constant (i.e. 1) for $5h^{-1}$ Gyr (since larger ages would red-shift the faint UV region into the blue photometric band and therefore contribute little). In this way one derives $\nu I_\nu \simeq 5 \times 10^{-13}$ wt cm^{-2} sterad^{-1}, which is comparable to the lower limit derived from galaxy counts in the B band (Fig. 12.1).

Partridge & Peebles computed several models making different assumptions about the star formation history of a typical galaxy. Their Model 1 assumes a constant star formation rate, corresponding to a nearly constant luminosity, and leads to the spectrum shown as a continuous curve stretching between 0.3 and 30 μm in Fig. 12.1, which falls nicely between the lower and upper limits. They, however, preferred other models with more drastic evolution because they felt a need to explain all of the helium in the universe as a result of stellar fuel consumption[1]; in addition, they assumed a cosmic density of stellar matter that is about an order of magnitude greater than some modern estimates (see Table 12.3). The dotted curve in the right half of Fig. 12.1 shows the background spectrum from their Model 5, which led them to predict that newly formed galaxies would be highly luminous and possibly detectable in the wave-length region between 5 and 15 μm.

Since then, there have been substantial advances in numerical modelling of galaxy evolution, including developments in stellar evolution theory, distinguishing between different sorts of galaxies and the inclusion of chemical evolution effects and radiation from AGNs and dust. These have been developed partly in the (so far largely vain) hope of deducing basic cosmological parameters from counts and red-shift surveys of galaxies. The background spectrum has been computed by, among others, Yoshii & Takahara (1988) and Franceschini *et al.* (1991) with similar outcomes to that from PP Model 1; the model spectrum by Franceschini *et al.*, which includes emission at around 10 μm from warm dust surrounding stars and AGNs, is shown in Fig. 12.1 as a bold curved segment between wave-lengths of 1.5 and 20 μm.

The IRAS satellite mission, launched in 1983, provided a more complete survey of disk and active galaxies containing dust than had been possible from optical observations. Follow-up measurements of red-shifts and other properties have led to significant results for cosmology and the discovery of many luminous star-forming and active galaxies enshrouded by dust. A typical spiral galaxy like our own radiates something like 25 per cent of its energy at wave-lengths between 10 and several hundred μm, coming from cool 'cirrus' clouds.

---

[1] Although Peebles had himself recently completed a calculation of primordial element synthesis including helium (Peebles 1966).

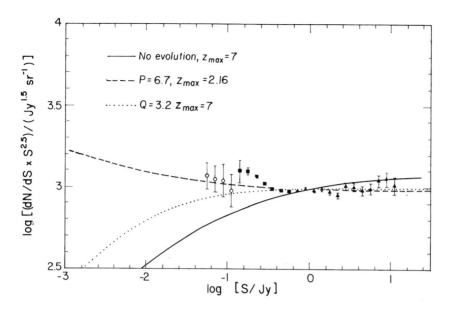

Fig. 12.2. IRAS differential source counts as a function of observed flux at 60 μm, normalized to a uniform population in Euclidean space. After S.J. Oliver, M. Rowan-Robinson & W. Saunders (1992), *MNRAS*, **256**, 15P.

An interesting feature of the IRAS galaxies is the evidence from red-shift surveys and from source counts as a function of observed flux that the population has undergone evolution (Fig. 12.2). This result is analogous to similar evidence from source counts of radio galaxies and quasars, as well as quasar red-shifts, and a correlation that has been observed between radio and infra-red luminosity suggests that the evolution could be similar in both cases. Typical simple models for such evolution include luminosity evolution according to

$$\frac{L(t)}{L(t_0)} = e^{QH_0(t_0-t)}; \quad z \leq z_{\max}, \tag{12.3}$$

where Q is some dimensionless number, or

$$\frac{L(t)}{L(t_0)} = (1+z)^q; \quad z \leq z_{\max}, \tag{12.4}$$

where z is the red-shift; and number-density evolution according to

$$\frac{\Phi(t)}{\Phi(t_0)} = (1+z)^P; \quad z \leq z_{\max}, \tag{12.5}$$

which could represent the effect of mergers. Some 'P' and 'Q' models (assuming an Einstein–de Sitter cosmology) are compared with the IRAS 60μm source counts in Fig. 12.2. Oliver, Rowan-Robinson and Saunders (1992; hereinafter ORS) calculated the diffuse background resulting from all three kinds of evolution, assuming the presence of two components, normal cirrus from galactic disks (the major contributor at most wave-lengths) and a warmer starburst component dominating between 60 and 120 μm, together with different values for the parameters. The bold curve in Fig. 12.1 shows the result for a 'Q' model with $Q = 3.2$, $H_0 = 50$ km s^{-1} Mpc^{-1} ($h = 0.5$), $z_{max} = 6.94$, while the higher broken-line curve represents a 'q' model with $q = 3.15$ (a number that has been adopted for luminous quasars), $z_{max} = 5.31$. The latter is seen to violate constraints more recently derived from the accurate black-body nature of the cosmic microwave background (the upward sloping line at the left end of Fig. 12.1), although there is less discrepancy with the tentative detection by Puget *et al.* shown by the black triangles.[1] The 'P' (number evolution) model shown in Fig. 12.2 was rejected by ORS themselves because the resulting diffuse background was too high. They concluded that, if number density evolution applies, then it must be cut off at $z_{max} < 2$, which could indicate a more recent peak of activity than for quasars and radio galaxies. In the optical and near infra-red (1μm), results from the Canada–France red-shift survey have been fitted to 'q' models with slightly smaller values of q between about 2 and 3 up to $z = 1$ and interpreted as resulting from a power-law decline in the star formation rate as $t^{-2.5}$ after a turn-on time of 2 to 3 Gyr (Lilly *et al.* 1996). The next section discusses some implications of the earlier models with regard to nuclear fuel consumption.

12.3   Luminosity evolution and nuclear fuel consumption

Some elementary considerations of the relationship between spent fuel in the Galaxy (including the amount in stellar interiors and white dwarfs) and its past luminosity evolution have been given in Section 7.3.4.

Using equations (7.17) to (7.20), we can estimate the fuel consumption as a function of the e-folding time-scale for the star-formation rate and thus check the corresponding implications of some of the evolutionary models discussed above (see Table 12.1, which is based on an age t_0 of 12 Gyr; L_0 is the present-day luminosity $L(t_0)$). In elliptical galaxies, the mass:light ratio could be considerably larger than the value of 1 gm/erg s^{-1} applying to the solar neighbourhood, and $<Z + \Delta Y>$ could also be higher, both as consequences of the presence of a greater number of white dwarfs. The top line of Table 12.1 implies, for example, a mass:light ratio of 12$\times$ solar for $<Z + \Delta Y> = 0.5$; however, our exponential formula is not an accurate representation of the model by Franceschini *et al.* for elliptical and lenticular (E/S0) galaxies, which is not severely constrained by these estimates. The table does, however, give a fairly good approximation to their model for disk galaxies, which is seen to give a very reasonable result.

The 'Q'-model of ORS corresponds to an e-folding time-scale of 6 Gyr and also leads to a reasonable result for a disk galaxy like our own. (E/S0 galaxies make only a minor contribution to the IRAS survey.) However, the 'q'-models, depending on a

---

[1] At the time of writing, there is a contradiction between the black triangles (detections) and the line (upper limits) around 0.5 mm wave-length which remains to be resolved.

Table 12.1. *Fuel consumption in models with* $L \propto e^{-\lambda t}$

$\left(\frac{M}{L_0}\right)_{cgs}$	$<Z + \Delta Y>$	λ^{-1} Gyr	$L_{init.}/L_0$
3		2.1	300^a
1		2.8	70
0.5		3.5	30
0.2		6	7.4^b
0.13		8.7	4.0^c
0.1		12	2.7
0.06		∞	1.0

^{a} Similar to Franceschini *et al.*'s model for E/S0 galaxies.
^{b} Similar to ORS's Q-model for IRAS galaxies.
^{c} Similar to Franceschini *et al.*'s model for disk galaxies.

Table 12.2. *Fuel consumption in models with* $L \propto (1+z)^{3.15}$ *up to* z_{max}

	$z_{max} = 2.16$	$z_{max} = 5.31$
$\left(\frac{M}{L_0}\right)_{cgs} \quad <Z + \Delta Y>$	0.31	1.09

power of $(1 + z)$, are more constrained. In this case, the fuel consumption is given by

$$<Z + \Delta Y> = 0.005 \frac{L_0}{M} \int_{z_{max}}^{0} (1 + z)^q \frac{dt}{dz} \, dz. \tag{12.6}$$

In Einstein–de Sitter cosmology, $t/t_0 = (1 + z)^{-3/2}$, so that Eq. (12.6) reduces to

$$<Z + \Delta Y> = 0.0075 \, t_0 \frac{L_0}{M} \left(\frac{(1 + z_{max})^{q-3/2} - 1}{q - 3/2} \right), \tag{12.7}$$

when t_0 is measured in Gyr.

Table 12.2 shows the results for two models by ORS; the diffuse background spectrum from the one with $z_{max} = 5.31$ is illustrated by a broken-line curve in Fig. 12.1. It is clear that the model with $z_{max} = 2.16$ gives a reasonable result for $(M/L_0)_{cgs} = 2$, say (i.e. 4 solar units), whereas larger values of z_{max} lead to excessive fuel consumption requirements, as well as violating recent background radiation limits in the sub-mm wave-length region. These arguments basically apply to disk galaxies; there could be a loophole in the case of starburst galaxies if, as has been suggested, they are newly forming ellipticals (cf. Kormendy & Sanders 1992).

Table 12.3. *Cosmic densities*

	$\frac{<\rho_0>}{10^{-32}}$ gm cm^{-3}	$<Z + \Delta Y>$	$(<\rho_0>)(<Z + \Delta Y>)$ 10^{-32} gm cm^{-3} $h = 0.5$	$h = 0.75$
E/S0 galaxies	$3.0h^2$	0.5	0.4	0.8
Spiral galaxies	$1.4h^2$	0.2	0.07	0.16
X-r gas in rich clusters	$0.71h^{0.7}$	0.2	0.09	0.12
Gas in poor cl. & groups	$0.26h^{0.7}$	0.2	0.03	0.04
Total	$5.2h^{1.6}$		0.6	1.1
ρ_b	18 to 30			
ρ_{crit}	$2000h^2$			

12.4   Abundance of processed material in the universe

Considerations of the smoothed-out density of processed material in the universe, which are inevitably rather rough, lead to somewhat similar conclusions to those that were reached in the foregoing. Because the energy of photons emitted at a red-shift z is degraded by a factor $(1 + z)$ by the time they are received, and their original energy is $0.007c^2$ per gm of material processed, there is a relation between the total intensity of diffuse background radiation emitted at a red-shift z and the average density of processed material:

$$\int_0^\infty I_v dv = \frac{0.007 <\rho_0(Z + \Delta Y)> c^3}{4\pi(1 + z)}, \tag{12.8}$$

where ρ_0 is the present-day density of gas and stars or their remnants, assuming dark baryonic matter to have primordial composition only. According to the model by Franceschini *et al.*, typical values of z contributing to the diffuse background vary quite strongly with wave-length, depending on whether the red-shift leads to a decrease or increase in intensity, but a typical value in the near infra-red where most of the energy resides is of the order of 2. Eq. (12.8) can then be re-written as

$$2.3 \int_{-\infty}^\infty v I_v d \log v = 5 \times 10^{-12} < \frac{3}{1 + z} > \frac{<\rho_0(Z + \Delta Y)>}{10^{-32} \text{ gm cm}^{-3}} \text{ wt cm}^{-2} \text{ sterad}^{-1} \tag{12.9}$$

and it will be assumed that $<3/(1 + z)> \simeq 1$.

The density of processed material can be estimated by taking the density of gas and stellar material deduced from the measured luminosity density, the luminosity functions of different sorts of galaxies and their expected mass:light ratios (Persic & Salucci 1992) and combining these with reasonable guesses as to their content of processed material. Estimates obtained in this way are given in Table 12.3; they correspond to an assumed overall mass:blue-light ratio of $4h^{0.6}$ solar units. It must be noted that the total density can have been underestimated by a factor of a few if there is a significant contribution from dwarf and low surface-brightness galaxies that have gone unnoticed in red-shift surveys (Bristow & Phillipps 1994), but the amount of processed material in such systems is likely to be relatively small.

From Table 12.3 it appears that $<\rho_0(Z+\Delta Y)>$ is close to 10^{-32} gm cm^{-3} ($\Omega_{\text{processed}} = 8 \times 10^{-4}h^{-0.5}$). The corresponding background light intensity integrated over all wavelengths is then between 3 and 5.5×10^{-12} $<3/(1+z)>$ wt cm^{-2} sterad^{-1}, from Eq. (12.9). Integration over the diffuse background spectra predicted in some of the models discussed above leads to the following numbers (in units of 10^{-12} wt cm^{-2} sterad^{-1}):

PP (1967) Model 1:	3.0
" " Model 5:	11
ORS (1992) 'Q'-model:	1.0
ORS 'q'-model, $z_{\text{max}} = 6.9$:	2.6

All of these models give acceptable values except for PP (1967) Model 5, which is not surprising given the large amount of processed material that that model was designed to produce. It does happen to fit one tentative detection in blue light shown as a triangle in Fig. 12.1. That detection was based on an interesting experiment with a dust cloud at a high Galactic latitude, which obscures the diffuse background but reflects light from the Milky Way (Mattila, Leinert & Schnur 1991). Confirmation or otherwise of that data point would obviously be of considerable interest.

12.5 **Starbursts and 'metal' production**

Blue and UV light from stellar populations is dominated by young, massive stars which also contribute the lion's share of the Z-elements ('metals'). Evolutionary synthesis computations show that between the Lyman limit (912 Å) and 2000 Å or so, the spectrum expressed as L_ν is more or less flat (Fig. 12.3). This leads to some rather simple considerations because I_ν retains its value (and flatness) when the light is red-shifted (cf. Eq. 12.1), because the degradation in energy is simply taken up by the factor $d\nu$ on integration (Lilly & Cowie 1987). Furthermore, the monochromatic luminosity L_ν is related to the total luminosity (approximately) by the simple expression

$$L_\nu \simeq L/\nu_{\text{Ly}} = 3 \times 10^{-16} L \text{ Hz}^{-1}. \tag{12.10}$$

When the light is dominated by massive stars, e.g. in starburst galaxies, the luminosity is related in turn to the rate of 'metal' production, since virtually all processed material is ejected in the form of 'metals' (and some helium). Thus there is a relationship between the mass going into nucleosynthesis and the corresponding energy emitted in a fixed frequency bandwidth (anywhere between 912 and about 2000 Å in the rest frame) given by

$$\epsilon_\nu \equiv \frac{L_\nu}{(\dot{M}Z)} = \frac{0.007c^2\beta}{\nu_{\text{Ly}}} \simeq 2000\beta \text{ erg Hz}^{-1} \text{ gm}^{-1}, \tag{12.11}$$

where β is a correction factor of the order of 2 to allow for the approximations made, notably the neglect of helium. Consequently, there is a relationship between the

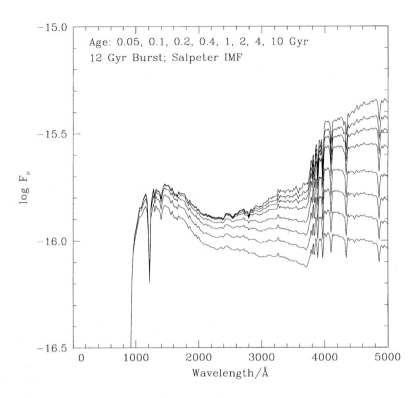

Fig. 12.3. Synthetic spectra for a region with constant star formation rate for the ages indicated, assuming a Salpeter IMF cut off at $75M_\odot$. After S.D.M. White, in C.S. Frenk *et al.* (eds.), *The Epoch of Galaxy Formation*, ©1989 Kluwer Academic Publishers, p. 15, Fig. 1, based on computations by Gustavo Bruzual. Courtesy Stephane Charlot. With kind permission from Kluwer Academic Publishers.

smoothed-out density of 'metals', $<\rho_0 Z>$, and that of monochromatic radiation, which is $4\pi/c \times$ its diffuse background intensity:

$$I_\nu = \frac{c}{4\pi} \epsilon_\nu <\rho_0 Z> . \qquad (12.12)$$

Taking $<Z> = 1$ per cent, $\epsilon_\nu = 5000$ and the total ρ_0 from Table 12.3 ($\Omega_Z = 2.6 \times 10^{-5} h^{-0.4}$), one gets

$$\nu I_\nu = \frac{3\mu\text{m}}{\lambda} 6.2 \times 10^{-14} h^{1.6} \text{ wt cm}^{-2} \text{ sterad}^{-1}. \qquad (12.13)$$

The upward sloping straight lines in the right half of Fig. 12.1 show the corresponding upper and lower limits derived by Songaila, Cowie & Lilly (1991). The background intensity expected from metal-producing starbursts should lie somewhere within this band, if not attenuated by dust. The black squares in the figure represent the background intensity from a population of faint blue galaxies that were once suspected to be dominated by starbursts, but subsequent measurements of red-shifts have shown

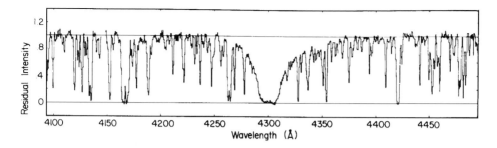

Fig. 12.4. Part of the spectrum of QSO 2344+124 (emission red-shift $z_{em} = 2.78$) showing a Lyman-α line with damping wings at a red-shift $z_{abs} = 2.54$ with column density $N(\text{H I}) = (2.7 \pm 0.5) \times 10^{20}$ cm^{-2} and [Si/H] $\simeq -1.7$. This low-resolution spectrum was taken with the ISIS spectrograph on the William Herschel Telescope at the Roque de los Muchachos Observatory, La Palma, Canary Is. on 12 Sept 1994. After K. Lipman 1995, Thesis, Cambridge University. Courtesy Max Pettini and Keith Lipman.

them to be mostly less than 1, so that this probably does not apply to the observed wave-length band. Indeed, at such red-shifts, a spectrum like that shown in Fig. 12.3 could be in danger of violating upper limits to the UV background (Pagel 1994b).

Blain & Longair (1993; see also Longair 1995) and Franceschini *et al.* (1994) have pointed out that star-bursting galaxies at large red-shifts are likely to be enshrouded in dust and detectable in the sub-mm waveband rather than the optical-UV, where it is evident from Fig. 12.1 that they make but a minor contribution to the diffuse background. Consequently the bulk of 'metal' production could actually come from such objects, rather than from galaxies detected in visible light. Blain & Longair have developed a series of models for the cosmic evolution of dusty starburst galaxies resulting from mergers, taking detailed account of the radiation properties of the dust, and the metal production from the embedded stars. They have found that they can just get consistency with the diffuse background upper limit longwards of 300 μm for a model that predicts $\Omega_Z = 2.5 \times 10^{-5} h^{-2}$; this is in fair agreement with our value of $2.6 \times 10^{-5} h^{-0.4}$. Their value for the total intensity $\int I_\nu d\nu$ is only of the order of 2×10^{-13} wt cm^{-2} sterad^{-1}, which reflects the large red-shifts involved ($z \leq 20$). Franceschini *et al.*, who do not appeal to mergers and consider somewhat lower red-shifts, predict a stronger far infra-red background spectrum resembling the full-drawn curve from ORS shown in Fig. 12.1.

12.6 Cosmic chemical evolution and high red-shift absorption-line systems

Absorption lines in the spectra of quasi-stellar objects (QSOs) showing emission lines at high red-shifts can be categorised into several distinct classes. The majority arise from Lyman-α absorption by clouds lying at various red-shifts along the line of sight to the QSO. These lines, extending from the Ly-α emission peak of the QSO down to the Ly-α

rest wave-length, are known collectively as the 'Lyman-α forest' (although they include some higher Lyman lines as well). The absorbing systems have very low metallicities (of order 0.01 solar or less) and could be some mixture of isolated hydrogen clouds with extended outer halos of galaxies. With increasing column density, one finds so-called metal-line systems with lines of Mg II and C IV, which appear to come from the halos of large galaxies. When the column density $N(\text{H I}) \geq 3 \times 10^{17}$ cm^{-2}, one has so-called Lyman limit systems in which the absorption edge at the Lyman limit appears in the continuum. For H I column densities above 2×10^{20} cm^{-2}, the Lyman-α line shows broad wings due to natural damping (Fig. 12.4) and these are known as damped Lyman-α systems.

Damped Ly-α absorbers are well suited to abundance analysis because the large column density ensures that O and N are essentially all in their neutral state while sulphur and metals are essentially all singly ionized, with column densities leading to readily measurable spectral features. Also the degree of depletion from the gas phase by the formation of dust is lower than in the local ISM. Some typical metallicities (based on zinc which is not seriously depleted even in the local ISM) are shown in Fig. 8.23. Another interesting feature of damped Ly-α absorbers is that their column densities (upward of $2M_\odot$ pc^{-2}) are comparable to those of the disks of spiral galaxies, of which they may well be precursors in which most of the mass is still in the form of gas. This idea is supported by estimates of Ω_{HI}, defined as their co-moving density in units of the present-day critical density $\rho_{\text{crit}} = 3H_0^2/8\pi G$ (Lanzetta *et al.* 1991). The argument is based on counts of systems with various column densities along the line of sight to many quasars in a suitable interval of red-shift centred on $z \simeq 2.5$, and their co-moving density is given by

$$\Omega_{\text{HI}} = \frac{H_0}{c}\frac{\mu m_{\text{H}}}{\rho_{\text{crit}}}\frac{\sum N(\text{H I})}{\Delta X} \tag{12.14}$$

$$= 0.7 \times 10^{-3}h^{-1} \text{ for a completely open universe} \tag{12.15}$$

$$= 1.3 \times 10^{-3}h^{-1} \text{ for an Einstein} - \text{de Sitter universe.} \tag{12.16}$$

Here ΔX is an interval of an 'absorption distance' X which is a dimensionless measure of the proper column density as a function of red-shift for a non-evolving population that maintains its density in co-moving coordinates. If the present volume density is ρ_0, then by the definition of X the column density is

$$N = X\rho_0 c/H_0 = \rho_0 \int_{t(z)}^{t_0} [a(t)]^{-3}c\,dt = \rho_0 c \int_z^0 (1+z)^3\frac{dt}{dz}\,dz, \tag{12.17}$$

where $a(t)$ is the expansion factor $(a(t_0) = 1)$. Since $H_0 t = (1+z)^{-1}$ for a completely open universe and $\frac{2}{3}(1+z)^{-3/2}$ for an Einstein–de Sitter universe, the corresponding absorption distances are given by

$$X(z) = \frac{1}{2}[(1+z)^2 - 1] \text{ (open), or } \frac{2}{3}[(1+z)^{3/2} - 1] \text{ (Einstein} - \text{de Sitter).} \tag{12.18}$$

The figures for Ω_{HI} in Eqs. (12.15) and (12.16) can be compared with the mass

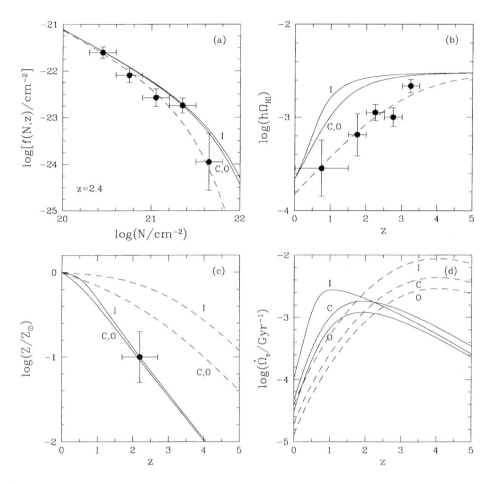

Fig. 12.5. Cosmic chemical evolution for an Einstein–de Sitter universe, assuming an initial H I density $\Omega_{HI} = 4 \times 10^{-3} h^{-1}$. Curves marked C, O, I represent GCE models assuming a closed box, outflow ($\eta = 0.5$) and inflow ($f = 0.5$) respectively. Broken-line curves (and points fitting them) are based on observed H I column densities, while continuous curves are based on the column densities corrected for obscuration. (a): Distribution function of column density at a red-shift of 2.4; (b) co-moving density of H I as a function of red-shift; (c) metallicity in solar units; (d) co-moving star formation rate. After Y.C. Pei & S.M. Fall (1995), *Ap. J.*, **454**, 69. Courtesy Michael Fall.

density of spiral galaxies given in Table 12.3: $\Omega_{spir} = 0.7 \times 10^{-3}$, which agrees quite well, although it is also the case that the frequency of high red-shift systems demands substantially higher cross-sections than those of present-day spirals. On the other hand, Ω_{HI} could be substantially larger if the largest column densities occur preferentially in dusty systems; these would have tended to be missed in surveys because of obscuration of the background QSO (Pei & Fall 1995). But then, Ω_{spir} may also have been underestimated by neglecting low surface-brightness galaxies.

Pei & Fall (1995) have calculated models of 'cosmic chemical evolution' assuming instantaneous recycling and constant yields. The basic equations are similar to those of Section 7.4.2, but now expressed in terms of the mean co-moving densities of stars, Ω_s, and gas (+ dust), Ω_g, measured in units of the present-day critical density. Treating the damped Ly-α systems as galaxies that can exchange matter with their surroundings (or with other phases such as ionized gas), the analogues of Eqs. (7.31), (7.35) are

$$\dot{\Omega}_g = \dot{\Omega}_F - \dot{\Omega}_E - \dot{\Omega}_s; \tag{12.19}$$

$$\Omega_g \dot{Z} = p\dot{\Omega}_s + (Z_F - Z)\dot{\Omega}_F, \tag{12.20}$$

where the dots denote differentiation with respect to proper time. Three kinds of models are considered, similar to some of those discussed in Chapter 8. These are

(1) The Simple (closed box) model, for which Eqs. (12.19), (12.20) lead to the solutions

$$\Omega_g = \Omega_{g,z=\infty} - \Omega_s; \tag{12.21}$$

$$Z = p\ln(\Omega_{g,\infty}/\Omega_g). \tag{12.22}$$

(2) An inflow model similar to that of Twarog (1980) with $\dot{\Omega}_F = f\dot{\Omega}_s$, $Z_F = 0$, leading to

$$\Omega_g = \Omega_{g,\infty} - (1-f)\Omega_s; \tag{12.23}$$

$$Z = (p/f)[1 - (\Omega_g/\Omega_{g,\infty})^{f/(1-f)}]. \tag{12.24}$$

(3) An outflow model similar to that of Hartwick (1976) with $\dot{\Omega}_E = \eta\dot{\Omega}_s$, leading to

$$\Omega_g = \Omega_{g,\infty} - (1+\eta)\Omega_s; \tag{12.25}$$

$$Z = [p/(1+\eta)]\ln(\Omega_{g,\infty}/\Omega_g). \tag{12.26}$$

Ω_g is deduced as a function of red-shift from the distribution of H I column densities as in Eq. (12.14) and the abundances then follow assuming a yield that fits overall solar abundance at the present time. The critical point lies in allowing for effects of dust obscuration on the distribution function of column densities (Fig. 12.5). Models that do not correct for absorption predict a maximum star formation rate at an early stage with but a slow growth in metallicity, giving rise to a 'G-dwarf problem' (Lanzetta, Wolfe & Turnshek 1995), whereas in Pei & Fall's models, star formation peaks at a red-shift near 1 at a metallicity of about one half solar, in quite good agreement with the model for the solar neighbourhood in Table 8.2 ($u = 1.6$) and with the results of the Canada–France red-shift survey. Background radiation from the models has been calculated by Fall, Charlot & Pei (1996). As can be expected from such sensible models as far as yields and metallicities are concerned, they fit within the well-established limits at most wave-lengths, but they predict a dip around 60μm which falls well below the lower limit from IRAS galaxy counts shown in Fig. 12.1. That gap could be filled by warm dust (perhaps in starburst galaxies) at somewhat smaller red-shifts than envisaged by Blain & Longair.

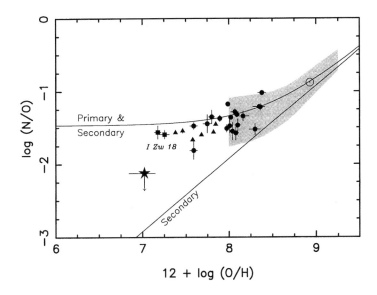

Fig. 12.6. Comparison of N/O *vs.* O/H in damped Ly-α absorbers (5-pointed stars), extragalactic H II regions and the Sun. After M. Pettini, K. Lipman & R.W. Hunstead (1995), *Ap. J.*, **451**, 100. Courtesy Max Pettini.

While evolutionary models for high red-shift absorption line systems are tempting, there may be need for some caution taking into account the existence of dispersion and abundance gradients at any given epoch. Phillipps & Edmunds (1996) have suggested that the average abundance observed along lines of sight could be significantly less than the mass-weighted abundance, and that the trends observed with red-shift could be due only partly to evolution. A possible clue to the answer to this question comes from studies of element:element ratios, which at the time of writing are still in their infancy. Fig. 12.6 shows results for the N/O ratio for some of these systems compared to 'primary' and 'secondary' lines proposed by Vila-Costas & Edmunds (1993). Dwarf emission-line galaxies seem to fit a 'primary + secondary' line (apart from the scatter discussed in Section 11.1.2), whereas the damped Ly-α systems seem to lie significantly lower. This difference can be understood if one assumes that these systems are being observed so soon after the first onset of star formation that there has not been enough time for primary nitrogen production by intermediate-mass stars (250 Myr for $4M_\odot$). Should the absence or extreme underabundance of nitrogen below a certain oxygen abundance be confirmed by further observations, then that would support the view that the variation in oxygen abundance is indeed largely due to the passage of time and not just a dispersion effect.

Clearly many issues in cosmic chemical evolution remain to be resolved. At what red-shift(s) did star formation begin, do the damped Ly-α systems represent a substantial part of the stellar material we have today and is it optical–near infra-red or far-infra-red–submm radiation that is more closely associated with nucleosynthesis? These are among the many fascinating questions in cosmology that remain to be tackled.

Notes to Chapter 12

An excellent introduction to cosmology, galaxy surveys, theories of structure formation and background radiation can be found in

A.R. Sandage, R.G. Kron & M.S. Longair, *The Deep Universe* (Saas-Fee lectures), Springer-Verlag 1995.

Many aspects of background radiation at all wave-lengths are discussed in

B. Rocca-Volmerange, J.M. Deharveng & J.T.T. Van (eds.), *The Early Observable Universe from Diffuse Backgrounds*, Ed. Frontières, Gif-sur-Yvette, 1991.

Problems

1. Justify Eq. (12.1).

2. How would the numbers in Table 12.2 be altered in the case of a completely open universe with $t = t_0/(1 + z)$?

3. If, on average, a total column density of 10^{20} H I atoms is found in the interval of red-shift 2.0 to 2.5 along one line of sight by counting damped Lyman-α systems, make an estimate of Ω_{HI} for the two universe models discussed in the text.

Appendix 1
Some historical landmarks

1675 Measurement of the speed of light (Ole Rømer).

1687 Publication of Newton's *Principia*.

1780 – 1800 W. Herschel studies Galactic structure by counting stars.

1784 Messier catalogue inspires Herschel's study of 'nebulae'. Herschel later speculated some gaseous 'proto-stars', others unresolved stellar systems over a million light years away. He also identified 'strata of nebulae' related to the Local Supercluster of galaxies.

1803 John Dalton's Atomic hypothesis.

1814 J. Fraunhofer locates and names 'Fraunhofer lines' A...L in solar spectrum. About the same time, Herschel discovers infra-red radiation from the Sun.

1816 William Prout's 'composite atoms' hypothesis.

1859 – 60 G. Kirchhoff and R. Bunsen discover spectral analysis and significance of Fraunhofer lines; Kirchhoff's law.

1868 J. Janssen and N. Lockyer independently discover helium from yellow emission line ('D3') in spectrum of prominence seen at eclipse.

1869 D.I. Mendeleev's Periodic Table of the elements.

1880s Stellar spectral classification. W. Huggins discovers spectroscopic distinction between ionised gaseous nebulae (mainly emission lines) and unresolved star clusters and galaxies (mainly absorption lines).

Development of statistical mechanics by Maxwell, Boltzmann *et al.*

1896 Radioactivity discovered by H. Becquerel.

1897 Electron discovered by J.J. Thomson.

1900 Planck's radiation formula.

1906 Rutherford shows age of Earth is about a Gyr or more from radioactive dating.

1905 Special Relativity

1911 Rutherford's model of nuclear atom.

Victor Hess discovers cosmic rays.

1913 Bohr's model of the hydrogen atom. Identification of atomic number with nuclear charge number (H. Moseley).

Hertzsprung–Russell (HR) diagram.

1916 General Relativity.

1917 Einstein develops first relativistic cosmological model and introduces concepts of transition probabilities and stimulated emission.

H. Shapley locates Galactic centre at centre of system of globular clusters, now thought to be 8.5 kpc away.

1919 Rutherford induces nuclear reaction $^{14}N(\alpha, p)^{17}O$.

Aston develops mass spectrometer

General relativity confirmed by eclipse observation.

1921 K. Lundmark introduces idea of a supernova, orders of magnitude brighter than an ordinary nova.

1923 M.N. Saha's ionization equation.

1924–7 Saha, R.H. Fowler, E.A. Milne and Cecilia Payne (later Payne-Gaposchkin) interpret stellar spectral sequence as temperature sequence with (mostly) constant chemical composition; crude abundances derived.

A.S. Eddington develops theory of radiative equilibrium (building on earlier work by A. Schuster and K. Schwarzschild) and applies it to internal constitution of stars, telling those who doubt the possibility of stellar energy generation by nuclear reactions to go and find a hotter place. He also pioneers physics of interstellar gas.

1925 E. Hubble uses cepheid period–luminosity law to demonstrate M31 is an external galaxy comparable to the Milky Way.

1926 Rise of quantum mechanics.

1928 B. Lindblad and J.H. Oort discover Galactic rotation. Theory of α radioactivity by quantum tunnelling (Gamow *et al.*).

1929 H.N. Russell analyses solar spectrum with theoretical transition probabilities and eye estimates of line intensities. Notes predominance of hydrogen (also deduced independently by Bengt Strömgren from stellar structure considerations) and otherwise similarity to meteorites rather than Earth's crust. M. Minnaert *et al.* introduce quantitative measurements of equivalent width, interpreted by the curve of growth developed by Minnaert, D.H. Menzel, and A. Unsöld.

Hubble announces velocity–distance law.

R. Atkinson and F. Houtermans apply Gamow's theory of potential barrier penetration by quantum tunneling to suggest how stars can release nuclear energy by synthesis of hydrogen into helium by an (unspecified) cyclic process.

1930 Neutrino hypothesis (Pauli).

1931 Discovery of deuterium (Urey), and of radio waves from the Galaxy (Jansky).

1932 Discovery of the neutron (Chadwick) and positron (Dirac, Anderson). First nuclear reaction induced in an accelerator ($^{7}Li(p, \alpha)$; Cockcroft & Walton). Baade and Zwicky suggest a neutron star may be created as residue of a supernova explosion.

W.H. McCrea develops models of atmospheres of A-type stars in radiative equilibrium with opacity due to H I.

1933 Atomic nucleus identified as an assemblage of protons and neutrons (Heisenberg).

1934 Russell explains spectra of carbon stars (types R,N,S) as consequence of reversal of the usual C/O ratio.

1935 Meson hypothesis (Yukawa).

1936 Compound nucleus theory (Bohr).

1938 H. Bethe and C. von Weizsäcker discover CNO cycle and p-p chain.
Discovery of nuclear fission (Hahn, Strassmann, Meitner, Frisch).

1939 Bengt Strömgren and A. Unsöld make quantitative analysis of solar abundances with fairly realistic model atmosphere, including H^- opacity source just discovered by R. Wildt.

1942 Unsöld makes quantitative analysis of a B0 star, τ Sco, getting similar abundances to solar.

1946 F. Hoyle puts forward idea of element synthesis in supernovae.
Solar ultra-violet spectrum observed in rocket flight.

1947 Discovery of π-meson (Powell).

1948 Heavy nuclei discovered in primary cosmic rays.
Gamow, R. Alpher and R. Herman develop 'Hot Big Bang' theory and suggest all elements created by neutron captures in early universe.
W. Baade distinguishes two stellar populations.

1949 Shell model of nuclear structure.

1951 P.W. Merrill discovers Tc I lines in R And (an S star), indicating nucleosynthesis within 10^6 years or less. E. Öpik and E.E. Salpeter discover 3-α reaction. J.W. Chamberlain and L.H. Aller discover that two classical 'subdwarfs' are deficient in metals by factors of order 100 relative to solar or local Galactic abundances, dispelling the idea of a universal cosmic abundance distribution.

1952 Baade revises Hubble's distance scale, so Big Bang theory not flawed by giving age of universe less than that of the Earth.
M. Schwarzschild, A. Sandage and others pioneer theory of stellar evolution without extensive mixing and apply it to HR diagrams of globular clusters getting ages of several Gyr, comparable to the new inverse Hubble constant.

1953 Hoyle successfully predicts existence of a 7.6 Mev resonance state of the carbon nucleus on grounds that otherwise little carbon would survive further processing into oxygen during stellar nucleosynthesis by helium burning, whereas in fact the C/O ratio is about 0.5.
Discovery of 'strange' particles.

1955 Salpeter introduces Initial Mass Function.

1956 Hans Suess and H.C. Urey compile 'Cosmic Abundance Distribution' from meteorites, Sun, stars and gaseous nebulae, revealing features correlated with nuclear properties.
Experimental discovery of neutrino (Reines & Cowan).
Parity violation in weak interactions (Lee, Yang, Wu *et al.*).

1957 G.R. and E.M. Burbidge, W.A. Fowler and F. Hoyle (B^2FH) publish influential article describing nuclear processes which generate essentially all nuclear species by reactions in stars or interstellar medium. Similar ideas put forward by A.G.W. Cameron in a Chalk River internal report.

1958 Recoilless γ-ray emission (Mössbauer).

1962 O.J. Eggen, D. Lynden-Bell and Sandage assemble dynamical and chemical evidence for formation of the Galaxy by free-fall collapse. The issue is still debated.

S. van den Bergh identifies the 'G-dwarf problem'.

Discovery of μ neutrino.

Discovery of cosmic X-ray background and Sco X-1 point source (R. Giacconi *et al.*).

1963 Discovery of quasars (M. Schmidt *et al.*).

1964 Hoyle and R.J. Tayler point out significance of additional neutrinos for helium synthesis in the Big Bang model.

Solar neutrino problem identified.

Quark model of hadrons (Gell-Mann, Zweig).

1965 Discovery by A.A. Penzias & R. Wilson of microwave background, predicted by Gamow *et al.* as consequence of hot Big Bang. Their ideas of Big Bang nucleosynthesis consequently resurrected up to ^{7}Li, with significant consequences for nuclear and high-energy particle physics.

1967 Discovery of first pulsar (i.e. neutron star) announced (A. Hewish, J. Bell *et al.*).

Electroweak model (Weinberg, Salam, Glashow).

T. Gold identifies spin-down of central pulsar as energy source for Crab nebula.

Discovery of γ-rays from Galaxy and diffuse background.

1971 Deuterium discovered in interstellar gas (Copernicus satellite) and quantitatively estimated in early Solar System, restricting baryonic density in Big Bang nucleosynthesis (BBNS) theory.

Trends in emission lines from H II regions in Scd galaxies as a function of galactocentric distance interpreted by L. Searle as a consequence of large-scale abundance gradients.

1972 Searle and W.L.W. Sargent identify two Zwicky blue compact galaxies as dwarf galaxies dominated by giant H II regions of low 'metallicity' but nearly normal helium abundance, supporting the existence of a universal 'floor' to helium abundance resulting from the Big Bang. Since then more such objects have been discovered, mostly in objective-prism surveys.

1975 Discovery of τ-lepton (Perl).

1979 Discovery of first gravitational lens (the double quasar 0957+561).

1982 F. and M. Spite discover Li in subdwarfs, confirming a prediction of BBNS theory.

1983 Discovery of weak bosons $W^\pm$ and Z^0 (Rubbia *et al.*).

1984 Estimates of primordial helium and deuterium abundance with Big Bang nucleosynthesis theory limit number of light neutrino families to 4 or fewer (Schramm *et al.*).

1987 Supernova 1987A tests theory of stellar nucleosynthesis. Neutrinos detected essentially as predicted.

1988 Estimates of primordial helium and deuterium with improved measurements of neutron half-life restrict number of light neutrino families to about 3. Similar, firmer result from LEP measurements at CERN of width of Z^0 resonance.

1990 COBE observations confirm black-body character of microwave background.

1992 COBE observations confirm presence of anisotropy of MWB at the 10^{-5} level.

Appendix 2 Some physical and astronomical constants

Gravitational constant	G	6.673×10^{-8}	$\text{cm}^3 \text{ gm}^{-1} \text{ s}^{-2}$
Speed of light	c	2.998×10^{10}	cm s^{-1}
Reduced Planck's constant	$\hbar$	1.0546×10^{-27}	erg s
Electronic charge	e	4.803×10^{-10}	esu
	e^2	1.440	MeV fm
Electron volt	eV	1.602×10^{-12}	erg
Freq. equiv. to 1 eV $= 10^8 \frac{e}{2\pi\hbar c}$		2.418×10^{14}	s^{-1}
Fine structure constant $e^2/(\hbar c)$	α	$1/137.036$	
Avogadro's number	N_A	6.022×10^{23}	mol^{-1}
Atomic mass unit	amu	1.661×10^{-24}	gm
Energy equiv. of ” ”	$m_u c^2$	931.494	MeV
Proton mass-energy	$m_p c^2$	938.272	MeV
Neutron mass-energy	$m_n c^2$	939.565	MeV
Electron mass	m_e	9.109×10^{-28}	gm
Electron mass-energy	$m_e c^2$	0.5110	MeV
Muon mass-energy	$m_\mu c^2$	105.66	MeV
Tauon mass-energy	$m_\tau c^2$	1777	MeV
Class. electron radius $e^2/(m_e c^2)$		2.818×10^{-13}	cm
Thomson sct. x-sec. $\frac{8\pi}{3}\left(\frac{e^2}{m_e c^2}\right)^2$	σ_T	6.653×10^{-25}	cm^2
Cl. damp. const. $2e^2\omega^2/(3m_e c^3)$	γ_cl	$8.9 \times 10^7 \left(\frac{0.5}{\lambda_{\mu\text{m}}}\right)^2$	s^{-1}
Radius of 1st Bohr orbit $\frac{\hbar}{Z m e^2}$	a_0	$0.529 \times 10^{-8}/Z$	cm
Boltzmann's constant	k	1.381×10^{-16}	erg K^{-1}
	kT	$0.08617\, T_6$	keV
Radiation density const. $\frac{\pi^2 k^4}{15(\hbar c)^3}$	a	7.564×10^{-15}	$\text{erg cm}^{-3} \text{ K}^{-4}$
Eqm photon density $\frac{2\zeta(3)}{\pi^2}\left(\frac{kT}{\hbar c}\right)^3$	n_γ	$2.03 \times 10^{28}\, T_9^3$	cm^{-3}
Quant. conc., protons $\left(\frac{m_p kT}{2\pi\hbar^2}\right)^{3/2}$		$5.95 \times 10^{33}\, T_9^{3/2}$	cm^{-3}
” ” electrons $\left(\frac{m_e kT}{2\pi\hbar^2}\right)^{3/2}$		$2.42 \times 10^{24}\, T_6^{3/2}$	cm^{-3}
Fermi coupling constant	g_F	1.4×10^{-49}	erg cm^3
Solar mass	$M_\odot$	1.989×10^{33}	gm
Solar radius	$R_\odot$	6.96×10^{10}	cm
Solar luminosity	$L_\odot$	3.84×10^{33}	erg s^{-1}
Solar effective temperature	$T_{\text{eff},\odot}$	5800	K
Solar surface gravity	$g_\odot$	2.74×10^4	cm s^{-2}

Solar mean density	$\bar{\rho}_{\odot}$	1.41	gm cm^{-3}
Year	yr	3.156×10^7	s
Astronomical unit	AU	1.496×10^{13}	cm
Light year	ly	9.46×10^{17}	cm
Parsec	pc	3.084×10^{18}	cm
Hubble constant	H_0	$100h$	km s^{-1} Mpc^{-1}
Hubble time	H_0^{-1}	$10^{10}h^{-1}$	yr

Appendix 3 Time-dependent perturbation theory and transition probabilities

A3.1 The Einstein coefficients

In a celebrated paper, Einstein (1917) analyzed the nature of atomic transitions in a radiation field and pointed out that, in order to satisfy the conditions of thermal equilibrium, one has to have not only a spontaneous transition probability per unit time A_{21} from an excited state '2' to a lower state '1' and an absorption probability $B_{12}J_v$ from '1' to '2', but also a stimulated emission probability $B_{21}J_v$ from state '2' to '1'. The latter can be more usefully thought of as negative absorption, which becomes dominant in masers and lasers.[1] Relations between the coefficients are found by considering detailed balancing in thermal equilibrium

$$N_2[A_{21} + B_{21}B_v(T)] = N_1B_{12}B_v(T) = \frac{g_1}{g_2}N_2e^{hv/kT}B_{12}B_v(T), \qquad (A3.1)$$

where N_1, N_2 are the population numbers in the two energy states with statistical weights g_1, g_2. Hence

$$B_v(T) = \frac{g_2A_{21}}{g_1B_{12}e^{hv/kT} - g_2B_{21}}. \qquad (A3.2)$$

Comparison of Eq. (A3.2) with Planck's law then gives the relations

$$g_1B_{12} = g_2B_{21}; \quad \frac{A_{21}}{B_{21}} = \frac{2hv^3}{c^2}, \qquad (A3.3)$$

which are always true because they refer to atomic constants.

To relate B_{12} to the absorption coefficient α_v, consider a unit volume of gas bathed in a radiation field J_v. The net rate at which energy is absorbed is then

$$(N_1B_{12} - N_2B_{21})J_vhv = N_14\pi J_v \int \alpha(\Delta v)\,d\Delta v, \qquad (A3.4)$$

---

[1] Einstein defined the B coefficients in terms of radiation energy density, but following E.A. Milne it is more usual to define them in terms of the angle-averaged specific intensity $J_v \equiv (c/4\pi)\rho_v$.

where Δv is the distance in frequency from the line centre. Hence

$$\alpha(\Delta v) = \frac{hv}{4\pi}\,\phi_{\Delta v}B_{12}\left(1 - \frac{N_2/g_2}{N_1/g_1}\right) \tag{A3.5}$$

where $\phi_{\Delta v}$ is a normalized profile function. In thermal equilibrium the factor in brackets reduces to the standard stimulated emission factor $[1-\exp(-hv/kT)]$, or more generally to $[1-\exp(-hv/kT_{\text{ex}})]$, where T_{ex} is an 'excitation temperature' obtained by force-fitting the two populations to a Boltzmann factor.

The corresponding result from classical electromagnetic theory is (in cgs units)

$$\int \alpha(\Delta v)\,d\Delta v = \frac{\pi e^2}{mc}. \tag{A3.6}$$

Comparison with Eq. (A3.5) then leads to the definition of a dimensionless 'oscillator strength'

$$f_{\text{abs}} = \frac{hv}{4\pi}B_{12}\Big/\frac{\pi e^2}{mc} = \frac{B_{12}mchv}{4\pi^2e^2} = \frac{g_2}{g_1}\frac{A_{21}}{3\gamma_{\text{cl}}} = -\frac{g_2}{g_1}f_{\text{em}}. \tag{A3.7}$$

A3.2 Time-dependent perturbation theory

In quantum mechanics, the state of an atom or nucleus is described by a complex wave function $\psi(\mathbf{r}_1, \mathbf{r}_2 \ldots, t)$ such that $\overline{\psi}\psi \equiv |\psi|^2$ is the probability density of finding particles in volume elements $d^3\mathbf{r}_i$ centred on $\mathbf{r}_i$ at time t. ψ satisfies the Schrödinger equation

$$H\psi = i\hbar\frac{\partial\psi}{\partial t} \tag{A3.8}$$

where H is the Hamiltonian operator $-(\hbar^2/2m)(\nabla_1^2 + \nabla_2^2 + \ldots) + V(\mathbf{r}_1, \mathbf{r}_2 \ldots)$.

Certain functions ψ_n are eigenfunctions of the operator H (when this is time-independent), i.e.

$$H\psi_n = E_n\psi_n \tag{A3.9}$$

where E_n is a number corresponding to a definite energy. In this case the solution of Eq. (A3.8) can be written as a product of space-dependent and time-dependent factors

$$\psi_n = u_n(\mathbf{r}_1 \ldots)\,e^{-i\omega_n t}, \tag{A3.10}$$

where $\omega_n \equiv E_n/\hbar$, and $|\psi|^2$ is time-independent, i.e. the system is in a stationary state satisfying the time-independent Schrödinger equation

$$Hu_n = E_n u_n. \tag{A3.11}$$

According to the postulates of QM, any ψ representing a physical state of the system can be expressed as a linear combination of energy eigenfunctions forming an infinite orthonormal set:

$$\psi = \sum_n a_n \psi_n = \sum_n a_n u_n e^{-i\omega_n t}, \tag{A3.12}$$

$|a_n|^2$ being the probability that a measurement of the state of the system will give the result that it is in state n. If the system is 'left alone', the $|a_n|$s are constant (strictly only true in the ground state), but if it is perturbed by altering V then they can change with time, i.e. the system undergoes a transition from one state to another. Eq. (A3.8) shows that the rate of change is closely related to the energy of the interaction, and one can find an expression for the Einstein–Milne absorption and stimulated emission coefficients from simple time-dependent perturbation theory.

Suppose that the atom (or nucleus) initially in an eigenstate '1' is subjected to a small time-dependent potential $V(t)$ on top of the unperturbed Hamiltonian H_0.[1] It is then possible to treat the coefficients a_n in Eq. (A3.12) as functions of time, with $|a_1|^2(\tau) \simeq 1$ being the probability that it is still in state '1' after a time τ and $|a_2|^2(\tau) \ll 1$ the probability that it has undergone a transition to another eigenstate '2'. Substituting in Schrödinger's equation (A3.8),

$$\sum_n [H_0 + V(t)] a_n \psi_n = i\hbar \left(\sum_n \dot{a}_n \psi_n + \sum_n a_n \dot{\psi}_n \right). \tag{A3.13}$$

The last term on the rhs cancels with $H_0 \sum a_n \psi_n$ by virtue of Eq. (A3.8) leaving

$$i\hbar \sum_n \dot{a}_n u_n e^{-i\omega_n t} = \sum_n V(t) a_n u_n e^{-i\omega_n t}. \tag{A3.14}$$

Multiplying through by $\bar{\psi}_2$ and integrating over space, we have

$$i\hbar \dot{a}_2 = \sum_n V_{2n}(t) e^{i\omega_{2n} t} a_n(t) \simeq V_{21}(t) e^{i\omega_{21} t} a_1(t), \tag{A3.15}$$

where $V_{2n} \equiv \langle u_2 | V(t) | u_n \rangle$ is a matrix element for the interaction. The first-order approximation solution to Eq. (A3.15) is

$$a_2(\tau) = \frac{1}{i\hbar} \int_0^\tau V_{21}(t) e^{i\omega_{21} t} dt \tag{A3.16}$$

since $a_1(t) \simeq 1$.

If the perturbation V_{21} is switched on suddenly at time $t = 0$ and is constant thereafter, Eq. (A3.16) can be integrated giving

$$a_2(\tau) = \frac{V_{21}}{\hbar \omega_{21}} (1 - e^{i\omega_{21}\tau}) \tag{A3.17}$$

and a transition probability

$$P_{1\to 2}(\tau) = \frac{|V_{21}|^2}{\hbar^2} \tau^2 \operatorname{sinc}^2(\omega_{21}\tau/2), \tag{A3.18}$$

---

[1] In the case of electromagnetic interaction, the smallness is measured by that of the fine-structure constant $\alpha = 1/137$.

where the function sinc x is defined as $\sin x/x$. This function is significantly large only for $|\omega_{12}\tau| \leq 2\pi$, i.e. for energy changes differing from zero by less than $2\pi\hbar/\tau$. Consequently, for transitions to a continuum which may be considered as a series of final states '2' with a density $\rho(E)$ per unit energy interval, the total transition probability is

$$P_{1\rightarrow \text{cont.}}(\tau) = \int_{E_1-2\pi\hbar/\tau}^{E_1+2\pi\hbar/\tau} \rho(E)\, dE \; \tau^2 \frac{|V_{21}|^2}{\hbar^2} \operatorname{sinc}^2\left(\frac{\omega_{21}\tau}{2}\right). \tag{A3.19}$$

Within this narrow range of energies, $\rho(E)$ and $|V_{21}|$ can be taken as constants outside the integral leading to

$$P_{1\rightarrow\text{cont.}} = \frac{|V_{21}|^2}{\hbar^2}\rho(E)\,\tau^2 \int_{E_1-2\pi\hbar/\tau}^{E_1+2\pi\hbar/\tau} \operatorname{sinc}^2\left(\frac{\omega_{21}\tau}{2}\right)\, dE. \tag{A3.20}$$

Now $dE = d(\hbar\omega) = (2\hbar/\tau)d(\omega\tau/2)$, and the variable $(\omega\tau/2)$ can be taken as going from $-\infty$ to ∞ since the contribution outside the relevant range $\pm\pi$ is negligible. Also, $\int_{-\infty}^{\infty} \operatorname{sinc}^2 x\, dx = \pi$. Hence the **Golden Rule** (cf. Eq. 2.91)

$$\lambda \equiv \frac{P_{1\rightarrow\text{cont.}}}{\tau} = \frac{2\pi}{\hbar}\, [|V_{21}|^2\rho(E)]_{E=E_1}. \tag{A3.21}$$

Suppose now that the system is subjected to an oscillating electromagnetic field with a representative Fourier component of the electric field $\mathbf{F}_0(\omega)\cos\omega t$. The predominant term in the interaction energy V is usually the electric dipole term 'E_1' , e.g. for an electron in an atom

$$V = -e\cos\omega t\, \mathbf{F}_0(\omega).\mathbf{r} = \cos\omega t\, \mathbf{F}_0.\mathbf{p}. \tag{A3.22}$$

There are also electric quadrupole 'E_2' terms of order $-ex_ix_j\partial F_i/\partial x_j \sim Fpr/\lambda$ and magnetic dipole 'M_1' terms of order $(\mathbf{r}\times\mathbf{v}).\mathbf{B}e/c \sim Fpv/c \sim Fpr/\lambda$ since $B_0 = F_0$. These provide smaller transition probabilities by factors of the order of $(r/\lambda)^2 \sim 10^{-8}$ in the optical region. However, when the dipole vanishes, they can give rise to 'forbidden' lines indicated by square brackets, e.g. [O III]. Still higher orders of transition are sometimes significant for nuclear γ-rays.

Substituting Eq. (A3.22) for V in Eq. (A3.15), we have

$$i\hbar\dot{a}_2 = \cos\omega t\, \mathbf{F}_0(\omega).\mathbf{p}_{21}e^{i\omega_{21}t}, \tag{A3.23}$$

where $\mathbf{p}_{21} \equiv e\sum_i <2|\mathbf{r}_i|1>$ is the 2,1 matrix element of the dipole moment and the sum is carried out over all electrons (or protons) present. Eq. (A3.23) leads to

$$\dot{a}_2(\omega) = \frac{1}{2i\hbar}\mathbf{F}_0.\mathbf{p}_{21}(e^{i\omega t} + e^{-i\omega t})\, e^{i\omega_{21}t}, \tag{A3.24}$$

which integrates up to

$$a_2(\tau,\omega) = -\frac{i}{2\hbar}\mathbf{F}_0(\omega).\mathbf{p}_{21} \left[\frac{e^{i(\omega_{21}+\omega)\tau}-1}{i(\omega_{21}+\omega)} + \frac{e^{i(\omega_{21}-\omega)\tau}-1}{i(\omega_{21}-\omega)} \right]. \tag{A3.25}$$

In the neighbourhood of a resonance, $|\omega| \simeq \omega_{21}$, the two terms on the rhs of Eq. (A3.25) correspond to stimulated emission (large for $\omega \simeq \omega_{12} = -\omega_{21} < 0$) and absorption (large for $\omega \simeq \omega_{21} > 0$) respectively. In the second case we recover a relation similar to Eq. (A3.18):

$$|a_2(\tau, \omega)|^2 = \frac{|\mathbf{F}_0 \cdot \mathbf{p}_{21}|^2}{4\hbar^2} \tau^2 \operatorname{sinc}^2 \left(\frac{\tau \Delta \omega}{2} \right). \tag{A3.26}$$

If continuous radiation is incident on the atom, Eq. (A3.26) must now be summed over all frequencies; lines are usually so narrow that J_ν can be taken as constant for $-\infty < \Delta\omega < \infty$. $|\mathbf{F}_0 \cdot \mathbf{p}_{21}|^2$ has to be expressed in terms of J_ν. To do this, note that for unpolarized isotropic radiation

$$\int_{\omega_1}^{\omega_2} J_\omega \, d\omega = \frac{c}{4\pi} \times \text{energy density} = \frac{c}{4\pi} \frac{3}{8\pi} \int_{\omega_1}^{\omega_2} F_0^2(\omega) \, d\omega, \tag{A3.27}$$

where $\frac{1}{2} F_0^2(\omega) d\omega$ is the mean squared electric field in the frequency range in any given direction, in particular that of $\mathbf{p}_{21}$.[1] Hence

$$|a_2(\tau)|^2 = \frac{8\pi^2}{3c\hbar^2} |p_{21}|^2 \, 2 J_\omega \tau \int_{-\infty}^{\infty} \operatorname{sinc}^2 \left(\frac{\tau \Delta\omega}{2} \right) d \left(\frac{\tau \Delta\omega}{2} \right) \tag{A3.28}$$

$$= \frac{8\pi^2}{3c\hbar^2} |p_{21}|^2 J_\nu \tau \equiv B_{12} J_\nu \tau. \tag{A3.29}$$

Calculation of A_{21} by quantum mechanics is much more difficult, but it can be found from B_{12} using Eq. (A3.3). Taking degeneracy into account,

$$\frac{c^2}{2h\nu^3} g_2 A_{21} = g_2 B_{21} = g_1 B_{12} = \frac{8\pi^2}{3c\hbar^2} \sum_{M_1, M_2} |\mathbf{p}_{21}|^2 \equiv \frac{8\pi^2}{3c\hbar^2} S_{12}, \tag{A3.30}$$

where $S_{12} \equiv S_{21}$ is called the line strength and the sum is taken over all individual (Zeeman) states of the energy levels 1 and 2. Because of the odd parity of $\mathbf{r}$, there is a selection rule that requires a change in parity between the spatial wave functions of states 1 and 2 for dipole radiation to occur; the opposite holds for E_2 and M_1.

---

[1] The total energy density is $3(\frac{1}{2} F_0^2 + \frac{1}{2} B_0^2)/8\pi = 3 F_0^2 / 8\pi$.

Appendix 4
Polytropic stellar models

A useful insight into the structure of nearly homogeneous stars or parts thereof can be gained from the study of so-called polytropic stellar models which depend on a combination of the principle of hydrostatic equilibrium with an assumed equation of state of the form

$$P = K\rho^{1+1/n} \tag{A4.1}$$

where n is called the polytropic index. Eq. (A4.1) applies precisely to certain cases, e.g. a perfect monatomic gas in adiabatic equilibrium ($n = 3/2$), ordinary degeneracy ($n = 3/2$ again) and relativistic degeneracy ($n = 3$), and it is assumed (with $n = 3$) in Eddington's Standard Model which gives quite a good approximation to the structure of main-sequence stars. (Note that for a perfect gas, Eq. (A4.1) makes $\rho \propto T^n$, which is often a good approximation in stellar interiors with $n \simeq 3$). The following description is based on that of Clayton (1968).

In developing the theory, it is convenient to express the density as a fraction of the central density ρ_c:

$$\rho = \rho_c \phi^n; \quad P = K\rho_c^{1+1/n} \phi^{n+1}, \tag{A4.2}$$

where ϕ is some suitable dimensionless function which diminishes monotonically from its value of 1 at the centre. The equation of hydrostatic equilibrium (Eq. 5.6) then gives (with Eq. 5.7)

$$\frac{1}{r^2} \frac{d}{dr} \left(\frac{r^2}{\rho} \frac{dP}{dr} \right) = -4\pi G\rho, \tag{A4.3}$$

or

$$(n+1)K\rho_c^{1/n} \frac{1}{r^2} \frac{d}{dr} \left(r^2 \frac{d\phi}{dr} \right) = -4\pi G\rho_c \phi^n. \tag{A4.4}$$

This equation is made dimensionless by introducing a unit of length. Let

$$r \equiv \xi a \equiv \xi \left[\frac{(n+1)K}{4\pi G\rho_c^{1-1/n}} \right]^{1/2}, \tag{A4.5}$$

Table A4.1. *Some constants of the Lane–Emden equation*

n	ξ_1	$-\xi_1^2 \left(\frac{d\phi}{d\xi}\right)_{\xi_1}$	$\rho_c/\bar{\rho}$
0.0	2.45	4.90	1.00
1.5	3.65	2.71	5.99
3.0	6.90	2.02	54.2
4.5	31.8	1.74	6190

so that Eq. (A4.4) reduces to the **Lane–Emden equation**

$$\frac{1}{\xi^2}\frac{d}{d\xi}\left(\xi^2\frac{d\phi}{d\xi}\right) = -\phi^n. \tag{A4.6}$$

Eq. (A4.6) is solved (which has to be done numerically for most values of n) using the inner boundary condition

$$\phi = 1; \quad \frac{d\phi}{d\xi} = 0, \quad \text{for } \xi = 0. \tag{A4.7}$$

For $n < 5$, ϕ vanishes at some value ξ_1 which represents the stellar surface. Some relevant quantities are given for a few polytropic indices in Table A4.1.

The stellar **radius** is

$$R = a\xi_1 = \left[\frac{(n+1)K}{4\pi G}\right]^{1/2}\rho_c^{-(n-1)/2n}\,\xi_1 \tag{A4.8}$$

and thus depends on both K and ρ_c (unless $n = 1$).

The **mass** is

$$M(\xi_1) = 4\pi a^3\rho_c\int_0^{\xi_1}\xi^2\phi^n d\xi = -4\pi a^3\rho_c\int_0^{\xi_1} d\left(\xi^2\frac{d\phi}{d\xi}\right) \tag{A4.9}$$

$$= -4\pi a^3\rho_c\left(\xi^2\frac{d\phi}{d\xi}\right)_{\xi_1} \tag{A4.10}$$

$$= -4\pi\left[\frac{(n+1)K}{4\pi G}\right]^{3/2}\rho_c^{(3-n)/2n}\left(\xi^2\frac{d\phi}{d\xi}\right)_{\xi_1}. \tag{A4.11}$$

The **ratio of central to mean density** is given by

$$\frac{\bar{\rho}}{\rho_c} = \frac{3}{4\pi}\frac{M}{R^3}\frac{1}{\rho_c} = -\frac{3}{\xi_1}\left(\frac{d\phi}{d\xi}\right)_{\xi_1} \tag{A4.12}$$

from Eqs. (A4.8), (A4.11). Thus this ratio depends only on n, which can be taken as a measure of the degree of central concentration.

When $n = 3$, ρ_c obligingly cancels out in Eq. (A4.11). For the case of **relativistic degeneracy**, we have, furthermore, from Eqs. (5.46), (5.49)

$$K = \frac{\pi \hbar c}{4}(3/\pi)^{1/3}(\mu_e m_H)^{-4/3} = 1.25 \times 10^{15} \mu_e^{-4/3}, \tag{A4.13}$$

which with Eq. (A4.11) and $\xi^2(d\phi/d\xi)_{\xi_1}$ from Table A4.1 gives the **Chandrasekhar–Landau limiting mass for white dwarfs**

$$M_{Ch} = \frac{5.83}{\mu_e^2} M_\odot. \tag{A4.14}$$

Another interesting application is **Eddington's Standard Model** (Eddington 1926). In this model it is assumed that gas pressure and radiation pressure contribute respective fractions β and $1 - \beta$ to the total pressure, with β constant in any one star. We then have

$$\beta P = P_g = \frac{\rho k T}{\mu m_H}; \tag{A4.15}$$

$$(1 - \beta)P = P_r = aT^4/3, \tag{A4.16}$$

whence

$$\frac{\beta^4}{1 - \beta} P^3 = \frac{3}{a}\left(\frac{k\rho}{\mu m_H}\right)^4, \tag{A4.17}$$

i.e.

$$P = K\rho^{4/3}, \tag{A4.18}$$

where

$$K = \left(\frac{k}{\mu m_H}\right)^{4/3}\left(\frac{3}{a}\right)^{1/3}\left(\frac{1-\beta}{\beta^4}\right)^{1/3} = \frac{2.67 \times 10^{15}}{\mu^{4/3}}\left(\frac{1-\beta}{\beta^4}\right)^{1/3}. \tag{A4.19}$$

The **mass** then follows from Eq. (A4.11)

$$M = \frac{18.3}{\mu^2}\frac{(1-\beta)^{1/2}}{\beta^2} M_\odot. \tag{A4.20}$$

Eq. (A4.20) (Eddington's quartic equation) determines $1 - \beta$ as a function of mass. Taking $\mu \simeq 0.6$, this means that, for the Sun, $1 - \beta \simeq 4 \times 10^{-4}$, whereas for $100M_\odot$, $1 - \beta \simeq 0.4$.

The **central pressure** follows from Eqs. (A4.2), (A4.11), (A4.12):

$$P_c = K\rho_c^{4/3} = K\left(\frac{\rho_c}{\bar{\rho}}\right)^{4/3}\left(\frac{3}{4\pi}\frac{M}{R^3}\right)^{4/3} \tag{A4.21}$$

and

$$K = \pi G\left[\frac{M}{-4\pi\xi^2(d\phi/d\xi)_{\xi_1}}\right]^{2/3}, \tag{A4.22}$$

whence

$$P_c = \frac{1}{16\pi} \left[\left(\frac{d\phi}{d\xi} \right)_{\xi_1} \right]^{-2} \frac{GM^2}{R^4} = 1.24 \times 10^{17} \left(\frac{M}{M_\odot} \right)^2 \left(\frac{R_\odot}{R} \right)^4 \text{ dyne cm}^{-2}.\text{(A4.23)}$$

This is several hundred times the lower limit estimated in Eq. (5.9).

The **central temperature** can be deduced from the central pressure and density:

$$T_c = \frac{\mu\beta m_H}{k} \frac{P_c}{\rho_c} = \frac{\mu\beta m_H}{k} P_c \frac{\bar{\rho}}{\rho_c} \left(\frac{4\pi R^3}{3M} \right) \tag{A4.24}$$

$$= 19.5 \times 10^6 \mu\beta \left(\frac{M}{M_\odot} \right) \left(\frac{R_\odot}{R} \right) \text{ K.} \tag{A4.25}$$

The radius can be eliminated by appealing to the homology transformation (Table 5.1) from which, approximately,

$$\frac{R}{R_\odot} \simeq \left(\frac{M}{M_\odot} \right)^{2/3}, \tag{A4.26}$$

whence

$$T_c \simeq 19.5 \times 10^6 \mu\beta \left(\frac{M}{M_\odot} \right)^{1/3} \text{ K} \tag{A4.27}$$

for main-sequence stars with $\mu = \mu_\odot$.

Finally, the **luminosity** can be derived from the radiative transfer equation Eq. (5.21)

$$\frac{dP_r}{dr} \equiv \frac{a}{3} \frac{dT^4}{dr} = \frac{\kappa\rho l(r)}{4\pi c r^2}. \tag{A4.28}$$

Dividing by the hydrostatic equilibrium condition Eq. (5.6),

$$\frac{dP_r}{dP} \equiv 1 - \beta = \frac{\kappa l(r)}{4\pi c G m(r)} = \frac{L}{4\pi c G M} \eta(r)\kappa(r), \tag{A4.29}$$

where

$$\eta(r) \equiv \frac{l(r)/m(r)}{L/M}. \tag{A4.30}$$

Consequently we have for the total luminosity

$$L = \frac{4\pi c G M(1 - \beta)}{\overline{\kappa\eta}} = 1.5 \times 10^{35} \frac{(\mu\beta)^4}{\overline{\kappa\eta}} \left(\frac{M}{M_\odot} \right)^3 \text{ erg s}^{-1}, \tag{A4.31}$$

from Eq. (A4.20), where $\overline{\kappa\eta}$ represents a suitable average over the star (κ increases from about 1 cm^2 gm^{-1} in the centre of the Sun to very large values in the envelope, whereas η decreases from large values in the centre to 1 at the surface, so that the product varies by a more limited amount; strictly it should be constant in order to make β constant). Thus we need to have $\overline{\kappa\eta} \simeq 5$ cm^2 gm^{-1} in order to reproduce the Sun's luminosity, which is right within a factor of a few.

See Arnett (1996) for applications of Eddington's Standard Model to more complicated situations.

Appendix 5 Dissipation and abundance gradients

Models of galaxy formation involving dissipative collapse (e.g. Larson 1974a) generically lead to a gradient in the mean stellar abundances (as well as a density gradient) caused by inflow of gas relative to stars. This can be understood analytically using the following simplified version of the 'concentration model' by Lynden-Bell (1975).

Consider star formation in a spherical galaxy of unit mass, consisting initially of gas. Stars are formed from the gas, which simultaneously contracts through the stars as a result of energy dissipation by cloud collisions etc. The abundance in the gas (assumed uniform) is still $z = \ln(1/g)$ as in the Simple model (Chapter 8), where g is the mass of gas remaining and z is in units of the yield.

Suppose that, at any time t, the mass of stars within a fixed mass coordinate m is $s(m, t)$ and that the gas is confined within a decreasing mass coordinate $m_a(t)$. We make the simplifying but reasonable assumption that, within $m_a(t)$, the radial distributions of stars and of the star formation rate are the same. Furthermore, we assume the *Ansatz*

$$m_a = g + s(m_a) = g^c \tag{A5.1}$$

where $c < 1$ is called the concentration index. To simplify the mathematics, we here take $c = 1/2$; for solutions with arbitrary c, see Lynden-Bell (1975).

With these assumptions, one can relate the rate of star formation within a fixed sphere of total mass m to the rate of consumption of gas in the system as a whole:

$$\frac{ds(m)}{dt} = -\frac{dg}{dt}\frac{s(m)}{s(m_a)} = -\frac{dg}{dt}\frac{s(m)}{g^{1/2} - g}; \quad m \le m_a(t); \tag{A5.2}$$

$$= -\frac{dg}{dt}; \quad m \ge m_a(t). \tag{A5.3}$$

These equations are easily integrated to give

$$s(m, t) = K(m)(1 - g^{1/2})^2; \quad m \le m_a = g^{1/2} = e^{-z/2}; \tag{A5.4}$$

$$= m - g; \quad m \ge m_a = g^{1/2} = e^{-z/2}. \tag{A5.5}$$

The integration constant $K(m)$ is fixed by equating (A5.4) to (A5.5) at $m = m_a(t) = [g(t)]^{1/2}$:

$$K(m) = \frac{m}{1 - m}. \tag{A5.6}$$

The mass of stars within the sphere m having metallicity less than z is then given by

$$s(m; < z) = \frac{m}{1-m}(1 - e^{-z/2})^2; \quad m \le e^{-z/2}; \quad z \le 2\ln\frac{1}{m}; \qquad (A5.7)$$

$$= m - e^{-z}; \qquad m \ge e^{-z/2}; \quad z \ge 2\ln\frac{1}{m}. \qquad (A5.8)$$

The differential distribution function of metallicities in a shell m to $m + dm$ is given by differentiating Eqs. (A5.7), (A5.8) with respect to both m and z:

$$\frac{\partial^2 s}{\partial m\,\partial z} = \frac{e^{-z/2} - e^{-z}}{(1-m)^2}; \quad z \le 2\ln\frac{1}{m}; \qquad (A5.9)$$

$$= 0; \qquad z > 2\ln\frac{1}{m}. \qquad (A5.10)$$

The mean metallicity of stars in the shell is then given by

$$\bar{z}\,(m \text{ to } m + dm) = \frac{\int_0^{2\ln\frac{1}{m}} z\frac{d^2 s}{dm\,dz}\,dz}{\int_0^{2\ln\frac{1}{m}} \frac{d^2 s}{dm\,dz}\,dz} = \frac{3 - 4m(1 - \ln m) + m^2(1 - 2\ln m)}{(1-m)^2}. \qquad (A5.11)$$

The ratio in Eq. (A5.11) tends to zero for $m = 1$ and to 3 for $m = 0$, i.e. a strong abundance gradient has been built up with the mean stellar metallicity at the centre exceeding the yield by a factor of 3. Cf. the numerical model by Larson shown in Fig. 11.6.

Appendix 6 Hints for problems

Chapter 2

1. Use Eq. (2.7).

2. $^3\text{He } \frac{1}{2}^+$; ^7Li, ^9Be, $^{11}\text{B } \frac{3}{2}^-$; ^{13}C, $^{15}\text{N } \frac{1}{2}^-$.

4. H→He 27.74; 6.93. He→C 7.27; 0.61; $^{12}\text{C}(\alpha, \gamma)$ 7.16; 0.45. $^{13}\text{C}(\alpha, n)$ 2.22; 0.13; $^{116}\text{Sn}(n, \gamma)$ 6.95; 0.059.

5. Momenta in CM system are $\pm m\mathbf{v}$.

6. The z−component of velocity will be more than w if the total speed $v \geq w$ and $\cos \theta \geq w/v$ where θ is the angle between $\mathbf{v}$ and the z−axis. Hence

$$N(\geq w) \quad \propto \quad \int_w^\infty v^2 e^{-mv^2/2kT} dv \int_0^{\arccos w/v} \sin \theta d\theta \tag{A6.1}$$

$$= \int_w^\infty v^2 \left(1 - \frac{w}{v}\right) e^{-mv^2/2kT} dv, \tag{A6.2}$$

$$\tag{A6.3}$$

and

$$\frac{dN}{dw} \propto -\int_w^\infty v e^{-mv^2/2kT} dv \propto e^{-mw^2/2kT}, \tag{A6.4}$$

i.e. a gaussian distribution with variance kT/m, the same as that of $v^{-2} dN/dv$. The variance of $w_1 - w_2$ is thus $kT(1/m_1 + 1/m_2)$ and this will also be that of its projection back into 3 dimensions $|\mathbf{v_1} - \mathbf{v_2}|$.

From Eq. (2.106), the distribution function of the centre-of-mass velocity will be a maxwellian with a mass of $m_1 + m_2$.

7. Take the starting point A at a fixed distance y_1 above the surface plane and the end-point B at a fixed distance $-y_2$ below it. The time is then $y_1 \sec i - ny_2 \sec r$, which has to be minimized subject to the subsidiary condition that $x_A - x_B = y_1 \tan i - y_2 \tan r$ is constant. Hence

$$\delta t = y_1 \sec i \tan i \, \delta i - ny_2 \sec r \tan r \, \delta r$$
$$-A(y_1 \sec^2 i \, \delta i - y_2 \sec^2 r \, \delta r) \tag{A6.5}$$
$$= 0 \quad \forall \, \delta i, \delta r. \tag{A6.6}$$

Equating coefficients,

$$\sec i \tan i \;=\; A \sec^2 i \tag{A6.7}$$

$$n \sec r \tan r \;=\; A \sec^2 r. \tag{A6.8}$$

8.

$$S = k \ln W \;=\; k \sum_i \left[\omega_i \ln \frac{\omega_i}{\omega_i - N_i} - N_i \ln \frac{N_i}{\omega_i - N_i} \right] \tag{A6.9}$$

$$\simeq\; k \sum_i \left[N_i \left(1 + \frac{\tilde{E}_i - \mu}{kT} \right) \right] \tag{A6.10}$$

$$=\; k \sum_i \left[N_i \left(1 + \frac{p_i^2}{2mkT} + \ln(un_Q/n) \right) \right]. \tag{A6.11}$$

$\sum_i N_i p_i^2$ is just $N \times$ the average square momentum $3mkT$.

9. Electrical neutrality gives

$$n_{e^-} - n_{e^+} = 8n_O = \frac{\rho}{2m_H} = 3.0 \times 10^{26} \text{ cm}^{-3}. \tag{A6.12}$$

The chemical potential condition gives

$$n_{e^+} n_{e^-} = 4n_Q^2 e^{-2m_e c^2/kT} = 1.9 \times 10^{53} \text{ cm}^{-6}. \tag{A6.13}$$

Hence $n_{e^+} = 3.1 \times 10^{26} \text{ cm}^{-3}$.

10. From Eqs. (2.55), (2.21), $E_0 = 1.7$ MeV. Since the reduced mass is $28/32 = 7/8$ of the mass of an α-particle, the laboratory energy of the αs should be $8/7 \times$ as much, i.e. 1.9 MeV.

11. For a gaussian, say $f(E) = (1/\sigma \sqrt{2\pi}) \exp[-(E - E_0)^2/2\sigma^2]$,

$$f(E_0)/f''(E_0) = \sigma^2. \tag{A6.14}$$

Hence from Eq. (2.58),

$$\sigma = \frac{\sqrt{2}}{3} kT \tau^{1/2} \tag{A6.15}$$

and the full $1/e$ width is

$$2\sigma \sqrt{2} = 4(E_0 kT/3)^{1/2}. \tag{A6.16}$$

12. ≤ 0.42 MeV, detectable by all the reactions except the first. In case (a), the deuteron and positron have equal and opposite momentum p. The K.E. of the positron is

$$T_{e^+} = \sqrt{m_e^2 c^4 + p^2 c^2} - m_e c^2. \tag{A6.17}$$

The K.E. of the deuteron is

$$T_d = p^2/4m_H. \tag{A6.18}$$

This makes $T_{e^+} = 0.42$ for $pc = 0.77$ MeV, so $T_d = 1.6 \times 10^{-4}$ MeV. In case (b), the kinetic energy of the neutrino is $pc = 0.42$ MeV, so the K.E. of the deuteron is 4.7×10^{-5} MeV.

13. In the first case, the result is obvious from Eq. (2.96). In the second case we have to go back to Eq. (2.95) with the following relations:

$$pc = \sqrt{T_e^2 + 2T_e m_e c^2}, \tag{A6.19}$$

$$c^2 p\,dp = (T_e + m_e c^2)dT_e, \tag{A6.20}$$

$$qc = \sqrt{(Q - T_e)^2 - m_\nu^2 c^4}, \tag{A6.21}$$

$$c^2 q\,dq = E_\nu dE_\nu = (Q - T_e)dE_f. \tag{A6.22}$$

Substituting in Eq. (2.95), we have

$$\frac{d\lambda}{dT_e} = \frac{d\lambda}{dp}\frac{dp}{dT_e} \propto (T_e + m_e c^2)pqE_\nu \tag{A6.23}$$

$$\propto (T_e + m_e c^2)(Q - T_e)\sqrt{T_e^2 + 2T_e m_e c^2}\sqrt{(Q - T_e)^2 - m_\nu^2 c^4}. \tag{A6.24}$$

The slope of the spectrum is $d^2\lambda/dT_e^2$. Differentiating Eq. (A6.24) we get the sum of a number of terms all but one of which vanish when $(Q - T_e) = m_\nu c^2$, but the remaining term has the factor
$$-(Q - T_e)/\sqrt{(Q - T_e)^2 - m_\nu^2 c^4} = -\infty.$$

Chapter 3

1. $W_{\text{exp}}/W_{\text{MEMMU}} = 2/\sqrt{\pi}$.
2. 1.7×10^{19} cm^{-2}.
3. This is best done graphically, using Table 3.3 ($a = 0.001$). The data are just barely compatible with the lines being on the linear part of the curve of growth, in which case the EW of D_1 is 200 mÅ and the column density is 2.0×10^{12} cm^{-2} from Eq. (3.38). In this case, the equivalent width of D_2 could be at most about 1 Doppler width, implying $b \geq 20$ km s^{-1}.
   The preferred doublet ratio is 0.21 dex, which corresponds to $\log(N\alpha_0\sqrt{\pi}) = 0.15$, $\log(W/\Delta\lambda_D) = 0.03$ for D_1, i.e. $b = 11$ km s^{-1} and $N = 3.0 \times 10^{12}$ cm^{-2} from Eqs. (3.26), (3.27).
   The smallest doublet ratio compatible with the data is 0.12 dex, with an EW of 260 mÅ for D_1. In this case, the curve gives $\log N\alpha_0\sqrt{\pi} = 0.6$, $\log W/\Delta\lambda_D = 0.32$, $b = 6.3$ km s^{-1}, $N = 1.5 \times 10^{13}$ cm^{-2}. So the final result is $\log N_{\text{cm}^{-2}} = 12.5^{+0.7}_{-0.2}$; $b \geq 6.3$ km s^{-1}.
4. From Fig. 3.12, the curve-of-growth shift $\log A = 0.7$. Hence from Eq. (3.58)

$$\frac{\text{FeI}_1/g_1}{H} = 2.6 \times 10^{-7}\kappa_{0.5\mu m} = 5.2 \times 10^{-9}P_e \tag{A6.25}$$

from Table 3.2. Applying Saha's equation, Eq. (3.7), we get

$$\log \frac{\text{Fe}^+}{\text{H}} = -4.55. \tag{A6.26}$$

Note that iron is sufficiently ionized in the solar atmosphere that the abundance of Fe I can be neglected and its partition function (or the ground-state statistical weight g_1) and the electron pressure cancel out.

5. Equating the legend to Fig. 3.13 with Eq. (3.59),

$$0.09(I - \chi) - 0.66 = [\text{Fe}^+/\text{H}] + 0.09(I - \chi - 0.75). \tag{A6.27}$$

Hence $[\text{Fe}/\text{H}] = -0.59$.

6. Following Russell (1934), we write for the fictitious undepleted pressures $P_O = P_H(O/H)$, $P_C = P_H(C/H)$ of oxygen and carbon

$$P_O = p_O + p_{CO} = p_O \left(1 + \frac{p_C}{K_{CO}}\right); \tag{A6.28}$$

$$P_C = p_C + p_{CO} = p_C \left(1 + \frac{p_O}{K_{CO}}\right). \tag{A6.29}$$

Substitution of p_C from the second equation into the first, and *vice-versa* for p_O, leads to two quadratic equations of which (3.73) and (3.74) are the solutions.

7. From Eq. (3.67)

$$5q_{21} = 1.72 \times 10^{-7}; \quad q_{21} = 3.45 \times 10^{-8} \text{ cm}^3 \text{ s}^{-1}. \tag{A6.30}$$

Hence

$$n_{\text{crit}} = 2 \times 10^5 \text{ cm}^{-3}. \tag{A6.31}$$

Chapter 4

1. Law of motion

$$\ddot{R} = \frac{1}{2}\frac{d}{dR}(\dot{R}^2) = -\frac{GM}{R^2} + \frac{1}{3}\Lambda R. \tag{A6.32}$$

Integrating,

$$3\dot{R}^2 = 8\pi G\rho R^2 + \Lambda R^2 + \text{const.} \tag{A6.33}$$

This is identical to Eq. (4.9) if the constant is identified as $-3kc^2$. If $\Lambda = 0$, then $-kc^2$ is $\dot{R}^2 - 2GM/R$, twice the total energy.

2. In adiabatic expansion,

$$T \propto \rho^{\gamma-1} \propto R^{-3(\gamma-1)}, \tag{A6.34}$$

i.e. R^{-1} for radiation and R^{-2} for a non-relativistic monatomic gas. Hence the present-day temperature is $2.73^2/1.2 \times 10^{13} = 6 \times 10^{-13}$ K.

3. Assuming $Y_P = 0.24$, the H I density is

$$n_H = 0.76 \times 3 \times 10^{-10} \times 20.3\, T^3 = 10^{-7}(1+z)^3 \qquad (A6.35)$$

and $n_e = n_p = 0.01 n_H$. Saha's equation then gives

$$10^{-11}(1+z)^3 = 2 \times 1.09 \times 10^{16}(1+z)^{3/2}10^{-25110/(1+z)}, \qquad (A6.36)$$

or

$$(1+z)^{3/2} = 2.2 \times 10^{27}\, 10^{-25110/(1+z)}, \qquad (A6.37)$$

which holds for $z \simeq 1100$, $T \simeq 3000$ K, $n_e \simeq 1$ cm^{-3}. Recombination time is thus of the order of 10^{12} s. From Eq. (4.26), the age of the universe is of the order of 10^{13} s, so that there will be some minor departure from Saha equilibrium due to expansion, but the effect on the parameters deduced here is negligible (cf. Jones 1977).

4. Assuming $g_* = 3.36$ today (i.e. all neutrinos massless),

$$1 + z_{eq} = \frac{1.88 \times 10^{-29} \Omega_0 h^2 c^2}{1.68 a (2.73)^4} = 2.4 \times 10^4 \Omega_0 h^2. \qquad (A6.38)$$

5.

$$\frac{D}{H} = \frac{3}{4} \times \frac{0.15}{1.15} \eta\, n_\gamma \times 2^{3/2}\left(\frac{2\pi\hbar^2}{m_p kT}\right)^{3/2} e^{2.224\ \mathrm{MeV}/kT} \qquad (A6.39)$$

$$= 3.0 \times 10^{-16}\, T_9^{3/2} e^{25.8/T_9} \qquad (A6.40)$$

$$= 4.8 \times 10^{-5}\ (T_9 = 1), \quad 3.4 \times 10^{-10}\ (T_9 = 2). \qquad (A6.41)$$

Also, the mass fraction $X_D = 2\,D/H$ since there is not yet a significant amount of helium at this stage.

6. From Eqs. (4.23), (4.24), an extra flavour of neutrinos increases g_* by 7/4. Hence from Eq. (4.50), Y_P is increased by 0.013.

7. Use $X + Y + Z = 1$, $Y/X = 4y$.

Chapter 5

2. Taking the electrostatic energy to be Ze^2/d, where $d \simeq (\rho/Am_H)^{1/3}$ is the average distance between ions, and the Fermi energy from Eq. (5.43) with $E_F = p_F^2/2m_e$, we have

	E.s. energy	Fermi energy	kT
Iron	7 eV	8 eV	0.025 eV
Carbon white dwarf	150 eV	35 keV	8.75 keV
Brown dwarf	125 eV	2.6 keV	0.44 keV

The energy density from electrostatic repulsion of ions is $(Ze)^2/d = (Ze)^2(\rho/Am_H)^{1/3}$ per ion and $(Ze)^2(\rho/Am_H)^{4/3}$ per unit volume, while the density of gravitational energy is

$$-\Omega \simeq GM\rho/R \simeq GM\rho/(M/\rho)^{1/3} = GM^{2/3}\rho^{4/3}. \tag{A6.42}$$

Thus both energy densities have the same dependence on ρ. The critical mass is then given by

$$M_{\text{plan}} \simeq \left(\frac{Z^4 e^2}{Gm_{\text{H}}^2}\right)^{3/2} m_{\text{H}}/\mu^2 \simeq 10^{-3} M_\odot \tag{A6.43}$$

for $Z = \mu = 1$. This is essentially just $\alpha^{3/2}$ times the Chandrasekhar mass (cf. Eq. 5.56).

3. From Saha's equation, $O^{7+}/O^{8+} = 0.63$. Lower states of ionization have lower populations, but are by no means negligible. Neglecting them nonetheless (as they will contribute some opacity), we have $O^{7+}/O \leq 0.39$. The opacity is then

$$10^{-19} \times 0.39 Z_{\text{O}}/A_{\text{O}} m_{\text{H}} \simeq 15 \text{ cm}^2 \text{ gm}^{-1} \gg \kappa_{\text{es}}. \tag{A6.44}$$

The mean separation between ions is $(\mu m_{\text{H}}/\rho)^{1/3} = 2.3 \times 10^{-9}$ cm, or about half the Bohr radius, so that no states of neutral hydrogen can exist.

4. Writing a monochromatic version of Eq. (3.15),

$$-\frac{dK_\nu}{\kappa_\nu \rho dz} = H_\nu \tag{A6.45}$$

and

$$-\frac{dK}{\bar{\kappa}\rho dz} = H = \int H_\nu d\nu. \tag{A6.46}$$

So

$$\frac{1}{\bar{\kappa}} = \frac{\int (1/\kappa_\nu)(dK_\nu/dz)d\nu}{dK/dz}. \tag{A6.47}$$

The Rosseland formula follows by dividing top and bottom by dT/dz and taking $K_\nu = J_\nu/3 = B_\nu(T)/3$.

5. Writing

$$B_\nu(T) \propto \nu^3 (e^{\nu\theta} - 1)^{-1}, \tag{A6.48}$$

where $\theta \equiv h/kT$, we have

$$\frac{dB_\nu}{d\theta} \propto \frac{\nu^4 e^{\nu\theta}}{(e^{\nu\theta} - 1)^2} \tag{A6.49}$$

and with $\kappa_\nu \propto \nu^{-3}\theta^{1/2}$

$$\frac{1}{\bar{\kappa}} = \frac{\int \nu^7 \theta^{-1/2} e^{\nu\theta} (e^{\nu\theta} - 1)^{-2} d\nu}{\int \nu^4 e^{\nu\theta} (e^{\nu\theta} - 1)^{-2} d\nu} = (kT)^{3.5} \frac{\int x^7 e^x (e^x - 1)^{-2} dx}{\int x^4 e^x (e^x - 1)^{-2} dx}. \tag{A6.50}$$

6. At the Eddington limit,

$$g = \frac{GM}{R^2} = \frac{\kappa \pi F}{c} = \frac{\kappa L}{4\pi R^2 c}, \qquad (A6.51)$$

whence

$$\frac{L_{\mathrm{Edd}}}{L_\odot} = \frac{4\pi c G M}{\kappa L_\odot} = 3.3 \times 10^4 \mu_e \frac{M}{M_\odot} \qquad (A6.52)$$

for electron scattering opacity.

7. $$P_c = \frac{G}{4\pi} \int \frac{m\,dm}{r^4} < \frac{G}{4\pi} \left(\frac{4\pi \rho_c}{3} \right)^{4/3} \int m^{-1/3} dm. \qquad (A6.53)$$

$$\frac{1-\beta}{\beta^4} = P^3 a \left(\frac{\mu m_{\mathrm{H}}}{k\rho} \right)^4 < 0.1 \mu^2 \left(\frac{M}{M_\odot} \right)^2 \qquad (A6.54)$$

(cf. Eq. A4.20). With $\mu = 0.6$, the upper limits to $1 - \beta$ are 0.03 ($M_\odot$), 0.56 ($20M_\odot$), 0.67 ($40M_\odot$). More precise, and considerably smaller, limits are derived from Eddington's Quartic Equation (Appendix 4).

8. -1.7.

10. If the metal abundance is $f \times$ some standard (e.g. solar), then κ/P_{ph} is replaced by $f\kappa/P_{\mathrm{ph}}$ in Eqs. (5.63), (5.64), equivalent to replacing M by $Mf^{-1/2}$ (for $a = 1$). So in Eq. (5.65) T_{eff} is multiplied by $Z^{-0.1}$.

11. Using B to denote $(2\pi \hbar^2 / m_{\mathrm{H}} kT)^{3/2}$,

$$\frac{{}^8\mathrm{Be}}{{}^4\mathrm{He}} = \frac{n_\alpha}{2^{3/2}} B e^{x/kT}; \qquad (A6.55)$$

$$\frac{{}^{12}\mathrm{C}^{**}}{{}^8\mathrm{Be}} = \frac{n_\alpha}{(8/3)^{3/2}} B e^{(7.275 - 7.654 - x)/kT}; \qquad (A6.56)$$

$$\frac{{}^{12}\mathrm{C}^{**}}{{}^4\mathrm{He}} = \left(\frac{3}{16} \right)^{3/2} \left(\frac{\rho}{4m_{\mathrm{H}}} \right)^2 B^2 e^{-0.379\,\mathrm{MeV}/kT} = 3.8 \times 10^{-29}. \qquad (A6.57)$$

Hence the volume density of ${}^{12}\mathrm{C}^{**}$ is

$$3.8 \times 10^{-29} (\rho / 4m_{\mathrm{H}}) = 0.57 \ \mathrm{cm}^{-3} \qquad (A6.58)$$

and the production rate of ground-state ${}^{12}\mathrm{C}$ is

$$0.57 \times 0.0037 \times 2.42 \times 10^{14} = 5 \times 10^{11} \ \mathrm{cm}^{-3}\,\mathrm{s}^{-1}. \qquad (A6.59)$$

12. Equating chemical potentials,

$$Z_i[m_p c^2 + kT \ln(n_p / 2n_{Q(\mathrm{p})})] + N_i[m_n c^2 + kT \ln(n_n / 2n_{Q(\mathrm{n})})]$$
$$= A_i[m_i c^2 + kT \ln(n_i / u_i n_{Q(\mathrm{i})})] \qquad (A6.60)$$

and $n_i \equiv Y_i \rho / m_{\mathrm{H}}$.

13. Use Eq. (5.114) with $Y_\alpha = 1/8$, $Y_p = Y_n = 1/4$. $T_9 \simeq 15$.

14. $$\frac{dL}{dt} \propto \frac{dL}{dM_c} \frac{dM_c}{dt} \propto M_c^{16.4} \propto L^{1.86}. \qquad (A6.61)$$

15. 1.9×10^{51} erg. $0.6GM^2/R = 3.2 \times 10^{50}$ erg.

Chapter 6

1. Taking $N_0 = 1$ initially, $N_0 = e^{-\sigma\tau}$ and we have

$$\frac{dN_1}{d\tau} + \sigma N_1 = \sigma e^{-\sigma\tau}; \quad N_1 = \sigma\tau e^{-\sigma\tau}; \tag{A6.62}$$

$$\frac{dN_2}{d\tau} + \sigma N_2 = \sigma^2\tau e^{-\sigma\tau}; \quad N_2 = \frac{1}{2}(\sigma\tau)^2 e^{-\sigma\tau}\dots \tag{A6.63}$$

leading to

$$N_k(\tau) = \frac{(\sigma\tau)^k}{k!}e^{-\sigma\tau}. \tag{A6.64}$$

With an exponential distribution of exposures,

$$N_k(\tau_0) = \frac{\sigma^k}{k!\tau_0}\int_0^\infty \tau^k e^{-\tau(\sigma+1/\tau_0)}d\tau \tag{A6.65}$$

$$= \frac{1}{\sigma\tau_0}\left(\frac{\sigma}{\sigma+1/\tau_0}\right)^{k+1}\frac{1}{k!}\int_0^\infty x^k e^{-x}dx \tag{A6.66}$$

$$= (\sigma\tau_0)^{-1}(1+1/\sigma\tau_0)^{-(k+1)}. \tag{A6.67}$$

2. 0.67.
3. In Eq. (2.95), $\lambda \propto \int_0^Q E^2(Q-E)^2 dE$.

Chapter 7

1. 8.26 to 1.
2.

	Salpeter		Eqs. (7.7)–(7.10)	
	$100M_\odot$	$50M_\odot$	$100M_\odot$	$50M_\odot$
Mean SN mass	$22.3M_\odot$	$18.7M_\odot$	$19.8M_\odot$	$17.6M_\odot$
Mass fr. above $10M_\odot$	.12	.093	.048	.040
SN rate per 100 yr ($\alpha \simeq 0.7$)	4.5	4.1	2.1	2.0

3. Oxygen yields are $0.0048/\alpha$ (Eq. 7.10) and $0.014/\alpha$ (Salpeter).

Chapter 8

1.

$$\frac{d}{dS}(gX_D) = -X_D - X_{D,E}\frac{E}{\psi} + X_{D,F}\frac{F}{\psi}. \tag{A6.68}$$

In the Simple model,

$$-g\frac{dX_D}{dg} = g\frac{dX_D}{ds} = -X_D\left(\frac{1}{\alpha}-1\right), \tag{A6.69}$$

leading to Eq. (8.55). The remaining fractions of primordial D are .67 and .2 respectively.

2. $u = \int_0^s ds'/g$ leads to Eq. (8.57). Using Eq. (7.36) with $dg/ds = -(1+\eta)$,

$$-g\frac{dz}{dg} = \frac{1}{1+\eta}, \tag{A6.70}$$

i.e. the same as the Simple model with p divided by $(1+\eta)$.

From Eqs. (8.56), (8.57),

$$z = (1+\eta)^{-1}\ln(1/g), \tag{A6.71}$$

and from Eq. (A6.68)

$$-g\frac{dX_D}{ds} = g\frac{dX_D}{dg}(1+\eta) = X_D\left(\frac{1}{\alpha}-1\right). \tag{A6.72}$$

Hence using Eq. (A6.71)

$$\frac{X_D}{X_{D,0}} = g^{(1/\alpha-1)/(1+\eta)} = e^{-z(1/\alpha-1)}. \tag{A6.73}$$

3. In Eq. (8.39), substitute z from Eq. (8.38) and differentiate wrt u.
4. Taking $\omega\Delta = 0.4$ (Table 8.1), $\dot{s}(u - \omega\Delta)/\dot{s}(u_{now}) = 0.87/0.75 = 1.16$ (Table 8.2). Hence $N_{Ia}/N_{II} = 0.21$.

Chapter 9

1. 0.52.
2. Since E is the kinetic energy, $p^2c^2 = (E + E_0)^2 - E_0^2$.
3. Number density is $\phi/c = 2.9 \times 10^{-10}$ cm^{-3}.
   Energy density is 0.46 eV cm^{-3}. Mean energy is 1.6 GeV.
4. 7×10^{-12}.
5. 5.3×10^{-12}.

Chapter 10

1. 50 yr.
2.

$$k\frac{e^{\lambda_{235}t} - 1}{e^{\lambda_{238}t} - 1} = \frac{207(t)/204 - 207(0)/204}{206(t)/204 - 206(0)/204}. \tag{A6.74}$$

The slope is 0.66 compared to 0.07 for Rb–Sr.

3. Eq. (10.31) is true if, for both species, $\frac{1}{2}\lambda^2 < (t- <\tau>)^2 > \ll 1$.
   For solar-system ^{232}Th, ^{238}U, $\Delta- <\tau> = 3.2$ Gyr. The error from uncertainties in $\ln K$ is of order 10 per cent, and that from neglecting the quadratic term in Eq. (10.30) is also of order 10 per cent, but is systematic in the sense that the 3.2 Gyr is an overestimate by about that amount.

4. For $\lambda_i \Delta \gg 1$, Eq. (10.15) reduces to $1/\lambda_i \Delta$. For a stable element it reduces to $1/(1-S)$. The corresponding limits for Eq. (10.21) are $1/\lambda_i \Delta$ and $1/2$.

Chapter 11

1. From Eq. (7.34),

$$\frac{d}{ds}(gZ) = p(1 - \eta^*) - Z(1 + \eta + \eta^* R^*/\alpha). \tag{A6.75}$$

Also,

$$\frac{dg}{ds} = -(1 + \eta + \eta^* R^*/\alpha), \tag{A6.76}$$

leading to Eq. (11.19). $R^* = \int_{10}^{100}(m - m_{rem})\phi(m)dm < 0.048$.

Chapter 12

1. The specific intensity is $(c/4\pi)\times$ the radiation energy density and all photons ever emitted are still here (in the absence of intergalactic absorption). Their energy is reduced by a factor of $(1 + z)$, which is taken up by the dv factor in any energy interval $I_v dv$, leaving I_v unchanged.
2. Eq. (12.7) now reads

$$\left(\frac{M}{L_0}\right)_{cgs} <Z + \Delta Y> = 0.005t_0 \left(\frac{(1 + z_{max})^{q-1} - 1}{q - 1}\right), \tag{A6.77}$$

leading to 0.38 (for $z_{max} = 2.16$) and 1.8 (for $z_{max} = 5.31$) if t_0 is taken as 15 Gyr.
3. From Eq. (12.18), $\Delta X = 1.62$ (open universe) or 0.90 (Einstein–de Sitter universe). Then from Eq. (12.14), $\Omega_{HI} = 6.5 \times 10^{-4}h^{-1}$ or $1.2 \times 10^{-3}h^{-1}$ respectively.

References

Alcock, C., Akerlof, C.W., Allsman, R.A. *et al.* 1993, *Nature*, **365**, 621.

Allen, C.W. 1973, *Astrophysical Quantities*, London: Athlone Press.

Aller, L.H. & Greenstein, J.L. 1960, *Ap. J. Suppl.*, **5**, 139.

Alloin, D., Collin-Souffrin, S., Joly, M. & Vigroux, L. 1979, *A & A*, **78**, 200.

Alpher, R.A., Follin, J.W. & Herman, R.C. 1953, *Phys. Rev.*, **92**, 1347.

Anders, E. 1988, in J.F. Kerridge & M.S. Matthews (eds.), *Meteorites and the Early Solar System*, University of Arizona Press, Tucson, p. 927.

Anders, E. & Grevesse, N. 1989, *Geochim. Cosmochim. Acta*, **53**, 197.

Anders, E. & Zinner, E. 1993, *Meteoritics*, **28**, 490.

Applegate, J.H. & Hogan, C. 1985, *Phys. Rev. D*, **31**, 3037.

Arimoto, N. 1989, in J.E. Beckman & B.E.J. Pagel (eds.), *Evolutionary Phenomena in Galaxies*, Cambridge University Press, p. 341.

Arnaud, K.A. 1996, in S.S. Holt & G. Sonneborn (eds.), *Cosmic Abundances*, Astr. Soc. Pacific Conf. Series, Vol. 99, p. 409.

Arnaud, M., Rothenflug, R., Boulade, O., Vigroux, L. & Vangioni-Flam, E. 1992, *A & A*, **254**, 49.

Arnett, D. 1996, *Supernovae and Nucleosynthesis*, Princeton University Press.

Arnett, W.D. 1978, *Ap. J.*, **219**, 1008.

Arnett, W.D. & Thielemann, F.-K. 1985, *Ap. J.*, **295**, 589, 604.

Aubourg, E., Bareyre, P., Bréhin, S. *et al.* 1993, *Nature*, **365**, 623.

Audouze, J. & Silk, J. 1989, *Ap. J. Lett.*, **342**, L5.

Axon, D.J., Staveley-Smith, L., Fosbury, R.A.E., Danziger, J., Boksenberg, A. & Davies, R.D. 1988, *MNRAS*, **231**, 1077.

Bahcall, J.N., Flynn, C. & Gould, A. 1992, *Ap. J.*, **389**, 234.

Bahcall, J.N., Huebner, W.F., Lubow, S.H., Parker, P.D. & Ulrich, R.K. 1982, *Rev. Mod. Phys.*, **54**, 767.

Balser, D.S., Bania, T.M., Brockway, C.J., Rood, R.T. & Wilson, T.L. 1994, *Ap. J.*, **430**, 667.

Bao, Z.Y. & Käppeler, F. 1987, *At. Data Nucl. Data Tables*, **36**, 411.

Barbuy, B. 1988, *A & A*, **191**, 121.

Barbuy, B. & Erdelyi-Mendez, M. 1989, *A & A*, **214**, 239.

Barnes, J.E. 1995, in C. Muñoz-Tuñon & F. Sánchez (eds.), *The Formation and Evolution of Galaxies*, Cambridge University Press, p. 399.

Bazan, G. & Mathews, W.G. 1990, *Ap. J.*, **354**, 644.

Beers, T.M., Preston, G.W., & Shectman, S.A. 1992, *Astr. J.*, **103**, 1987.

Beers, T.M. & Sommer-Larsen, J. 1995, *Ap. J. Suppl.*, **96**, 175.

Belley, J. & Roy, J.-R. 1991, *Ap. J. Suppl.*, **78**, 61.

Bender, R. 1992, in B. Barbuy & A. Renzini (eds.), IAU Symp. no. 149: *The Stellar Populations of Galaxies*, Kluwer, Dordrecht, p. 267.

Bender, R., Burstein, D. & Faber, S.M. 1992, *Ap. J.*, **399**, 462.

Bender, R., Burstein, D. & Faber, S.M. 1993, *Ap. J.*, **411**, 153.

Bessell, M.S., Sutherland, R. & Ruan, K. 1991, *Ap. J. Lett.*, **383**, L71.

Bica, E. 1988, *A & A*, **195**, 76.

Bica, E. & Alloin, D. 1986, *A & A*, **162**, 21.

Bica, E. & Alloin, D. 1987, *A & A*, **186**, 49.

Bienaimé, O., Robin, A.S. & Crézé, M. 1987, *A & A*, **180**, 94.

Binney, J. & Tremaine, S. 1987, *Galactic Dynamics*, Princeton University Press.

Black, D.C. 1971, *Nature Phys. Sci.*, **234**, 148.

Blackwell, D.E., Booth, A.J., Haddock, D.J., Petford, A.D. & Leggett, S.K. 1986, *MNRAS*, **220**, 549.

Blain, A.W., & Longair, M.S. 1993, *MNRAS*, **265**, L21.

Bloemen, H., Wijnans, R., Bennett, K. *et al.* 1994, *A & A*, **281**, L5.

Blumenthal, G.R., Faber, S.M., Primack, J.R. & Rees, M.J. 1984, *Nature*, **311**, 517.

Boesgaard, A. & King, J.R. 1993, *Astr. J.*, **106**, 2309.

Boesgaard, A. & Steigman, G. 1985, *Ann. Rev. Astr. Astrophys.*, **23**, 319.

Boesgaard, A. & Tripicco, M.J. 1986, *Ap. J. Lett.*, **302**, L49.

Bressan, A., Chiosi, C. & Fagotto, F. 1994, *Ap. J. Suppl.*, **94**, 63.

Bristow, P.D. & Phillipps, S. 1994, *MNRAS*, **267**, 13.

Brocato, E., Matteucci, F., Mazzitelli, I. & Tornambé, A. 1990, *Ap. J.*, **349**, 458.

Brown, G.E., Bruenn, S.W. & Wheeler, J.C. 1992, *Comments Astrophys.*, **16**, 153.

Bruzual, A. & Charlot, S. 1993, *Ap. J.*, **405**, 538.

Burbidge, E.M., Burbidge, G.R., Fowler, W.A. & Hoyle, F. 1957, *Rev. Mod. Phys.*, **29**, 547.

Burkert, A. & Hensler, G. 1989, in J.E. Beckman & B.E.J. Pagel (eds.), *Evolutionary Phenomena in Galaxies*, Cambridge University Press, p. 230.

Burkert, A., Truran, J.W. & Hensler, G. 1992, *Ap. J.*, **391**, 651.

Butcher, H.R. 1987, *Nature*, **328**, 127.

Buzzoni, A. 1990, *Ap. J. Suppl.*, **71**, 817.

Buzzoni, A., Fusi-Pecci, F., Buonanno, R. & Corsi, C.E. 1983, *A & A*, **128**, 94.

Cameron, A.G.W. 1957, Atomic Energy of Canada Ltd, CRL-41; *Pub. Astr. Soc. Pacific*, **69**, 201.

Cameron, A.G.W. 1982a, in C.A. Barnes, D.D. Clayton & D.N. Schramm (eds.), *Essays in Nuclear Astrophysics*, Cambridge University Press, p. 23.

Cameron, A.G.W. 1982b, *Astrophys. Space Sci.*, **82**, 123.

Cameron, A.G.W. 1993, in E.H. Levy, E.H. Lunine & M. Matthews (eds.), *Protostars and Planets III*, University of Arizona Press, Tucson, p. 47.

Cameron, A.G.W. & Fowler, W.A. 1971, *Ap. J.*, **164**, 111.

Caputo, F., Martínez Roger, C. & Páez, E. 1987, *A & A*, **183**, 228.

Carigi, L. 1994, *Ap. J.*, **424**, 181.

Carney, B.W. 1983, in P.A. Shaver, D. Kunth & K. Kjär (eds.), *Primordial Helium*, ESO, Garching, p. 179.

Carney, B.W., Laird, J.B., Latham, D.W. & Aguilar, L.A. 1996, *Astr. J.*, **112**, 668.

Carney, B.W., Laird, J.B., Latham, D.W. & Kurucz, R.L. 1987, *Astr. J.*, **94**, 1066.

Carney, B.W., Latham, D.W. & Laird, J.B. 1990, *Astr. J.*, **99**, 572.

Carr, B.J. & Rees, M.J. 1979, *Nature*, **278**, 605.

Catchpole, R., Pagel, B.E.J. & Powell, A.L.T. 1967, *MNRAS*, **136**, 403.

Caughlan, G.R. & Fowler, W.A. 1988, *At. Data Nucl. Data Tables*, **40**, 283.

Cayrel, R. & Jugaku, J. 1963, *Ann. d'Ap.*, **26**, 495.

Chaboyer, B., Demarque, P. & Sarajedini, A. 1996, *Ap. J.*, **459**, 558.

Chamberlain, J.W. & Aller, L.H. 1951, *Ap. J.*, **114**, 52.

Chamcham, K., Pitts, E. & Tayler, R.J. 1993, *MNRAS*, **263**, 967.

Chen, B., Dobaczewski, J., Kratz, K.-L., Langanke, K., Pfeiffer, B., Thielemann, F.-K. & Vogel, P. 1995, *Phys. Lett. B*, **355**, 37.

Chevalier, R.A. 1976, *Nature*, **260**, 689.

Christy, R.F. 1966, *Ann. Rev. Astr. Astrophys.*, **4**, 353.

Ciotti, L., D'Ercole, A., Pellegrini, S. & Renzini, A. 1991, *Ap. J.*, **376**, 380.

Clarke, C.J. 1989, *MNRAS*, **238**, 283.

Clarke, C.J. 1991, *MNRAS*, **249**, 704.

Clayton, D.D. 1964, *Ap. J.*, **139**, 637.

Clayton, D.D. 1968, 1984, *Principles of Stellar Evolution and Nucleosynthesis*, McGraw-Hill and University of Chicago Press.

Clayton, D.D. 1982, *Quart. J. RAS*, **23**, 174.

Clayton, D.D. 1985a, in Arnett, W.D. & Truran, J.W. (eds.), *Nucleosynthesis: Challenges and New Developments*, University of Chicago Press, p. 65.

Clayton, D.D. 1985b, *Ap. J.*, **285**, 411.

Clayton, D.D. 1988, *MNRAS*, **234**, 1.

Clayton, D.D., Fowler, W.A., Hull, T.E. & Zimmerman, B.A. 1961, *Annals of Phys.*, **12**, 331.

Clayton, D.D. & Ward, R.A. 1974, *Ap. J.*, **193**, 397.

Conti, P.S., Greenstein, J.L., Spinrad, H., Wallerstein, G. & Vardya, M.S. 1967, *Ap. J.*, **148**, 105.

Copi, C., Schramm, D.N. & Turner, M.S. 1995, *Science*, **267**, 192.

Cord, M.S., Peterson, J.D., Lojko, M.S. & Haas, R.H. 1968, *Microwave Spectral Tables V. Spectral Line Listing*, U.S. Govt. Printing Office, Washington.

Cowan, J.J., Thielemann, F.-K. & Truran, J.W. 1987, *Ap. J.*, **323**, 523.

Cowan, J.J., Thielemann, F.-K. & Truran, J.W. 1991a, *Phys. Rep.*, **208**, 267.

Cowan, J.J., Thielemann, F.-K. & Truran, J.W. 1991b, *Ann. Rev. Astr. Astrophys.*, **29**, 447.

Cowan, R.D. 1981, *The Theory of Atomic Structure and Spectra*, University of California Press, Berkeley.

Cowley, C.R. 1995, *An Introduction to Cosmochemistry*, Cambridge University Press.

Cox, D.P. 1972, *Ap. J.*, **178**, 143.

Cox, D.P. & Smith, B.W. 1976, *Ap. J.*, **203**, 361.

Crane, P. (ed.) 1995, *The Light Element Abundances*, Springer-Verlag.

Cunha, K. & Lambert, D.L. 1992, *Ap. J.*, **399**, 586.

Cunha, K. & Lambert, D.L. 1994, *Ap. J.*, **426**, 170.

Dame, T.M. 1993, in S.S. Holt & F. Verter (eds.), *Back to the Galaxy*, Amer. Inst. Phys. Publ., p. 267.

da Silva, L., de la Reza, R., & Magalhães, S.D. 1990, in E. Vangioni-Flam, M. Cassé, J. Audouze & J.T.T. Van (eds.), *Astrophysical Ages and Dating Methods*, Ed. Frontières, Gif-sur-Yvette, p. 419.

Dekel, A. & Silk. J. 1986, *Ap. J.*, **303**, 39.

Delbouille, L., Roland, G. & Neven, L. 1973, *Spectrophotometric Atlas of the Solar Spectrum*, Liège.

Deliyannis, C.P., Demarque, P. & Kawaler, S. 1990, *Ap. J. Suppl.*, **73**, 21.

Dimopoulos, S., Emailzadeh, R., Hall, L.J. & Starkman, G.D. 1988, *Phys. Rev. Lett.*, **60**, 7

Dopita, M. A. 1985, *Ap. J. Lett.*, **295**, L5.

Dopita, M.A. 1996, in J.T.T. Van, L.M. Cernikier, H.C. Trung & S. Vauclair (eds.), *2me Rencontre du Vietnam: The Sun and Beyond*, Ed. Frontières, Gif-sur-Yvette.

Dopita, M.A., D'Odorico, S. & Benvenuti, P. 1980, *Ap. J.*, **236**, 628.

Dorman, B., Lee, Y.-W. & VandenBerg, D. 1991, *Ap. J.*, **366**, 115.

Dorman, B., VandenBerg, D.A. & Laskarides, P.G. 1989, *Ap. J.*, **343**, 750.

Duncan, D., Lambert, D.L. & Lemke, M. 1992, *Ap. J.*, **401**, 584.

Dwek, E. & Slavin, J. 1994, *Ap. J.*, **436**, 696.

Eddington, A.S. 1926, *The Internal Constitution of the Stars*, Cambridge University Press.

Edmunds, M.G. 1990, *MNRAS*, **246**, 678.

Edmunds, M.G. 1992, in M.G. Edmunds & R.J. Terlevich (eds.), *Elements and the Cosmos*, Cambridge University Press, p. 289.

Edmunds, M.G. & Greenhow, R.M. 1995, *MNRAS*, **272**, 241.

Edmunds, M.G., Greenhow, R.G., Johnson, D., Kluckers, V. & Vila, B.M. 1991, *MNRAS*, **251**, 33P.

Edmunds, M.G. & Pagel, B.E.J. 1978, *MNRAS*, **185**, 77P.

Edmunds, M.G. & Pagel, B.E.J. 1984, *MNRAS*, **211**, 507.

Edvardsson, B., Andersen, J., Gustafsson, B., Lambert, D.L., Nissen, P.E. & Tomkin, J. 1993, *A & A*, **275**, 101.

Edvardsson, B., Gustafsson, B., Johansson, S.G., Kiselman, D., Lambert, D.L., Nissen, P.E. & Gilmore, G. 1994, *A & A*, **290**, 176.

Eggen, O.J., Lynden-Bell, D. & Sandage, A.R. 1962, *Ap. J.*, **136**, 748.

Einstein, A. 1917, *Phys. Zs.*, **18**, 121.

Epstein, R.I., Lattimer, J.M. & Schramm, D.N. 1976, *Nature*, **263**, 198.

Faber, S.M. 1973, *Ap. J.*, **179**, 731.

Faber, S.M., Burstein, D. & Dressler, A. 1977, *Astr. J.*, **82**, 941.

Faber, S.M. & Jackson, R.E. 1976, *Ap. J.*, **204**, 668.

Faber, S.M. & Lin, D. 1983, *Ap. J. Lett.*, **266**, L17.

Fall, S.M., Charlot, S. & Pei, Y.C. 1996, *Ap. J. Lett.*, **464**, L43.

Faulkner, J. & Iben, I., Jr. 1966, *Ap. J.*, **144**, 995.

Feltzing, S. & Gustafsson, B. 1994, *Ap. J.*, **423**, 68.

Fermi, E. 1949, *Phys. Rev.*, **75**, 1169.

Fermi, E. 1954, *Ap. J.*, **119**, 1.

Fields, B.D., Olive, K.A. & Schramm, D.N. 1995, *Ap. J.*, **439**, 854.

Fish, R.A. 1964, *Ap. J.*, **139**, 284.

Fitch, M. & Silkey, M. 1991, *Ap. J.*, **366**, 107.

Fleming, T.A., Liebert, J. & Green, R.F. 1986, *Ap. J.*, **308**, 176.

Flynn, C. & Fuchs, B. 1994, *MNRAS*, **270**, 471.

Fowler, W.A. 1987, *Quart. J. RAS*, **28**, 87.

Fowler, W.A., Caughlan, G.R. & Zimmerman, B.A. 1967, *Ann. Rev. Astr. Astrophys.*, **5**, 525.

Fowler, W.A., Caughlan, G.R. & Zimmerman, B.A. 1975, *Ann. Rev. Astr. Astrophys.*, **13**, 69.

Franceschini, A., Mazzei, P., De Zotti, G. & Danese, L. 1994, *Ap. J.*, **427**, 140.

Franceschini, A., Toffolatti, L., Mazzei, P., Danese, L. & De Zotti, G. 1991, *A & A Suppl.*, **89**, 285.

François, P., Spite, M. & Spite, F. 1993, *A & A*, **274**, 821.

Franx, M. & Illingworth, G. 1990, *Ap. J. Lett.*, **359**, L41.

Freeman, K.C. 1990, in R. Wielen (ed.), *Dynamics and Interactions of Galaxies*, Springer-Verlag, p. 36.

Friel, E.D. & Janes, K.A. 1993, *A & A*, **267**, 75.

Gallino, R. 1989, in H.R. Johnson & B. Zuckerman (eds.), *Evolution of Peculiar Red Giant Stars*, Cambridge University Press, p. 176.

Garnett, D.R. 1990, *Ap. J.*, **363**, 142.

Garnett, D.R., Skillman, E.D., Dufour, R.J., Peimbert, M., Torres-Peimbert, S., Shields, G.A., Terlevich, R.J. & Terlevich, E. 1995, *Ap. J.*, **443**, 64.

Gasson, R.E.M. & Pagel, B.E.J. 1966, *Observatory*, **86**, 196.

Geisler, D. & Friel, E.D. 1992, *Astr. J.*, **104**, 128.

Geiss, J. 1993, in N. Prantzos, E. Vangioni-Flam & M. Cassé (eds.), *Origin and Evolution of the Elements*, Cambridge University Press, p. 89.

Geiss, J. & Reeves, H. 1972, *A & A*, **18**, 126.

Gerola, H. & Seiden, P.E. 1978, *Ap. J.*, **223**, 129.

Gies, D.R. & Lambert, D.L. 1992, *Ap. J.*, **387**, 673.

Gilmore, G., Edvardsson, B. & Nissen, P.E. 1991, *Ap. J.*, **378**, 17.

Gilmore, G., Gustafsson, B., Edvardsson, B. & Nissen, P.E. 1992, *Nature*, **357**, 379.

Gilmore, G. & Reid, I.N. 1983, *MNRAS*, **202**, 1025.

Gilmore, G. & Wyse, R.F.G. 1985, *Astr. J.*, **90**, 2015.

Gilmore, G. & Wyse, R.F.G. 1991, *Ap. J. Lett.*, **367**, L55.

Gilmore, G., Wyse, R.F.G. & Jones, J.B. 1995, *Astr. J.*, **109**, 1095.

Gilmore, G., Wyse, R.F.G. & Kuijken, K. 1989, *Ann. Rev. Astr. Astrophys.*, **27**, 555.

Gloeckler, G. & Geiss, J. 1996, *Nature*, **381**, 210.

Götz, M. & Köppen, J. 1992, *A & A*, **260**, 455.

Goldstein, M.A., Fisk, L.A. & Ramaty, R. 1970, *Phys. Rev. Lett.*, **25**, 832.

Gordy, W. & Cook, R.L. 1984, *Microwave Molecular Spectra*, Wiley.

Gorgas, J., Efstathiou, G. & Aragón Salamanca, A. 1990, *MNRAS*, **245**, 217.

Gough, D.O. 1994, *Phil. Trans. R. Soc. London, A*, **346**, 39.

Gratton, R., Carretta, E., Matteucci, F. & Sneden, C. 1997, preprint.

Gray, D.F. 1976, *The Observation and Analysis of Stellar Photospheres*, John Wiley & Sons.

Grenon, M. 1989, *Ap. Sp. Sci.*, **156**, 29.

Grenon, M. 1990, in B.J. Jarvis & D.M. Terndrup (eds.), *Bulges of Galaxies*, ESO Conf. & Workshop Proc. no. 5, ESO, Garching, p. 143.

Grevesse, N., Noels, A. & Sauval, A.J. 1996, in S.S. Holt & G. Sonneborn (eds.), *Cosmic Abundances*, Astr. Soc. Pacific Conf. Series, p. 117.

Guiderdoni, B. & Rocca-Volmerange, B. 1987, *A & A*, **186**, 1.

Gunn, J.E. & Peterson, B.A. 1965, *Ap. J.*, **142**, 1633.

Gustafsson, B. 1980, in P.E. Nissen & K. Kjär (eds.), Workshop on *Methods of Abundance Determination for Stars*, ESO, Geneva.

Gustafsson, B. 1989, *Ann. Rev. Astr. Ap.*, **27**, 701.

Harris, M.J., Fowler, W.A., Caughlan, G.R. & Zimmerman, B.A. 1983, *Ann. Rev. Astr. Ap.*, **21**, 165.

Harrison, G.R. 1939, *MIT Wavelength Tables*, John Wiley & Sons.

Hartwick, F.D.A. 1976, *Ap. J.*, **209**, 418.

Hartwick, F.D.A. 1987, in G. Gilmore & R.F. Carswell (eds.), *The Galaxy*, Reidel, p. 281.

Hayashi, C. 1950, *Prog. Theor. Phys.* (Japan), **5**, 224.

Hayashi, C. 1961, *Pub. Astr. Soc. Japan*, **13**, 450.

Hayashi, C. 1966, *Ann. Rev. Astr. Ap.*, **4**, 171.

Heasley, J.N. & Milkey, R.W. 1978, *Ap. J.*, **221**, 677.

Henry, R.B.C. & Howard, J.W. 1995, *Ap. J.*, **438**, 170.

Hensler, G. & Burkert, A. 1990, *Astrophys. Space Sci.*, **170**, 231, and **171**, 149.

Herzberg, G. 1945, *Molecular Spectra and Molecular Structure II: Infra-red and Raman Spectra of Polyatomic Molecules*, Van Nostrand, Princeton.

Herzberg, G. 1950, *Molecular Spectra and Molecular Structure I: Spectra of Diatomic Molecules*, Van Nostrand, Princeton.

Herzberg, G. 1966, *Molecular Spectra and Molecular Structure III: Electronic Spectra and Electronic Structure of Polyatomic Molecules*, Van Nostrand, New York.

Herzberg, G. 1971, *The Spectra and Structures of Simple Free Radicals*, Cornell University Press, Ithaca.

Hohenberg, C.M., Podosek, F.A. & Reynolds, J.H. 1967, *Science*, **156**, 233.

Holweger, H. & Müller, E.A. 1974, *Solar Phys.*, **39**, 19.

Howard, W.M., Meyer, B.S. & Woosley, S.E. 1991, *Ap. J. Lett.*, **373**, L5.

Hoyle, F. 1946, *MNRAS*, **106**, 343.

Hoyle, F. 1953, *Ap. J.*, **118**, 512.

Hoyle, F. 1954, *Ap. J. Suppl.*, **1**, 121.

Hoyle, F. & Fowler, W.A. 1960, *Ap. J.*, **132**, 565.

Hoyle, F. & Tayler, R.J. 1964, *Nature*, **203**, 1108.

Hübner, W.F., Mertz, A.L., McGee, N.H. & Argo, M.F. 1977, Los Alamos Rep. LA-6760-M.

Ibata, R.A., Gilmore, G. & Irwin, M.J. 1994, *Nature*, **370**, 194.

Iben, I., Jr. 1965, *Ap. J.*, **141**, 993.

Iben, I., Jr. 1967, *Ann. Rev. Astr. Astrophys.*, **5**, 571.

Iben, I., Jr. 1985, *Ap. J. Suppl.*, **58**, 661.

Iben, I., Jr. 1991, *Ap. J. Suppl.*, **76**, 55.

Iben, I., Jr. 1993, *Ap. J.*, **415**, 767.

Iben, I., Jr. & Renzini, A. 1983, *Ann. Rev. Astr. Astrophys.*, **21**, 271.

Iben, I., Jr. & Renzini, A. 1984, *Phys. Rep.*, **105**, 329.

Iben, I., Jr. & Tutukov, A.V. 1984a, *Ap. J.*, **282**, 615.

Iben, I., Jr. & Tutukov, A.V. 1984b, *Ap. J. Suppl.*, **54**, 335.

Iben, I., Jr. & Tutukov, A.V. 1987, *Ap. J.*, **313**, 727.

Int. Astr. Union, *Transactions*, publ. triennially.

Izotov, Y.I., Thuan, T.X. & Lipovetsky, V.A. 1997, *Ap. J. Suppl.*, **108**, 1.

Johnson, H.R. & Zuckerman, B. (eds.) 1989, *Evolution of Peculiar Red Giant Stars*, Cambridge University Press.

Jones, B.J.T. 1977, *MNRAS*, **180**, 151.

Jura, M. 1989, in H.R. Johnson & B. Zuckerman (eds.), *Evolution of Peculiar Red Giant Stars*, Cambridge University Press, p. 348.

Käppeler, F., Beer, H. & Wisshak, K. 1989, *Rep. Prog. Phys.*, **52**, 945.

Käppeler, F., Gallino, R., Busso, M. Picchio, G. & Raiteri, C.M. 1990, *Ap. J.*, **354**, 630.

Kaye, G.W.C. & Laby, T.H. 1995, *Tables of Physical and Chemical Constants*, 16th edition, Longman.

Kennicutt, R.C., Jr., 1983, *Ap. J.*, **272**, 54.

Kennicutt, R.C., Jr., 1989, *Ap. J.*, **344**, 685.

Kennicutt, R.C., Jr., Tamblyn, P. & Congdon, C.W. 1994, *Ap. J.*, **435**, 22.

Kilian, J. 1992, *Astr. Ap.*, **262**, 171.

Kingsburgh, R.L. & Barlow, M.J. 1994, *MNRAS*, **271**, 257.

Kippenhahn, R. & Weigert, A. 1990, *Stellar Structure and Evolution*, Springer-Verlag.

Kiselman, D. & Carlsson, M. 1994, in P. Crane (ed.), *The Light Element Abundances*, Springer-Verlag, p. 372.

Kolb, E.W. & Turner, M.S. 1990, *The Early Universe*, Addison-Wesley Press

Koonin, S.E., Tombrello, T.A. & Fox, G. 1974, *Nucl. Phys. A*, **220**, 221.

Köppen, J. & Arimoto, N. 1990, *A & A*, **240**, 22.

Köppen, J. & Arimoto, N. 1991, *A & A Suppl.*, **87**, 109, and *Erratum*, *A & A Suppl*, **89**, 420.

Kormendy, J. & Sanders, D.B. 1992, *Ap. J. Lett.*, **390**, L53.

Kraft, R.P. 1994, *Pub. Astr. Soc. Pacific*, **106**, 553.

Krane, K.S. 1988, *Introductory Nuclear Physics*, Wiley.

Kroupa, P., Tout, C.A. & Gilmore, G. 1991, *MNRAS*, **251**, 293.

Kudritsky, R.F. & Hummer, D.G. 1990, *Ann. Rev. Astr. Astrophys.*, **28**, 303.

Kuhn, H.G. 1971, *Atomic Spectra*, Longman.

Kuijken, K. & Gilmore, G. 1989, *MNRAS*, **239**, 605.

Kulkarni, S.R., Blitz, L. & Heiles, C. 1982, *Ap. J. Lett.*, **259**, L63.

Kulkarni, S.R. & Heiles, C. 1987, in D. Hollenbach & H. Thronson (eds.), *Interstellar Processes*, Reidel, p. 87.

Kumai, Y. & Tosa, M. 1992, *A & A*, **257**, 511.

Kunth, D., Lequeux, J., Sargent, W.L.W. & Viallefond, F. 1994, *A & A*, **282**, 709.

Kunth, D. & Sargent, W.L.W. 1983, *Ap. J.*, **273**, 81.

Kurki-Suonio, H., Matzner, R., Olive, K. & Schramm, D.N. 1990, *Ap. J.*, **353**, 406.

Kurucz, R.L. 1979, *Ap. J. Suppl.*, **40**, 1.

Kurucz, R.L. & Peytremann, E. 1975, *A Table of Semiempirical gf-Values*, Smithsonian Ap. Obs. Special Rep., **362**.

Lacey, C.G. & Fall, S.M. 1985, *Ap. J.*, **290**, 154.

Laird, J.B., Rupen, M.P., Carney, B.W. & Latham, D.W. 1988, *Astr. J.*, **96**, 1908.

Lambert, D.L. 1989, in C.F. Waddington (ed.), *Cosmic Abundances of Matter*, New York: Amer. Inst. Phys., p. 168.

Lambert, D.L. 1992, *A & A Rev.*, **3**, 201.

Lambert, D.L., Heath, J.E. & Edvardsson, B. 1991, *MNRAS*, **253**, 610.

Lambert, D.L., Smith, V.V., Busso, M., Gallino, R. & Straniero, O. 1995, *Ap. J.*, **450**, 502.

Lanzetta, K.M., Wolfe, A.M. & Turnshek, D.A. 1995, *Ap. J.*, **440**, 435.

Lanzetta, K.M., Wolfe, A.M., Turnshek, D.A., Lu, L., McMahon, R.G. & Hazard, C. 1991, *Ap. J. Suppl.*, **77**, 1.

Larson, R.B. 1972, *Nature Phys. Sci.*, **236**, 7.

Larson, R.B. 1974a, *MNRAS*, **166**, 585.

Larson, R.B. 1974b, *MNRAS*, **169**, 229.

Larson, R.B. 1976, *MNRAS*, **176**, 31.

Larson, R.B. 1986, *MNRAS*, **218**, 409.

Larson, R.B. & Dinerstein, H. 1975, *Pub. Astr. Soc. Pacific*, **87**, 911.

Larson, R.B. & Tinsley, B.M. 1978, *Ap. J.*, **219**, 46.

Laurent, C. 1983, in P.A. Shaver, D. Kunth & K. Kjär (eds.), *Primordial Helium*, ESO, Garching, p. 335.

Lawler, J.E., Whaling, W. & Grevesse, N. 1990, *Nature*, **346**, 635.

Lee, T., Papanastassiou, D.A. & Wasserburg, G. 1977, *Ap. J. Lett.*, **211**, L107.

Lemoine, M., Ferlet, R. & Vidal-Madjar, A. 1994, *A & A*, **298**, 879.

Lequeux, J., Peimbert, M., Rayo, J.F., Serrano, A. & Torres-Peimbert, S. 1979, *A & A*, **80**, 155.

Lewis, J.R. & Freeman, K.C. 1989, *Astr. J.*, **97**, 139.

Lewis, R.S., Tang, M., Wacker, J.F., Anders, E. & Steel, E. 1987, *Nature*, **326**, 160.

Lilly, S.J. & Cowie, L.L. 1987, in C.GG. Wynn-Williams & E.E. Becklin (eds.), *Infrared Astronomy with Arrays*, University of Hawaii, Honolulu, p. 473.

Lilly, S.J., Le Fèvre, O., Hammer, F. & Crampton, D. 1996, *Ap. J. Lett.*, **460**, L1.

Linsky, J.L., Brown, A., Gayley, K., Diplas, A., Savage, B.D., Ayres, T.R., Landsman, W., Shore, S.N. & Heap, S.R. 1993, *Ap. J.*, **402**, 694.

Lipman, K. 1995, Thesis, Cambridge University.

Longair, M.S. 1995, in A.R. Sandage, R.G. Kron & M.S. Longair, *The Deep Universe* (Saas-Fee Lectures), Springer-Verlag.

Low, C. & Lynden-Bell, D. 1976, *MNRAS*, **176**, 367.

Lynden-Bell, D. 1973, in G. Contopulos, M. Hénon & D. Lynden-Bell, *Dynamical Structure and Evolution of Stellar Systems*, Geneva Observatory, p. 132.

Lynden-Bell, D. 1975, *Vistas in Astr.*, **19**, 299.

Lynden-Bell, D. 1992, in M.G. Edmunds & R.J. Terlevich (eds.), *Elements and the Cosmos*, Cambridge University Press, p. 270.

Lynden-Bell, D. & Gilmore, G. (eds.) 1990, *Baryonic Dark Matter*, Kluwer, Dordrecht

Maciel, W.J. 1992, in M.G. Edmunds & R.J. Terlevich (eds.), *Elements and the Cosmos*, Cambridge University Press, p. 210.

Maeder, A. 1990, *A & A Suppl.*, **84**, 139.

Maeder, A. 1992, *A & A*, **264**, 105.

Maeder, A. 1993, *A & A*, **268**, 833.

Maeder, A. & Meynet, G. 1987, *A & A*, **182**, 243.

Maeder, A. & Meynet, G. 1989, *A & A*, **210**, 155.

Malaney, R.A. & Butler, M.N. 1993, *Ap. J. Lett.*, **407**, L73.

Malaney, R.A. & Fowler, W.A. 1988, *Ap. J.*, **333**, 14.

Malaney, R.A. & Fowler, W.A. 1989a, *Ap. J. Lett.*, **345**, L5.

Malaney, R.A. & Fowler, W.A. 1989b, *MNRAS*, **237**, 67.

Malaney, R.A., Mathews, G.J. & Dearborn, D. 1989, *Ap. J.*, **345**, 169.

Malinie, G., Hartmann, D.H. & Mathews, W.G. 1991, *Ap. J.*, **376**, 520.

Mandl, F. 1992, *Quantum Mechanics*, J. Wiley & Sons, Chichester.

Marigo, P., Bressan, A. & Chiosi, C. 1996, *A & A*, **313**, 545.

Martin, C. & Bowyer, S. 1989, *Ap. J.*, **338**, 677.

Martin, W.C. 1992, in P.L. Smith & W.L. Wiese (eds.), *Atomic and Molecular Data for Space Astronomy: Needs, Analysis and Availability*, Springer-Verlag.

Mather, J.C., Cheng, E.S., Cottingham, D.A. et al. 1994, *Ap. J.*, **420**, 439.

Mathews, G.J., Bazan, G. & Cowan, J.J. 1992, *Ap. J.*, **391**, 719.

Mathews, G.J. & Cowan, J.J. 1990, *Nature*, **345**, 491.

Mathews, G.J. & Schramm, D.N. 1988, *Ap. J. Lett.*, **324**, L67.

Mathews, W.B. & Baker, J.C. 1971, *Ap. J.*, **170**, 241.

Matteucci, F. 1991, in I.J. Danziger & K. Kjär (eds.), ESO/EIPC Workshop: *SN 1987A and Other Supernovae*, ESO, Garching, p. 703.

Matteucci, F. 1992, *Ap. J.*, **397**, 32.

Matteucci, F. 1994, *A & A*, **288**, 57.

Matteucci, F., D'Antona, F. & Timmes, F.X. 1995, *A & A*, **303**, 460.

Matteucci, F. & François, P. 1989, *MNRAS*, **239**, 885.

Matteucci, F. & Greggio, L. 1986, *A & A*, **154**, 279.

Matteucci, F. & Tornambé, A. 1987, *A & A*, **185**, 51.

Matteucci, F. & Tosi, M. 1985, *MNRAS*, **217**, 391.

Mattila, K., Leinert, Ch. & Schnur, G. 1991, in B. Rocca-Volmerange, J.M. Deharveng & J.T.T. Van (eds.), *The Early Observable Universe from Diffuse Backgrounds*, Ed. Frontières, Gif-sur-Yvette, p. 133.

Mayor, M. & Vigroux, L. 1981, *A & A*, **98**, 1.

McClure, R.D., Vandenberg, D.A., Bell, R.A., Hesser, J.E. & Stetson, P.B. 1987, *Astr. J.*, **93**, 1144.

McGaugh, S. 1991, *Ap. J.*, **380**, 140.

McWilliam, A. & Rich, R.M. 1994, *Ap. J. Suppl.*, **91**, 749.

Meggers, W.F., Corliss, C.H. & Scribner, B.F. 1975, *Tables of Spectral Line Intensities*, NBS Monograph **145**, Govt. Pr. Off., Washington.

Melnick, J. 1987, in T.X. Thuan, T. Montmerle & J.T.T. Van (eds.), *Starbursts and Galaxy Evolution*, Ed. Frontières, Gif-sur-Yvette, p. 215.

Melnick, J., Haydari-Malayeri, M. & Leisy, P. 1992, *A & A*, **253**, 16.

Meneguzzi, M., Audouze, J. & Reeves, H. 1971, *A & A*, **15**, 337.

Meneguzzi, M. & Reeves, H. 1975, *A & A*, **40**, 99.

Merrill, P.W. 1952, *Science*, **115**, 484.

Meyer, B.S. & Schramm, D.N. 1986, *Ap. J.*, **311**, 406.

Meyer, J.-P. 1989, in C.J. Waddington (ed.), *Cosmic Abundances of Matter*, Amer. Inst. Phys., p. 245.

Meylan, T., Furenlid, I., Wiggs, M.S. & Kurucz, R.L. 1993, *Ap. J. Suppl.*, **85**, 163.

Meynet, G., Maeder, A., Schaller, G., Schaerer, D. & Charbonnel, C. 1994, *A & A Suppl.*, **103**, 97.

Mezger, P. & Wink, J.E. 1983, in P.A. Shaver, D. Kunth & K. Kjär (eds.), *Primordial Helium*, ESO, Garching, p. 281.

Mihalas, D. 1970, *Stellar Atmospheres*, 1st edition, W.H. Freeman & Co., San Francisco.

Mihalas, D. 1978, *Stellar Atmospheres*, 2nd edition, W.H. Freeman & Co., San Francisco.

Mihalas, D. & Binney, J. 1981, *Galactic Astronomy*, W.H. Freeman & Co., San Francisco.

Mihara, K. & Takahara, F. 1994, *Publ. Astr. Soc. Japan*, **46**, 447.

Moore, C.E. 1962, *An Ultraviolet Multiplet Table*, NBS Circ. no. 488, Nat. Bur. Stand., Washington.

Moore, C.E. 1972, *A Multiplet Table of Astrophysical Interest*, revised ed. (*RMT*), *NSRDS-NBS* **40**, Nat. Bur. Stand., Washington.

Moore, C.E., Minnaert, M. & Houtgast, J. 1966, *The Solar Spectrum 2935 Å to 8770 Å*, *Natl. Bur. Stand. U.S. Monograph* **61**, Washington.

Morell, O., Källander, D. & Butcher, H.R. 1992, *A & A*, **259**, 543.

Morrison, H.L., Flynn, C. & Freeman, K.C. 1990, *Astr. J.*, **100**, 1191.

Morton, D.C. 1991, *Ap. J. Suppl.*, **77**, 119.

Mould, J. 1978, *Ap. J.*, **220**, 434.

Mushotzky, R.F., Loewenstein, M., Arnaud, K.A. *et al.* 1996, *Ap. J.*, **466**, 686.

Neese, C. & Yoss, K. 1988, *Astr. J.*, **95**, 463.

Nissen, P.E. & Edvardsson, B. 1992, *A & A*, **261**, 255.

Nissen, P.E. & Schuster, W.J. 1991, *A & A*, **251**, 457.

Nomoto, K. & Hashimoto, M. 1988, *Phys. Rep.*, **163**, 13.

Nomoto, K. & Kondo, Y. 1991, *Ap. J. Lett.*, **367**, L19.

Nomoto, K., Thielemann, F.-K. & Yokoi, K. 1984, *Ap. J.*, **286**, 644.

Norris, J. 1994, *Ap. J.*, **431**, 645.

Norris, J., Bessell, M.S. & Pickles, A.J. 1985, *Ap. J. Suppl.*, **58**, 463.

Norris, J., Ryan, S.G., & Stringfellow, G.S. 1994, *Ap. J.*, **423**, 386.

O'Dell, C.R., Peimbert, M. & Kinman, T.D. 1964, *Ap. J.*, **140**, 119.

Olive, K.A. & Steigman, G. 1995, *Ap. J. Suppl.*, **97**, 49.

Oliver, S.J., Rowan-Robinson, M. & Saunders, W. 1992, *MNRAS*, **256**, 15P.

Öpik, E.J. 1951, *Proc. R. Irish Acad. A*, **54**, 49.

Osterbrock, D.E. 1988, *Pub. Astr. Soc. Pacific*, **100**, 412.

Osterbrock, D.E. 1989, *Astrophysics of Gaseous Nebulae and Active Galactic Nuclei*, University Science Books, Mill Valley, Calif.

Ostriker, J.B. & Thuan, T.X. 1975, *Ap. J.*, **202**, 353.

Ostriker, J.B. & Tinsley, B.M. 1975, *Ap. J. Lett.*, **201**, L51.

Ott, U. 1993, *Nature*, **364**, 25.

Pagel, B.E.J. 1965, *Roy. Obs. Bull.* no. 104.

Pagel, B.E.J. 1977, in *Highlights of Astronomy*, E.A. Müller (ed.), **4**, pt II, Int. Astr. Union, p. 119.

Pagel, B.E.J. 1981, in S.M. Fall & D. Lynden-Bell (eds.), *The Structure and Evolution of Normal Galaxies*, Cambridge University Press, p. 211.

Pagel, B.E.J. 1989a, *Rev. Mex. Astr. Astrofis.*, **18**, 153.

Pagel, B.E.J. 1989b, in J.E. Beckman & B.E.J. Pagel (eds.), *Evolutionary Phenomena in Galaxies*, Cambridge University Press, p. 201.

Pagel, B.E.J. 1991, in H. Oberhummer (ed.), *Nuclei in the Cosmos*, Springer-Verlag, p. 89.

Pagel, B.E.J. 1992a, in D. Alloin & G. Stasinska (eds.), *The Feedback of Chemical Evolution on the Stellar Content of Galaxies*, Paris: Observatoire, p. 87.

Pagel, B.E.J. 1992b, in B. Barbuy & A. Renzini (eds.), IAU Symp. 149: *The Stellar Populations of Galaxies*, Kluwer, p. 133.

Pagel, B.E.J. 1993, in N. Prantzos, E. Vangioni-Flam & M. Cassé (eds.), *Origin and Evolution of the Elements*, Cambridge University Press, p. 496.

Pagel, B.E.J. 1994a, in D. Lynden-Bell (ed.), *Cosmical Magnetism*, Kluwer, Dordrecht, p. 113.

Pagel, B.E.J. 1994b, in Th. Montmerle, Ch. J. Lada, I.F. Mirabel & J.T.T. Van (eds.), *The Cold Universe*, Ed. Frontières, Gif-sur-Yvette, p. 395.

Pagel, B.E.J. 1995, in P. Crane (ed.), *The Light Element Abundances*, Springer-Verlag, p. 155.

Pagel, B.E.J. 1997, in G. Münch & A. Mampaso (eds.), *The Universe at Large*, Cambridge University Press.

Pagel, B.E.J. & Edmunds, M.G. 1981, *Ann. Rev. Astr. Ap.*, **19**, 77.

Pagel, B.E.J., Edmunds, M.G., Fosbury, R.A.E. & Webster, B.L. 1978, *MNRAS*, **184**, 569.

Pagel, B.E.J. & Kazlauskas, A. 1992, *MNRAS*, **256**, 49P.

Pagel, B.E.J. & Patchett, B.E. 1975, *MNRAS*, **172**, 13.

Pagel, B.E.J., Simonson, E., Terlevich, R.J. & Edmunds, M.G. 1992, *MNRAS*, **255**, 325.

Pagel, B.E.J. & Tautvaišienė, G. 1995, *MNRAS*, **276**, 505.

Pagel, B.E.J. & Tautvaišienė, G. 1997, *MNRAS*, in press.

Pantelaki, I. & Clayton, D.D. 1987, in T.X. Thuan, T. Montmerle & J.T.T. Van (eds.), *Starbursts and Galaxy Evolution*, Ed. Frontières, Gif-sur-Yvette, p. 145.

Parker, J.W. & Garmany, C.D. 1993, *Astr. J.*, **106**, 1471.

Partridge, R.B. & Peebles, P.J.E. 1967, *Ap. J.*, **148**, 377.

Peebles, P.J.E. 1966, *Ap. J.*, **146**, 542.

Pei, Y.C. & Fall, S.M. 1995, *Ap. J.*, **454**, 69.

Peimbert, M. 1983, in P.A. Shaver, D. Kunth & K. Kjär (eds.), *Primordial Helium*, ESO, Garching, p. 267.

Peimbert, M., Colin, P. & Sarmiento, A. 1994, in G. Tenorio-Tagle (ed.), *Violent Star Formation from 30 Doradus to QSOs*, Cambridge University Press, p. 79.

Peimbert, M. & Serrano, A. 1982, *MNRAS*, **198**, 563.

Peimbert, M. & Torres-Peimbert, S. 1974, *Ap. J.*, **193**, 327.

Peimbert, M. & Torres-Peimbert, S. 1976, *Ap. J.*, **203**, 581.

Perkins, D.H. 1982, *Introduction to High Energy Physics*, 2nd edition, Addison-Wesley Press, p. 24.

Perkins, D.H. 1987, *Introduction to High Energy Physics*, 3rd edition, Addison-Wesley Press.

Perryman, M.A.C., Lindegren, L., Kovalevsky, J. *et al.* 1995, *A & A*, **304**, 69.

Persic, M. & Salucci, P. 1992, *MNRAS*, **258**, 14P.

Pettini, M. & Lipman, K. 1995, *A & A*, **313**, 792.

Pettini, M., Lipman, K. & Hunstead, R.W. 1995, *Ap. J.*, **451**, 100.

Pettini, M., Smith, L.J., Hunstead, R.W. & King, D.L. 1994, *Ap. J.*, **426**, 79.

Pey, Y. & Fall, S.M. 1995, *Ap. J.*, **454**, 69.

Phillipps, S. & Edmunds, M.G. 1996, *MNRAS*, **281**, 362.

Phillips, A.C. 1994, *The Physics of Stars*, John Wiley & Sons, Chichester.

Pickles, A. 1985, *Ap. J.*, **296**, 340.

Pilyugin, L. 1993, *A & A*, **277**, 42.

Pilyugin, L. 1996, *A & A*, **310**, 751.

Pilyugin, L. & Edmunds, M.G. 1996, *A & A*, **313**, 792.

Pitts, E. & Tayler, R.J. 1989, *MNRAS*, **240**, 373.

Podosek, F.A. & Swindle, T.D. 1988, in J.F. Kerridge & M.S. Matthews (eds.), *Meteorites and the Early Solar System*, University of Arizona Press, Tucson, p. 47.

Pottasch, S.R. 1984, *Planetary Nebulae*, Reidel.

Powell, A.L.T. 1969, *Roy. Obs. Bull.*, no. 152.

Prantzos, N. 1994a, *A & A*, **284**, 477.

Prantzos, N. 1994b, in M. Busso, R. Gallino & C.M. Raiteri (eds.), *Nuclei in the Cosmos III*, Amer. Inst. Phys. Conf. Proc. no. 327, p. 531.

Prantzos, N., Cassé, M. & Vangioni-Flam, E. 1993, *Ap. J.*, **403**, 630.

Prantzos, N., Vangioni-Flam, E. & Chauveau, 1994, *A & A*, **285**, 132.

Puget, J.-L., Abergel, A., Bernard, J.-P. *et al.* 1996, *A & A Lett.*, **308**, L5.

Raiteri, C.M., Gallino, R. & Busso, M. 1992, *Ap. J.*, **387**, 263.

Raiteri, C.M., Gallino, R., Busso, M., Neuberger, D. & Käppeler, F. 1993, *Ap. J.*, **419**, 207.

Ramaty, R., Reeves, H., Lingenfelter, R.E. & Kozlovsky, B. 1996, in *Nuclei in the Cosmos*, Conference proceedings.

Read, S.M. & Viola, V.E. 1984, *At. Data Nucl. Data Tables*, **31**, 359.

Rebolo, R., Molaro, P. & Beckman, J. 1988, *A & A*, **192**, 192.

Rees, M.J. 1976, *MNRAS*, **176**, 483.

Reeves, H. 1990, in F. Sánchez, M. Collados & R. Rebolo (eds.), *Observational and Physical Cosmology*, Cambridge University Press, p. 73.

Reeves, H. 1991, *A & A*, **244**, 294.

Reeves, H., Fowler, W.A. & Hoyle, F. 1970, *Nature*, **226**, 727.

Reeves, H. & Meyer, J.-P. 1978, *Ap. J.*, **226**, 613.

Refsdal, S. & Weigert, A. 1970, *A & A*, **6**, 426.

Reisenegger, A. 1996, in S.S. Holt & G. Sonneborn (eds.), *Cosmic Abundances*, Astr. Soc. Pacific Conf. Series, Vol. 99, p. 405.

Renzini, A., Ciotti, L. & Pellegrini, S. 1993, in I.J. Danziger, W.W. Zeilinger & K. Kjär (eds.), ESO Conference & Workshop Proc. no. 45: *Structure, Dynamics and Chemical Evolution of Elliptical Galaxies*, Garching, p. 443.

Renzini, A., Greggio, L., Ritossa, C. & Ferrario, L. 1992, *Ap. J.*, **400**, 280.

Renzini, A. & Voli, M. 1981, *A & A*, **94**, 175.

Richtler, T., de Boer, K.S. & Sagar, R. 1991, ESO *Messenger*, no. 64, p. 50.

Rocca-Volmerange, B., Deharveng, J.M. & J.T.T. Van (eds.) 1991, *The Early Observable Universe from Diffuse Backgrounds*, Ed. Frontières, Gif-sur-Yvette.

Rocha-Pinto, H.J. & Maciel, W.J. 1996, *MNRAS*, **279**, 447.

Rogers, F.J. & Iglesias, C.A. 1992, *Ap. J. Suppl.*, **79**, 507.

Rolfs, C.E. & Rodney, W.S. 1988, *Cauldrons in the Cosmos*, University of Chicago Press.

Rolleston, W.R.J., Brown, P.J.F., Dufton, P.L. & Fitzsimmons, A. 1993, *A & A*, **270**, 107.

Rood, R.T., Bania, T.M., Wilson, T.L. & Balser, D.S. 1995, in P. Crane (ed.), *The Light Element Abundances*, Springer-Verlag, p. 201.

Rose, J. 1985, *Astr. J.*, **90**, 1927.

Rubin, R.H., Simpson, J.P., Haas, M.R. & Erickson, E.F. 1991, *Ap. J.*, **374**, 564.

Russell, H.N. 1934, *Ap. J.*, **79**, 317.

Rutherford, E. 1929, *Nature*, **123**, 313.

Ryan, S.G. & Norris, J. 1991, *Astr. J.*, **101**, 1835.

Ryan, S.G., Norris, J. & Bessell, M.S. 1991, *Astr. J.*, **102**, 303.

Ryan, S.G., Norris, J., Bessell, M.S. & Deliyannis, C. 1992, *Ap. J.*, **388**, 184.

Sackmann, I.-J. & Boothroyd, A.I. 1992, *Ap. J. Lett.*, **392**, L71.

Salpeter, E.E. 1952, *Ap. J.*, **115**, 326.

Salpeter, E.E. 1955, *Ap. J.*, **121**, 161.

Sandage, A.R. 1990, *J.R.A.S. Canada*, **84**, 70.

Sandage, A.R. & Fouts, G. 1987, *Astr. J.*, **93**, 74.

Scalo, J.M. 1986, *Fund. Cosmic Phys.*, **11**, 1.

Schaller, G., Schaerer, D., Meynet, G. & Maeder, A. 1992, *A & A Suppl.*, **96**, 269.

Schmidt, A.A., Copetti, M.V.F., Alloin, D. & Jablonka, J. 1991, *MNRAS*, **249**, 766.

Schmidt, M. 1959, *Ap. J.*, **129**, 243.

Schmidt, M. 1963, *Ap. J.*, **137**, 758.

Schönberg, M. & Chandrasekhar, S. 1942, *Ap. J.*, **96**, 161.

Schramm, D.N. 1974, *Ann. Rev. Astr. Astrophys.*, **12**, 383.

Schramm, D.N. 1995, in P. Crane (ed.), *The Light Element Abundances*, Springer-Verlag, p. 51

Schramm, D.N. & Wasserburg, G.J. 1970, *Ap. J.*, **162**, 57.

Searle, L. & Sargent, W.L.W. 1972, *Ap. J.*, **173**, 25.

Searle, L. & Zinn, R. 1978, *Ap. J.*, **225**, 357.

Seaton, M.J. 1987, *J. Phys. B Atom. Molec. Phys.*, **20**, 6363.

Seaton, M.J. 1996, *The Opacity Project*, Vol I: *Selected Research Papers, Atomic Data Tables for He to Si*; Vol II: *Atomic Data Tables for S to Fe, Photo-ionization Cross-section Graphs*, Inst. of Phys. Publ., Bristol.

Seaton, M.J., Yan, Y., Mihalas, D. & Pradhan, A.K. 1994, *MNRAS*, **266**, 805.

Seeger, P.A., Fowler, W.A. & Clayton, D.D. 1965, *Ap. J. Suppl.*, **11**, 121.

Shaver, P.A., Kunth, D. & Kjär, K. (eds.) 1983, *Primordial Helium*, ESO, Garching.

Shaver, P.A., McGee, R.X., Danks, A.C. & Pottasch, S.R. 1983, *MNRAS*, **204**, 53.

Shields, G.A. 1990, *Ann. Rev. Astr. Ap.*, **28**, 303.

Silk, J. & Schramm, D.N. 1992, *Ap. J. Lett.*, **393**, L9.

Skillman, E.D., Kennicutt, R.C. & Hodge, P.W. 1989, *Ap. J.*, **347**, 875.

Skillman, E.D., Terlevich, R.J., Kennicutt, R.C., Jr., Garnett, D. & Terlevich, E. 1994, *Ap. J.*, **431**, 172.

Smith, M.S., Kawano, L.H. & Malaney, R.A. 1993, *Ap. J. Suppl.*, **85**, 219.

Smith, V.V. & Lambert, D.L. 1990, *Ap. J. Suppl.*, **72**, 387.

Smith, V.V., Lambert, D.L. & Nissen, P.E. 1993, *Ap. J.*, **408**, 262.

Sneden, C., McWilliam, A., Preston, G.W., Cowan, J.J., Burris, D.L. & Armosky, B.J. 1996, *Ap. J.*, **467**, 819.

Soderblom, D.R., Duncan, D.K. & Johnson, D.R.H. 1991, *Ap. J.*, **375**, 722.

Sommer-Larsen, J. 1991a, *MNRAS*, **249**, 356.

Sommer-Larsen, J. 1991b, *MNRAS*, **250**, 356.

Sommer-Larsen, J. & Yoshii, Y. 1990, *MNRAS*, **243**, 468.

Songaila, A., Cowie, L.L. & Lilly, S.J. 1991, *Ap. J.*, **348**, 371.

Spinrad, H. & Taylor, B. 1969, *Ap. J. Suppl.*, **22**, 445.

Spite, F. & Spite, M. 1992, *Nature*, **297**, 483.

Spite, F. & Spite, M. 1986, *A & A*, **163**, 140.

Spitzer, L. & Jenkins, E.B. 1975, *Ann. Rev. Astr. Astrophys.*, **13**, 133.

Steigman, G. & Walker, T.P. 1992, *Ap. J. Lett.*, **385**, L13.

Straniero, O., Gallino, R., Busso, M. *et al.* 1995, *Ap. J. Lett.*, **440**, L85.

Strauss, M.A. & Willik, J.A. 1995, *Phys. Rep.*, **261**, 272.

Strömgren, B. 1987, in G. Gilmore & R.F. Carswell (eds.), *The Galaxy*, Reidel, p. 229.

Suess, H.E. & Urey, H.C. 1956, *Rev. Mod. Phys.*, **28**, 53.

Sweigart, A.V. & Mengel, J.G. 1979, *Ap. J.*, **229**, 624.

Takahashi, K., Witti, J. & Janka, H.-Th. 1994, *A & A*, **286**, 857.

Talbot, R.J. & Arnett, W.D. 1971, *Ap. J.*, **170**, 409.

Talbot, R.J. & Arnett, W.D. 1973a, *Ap. J.*, **186**, 51.

Talbot, R.J. & Arnett, W.D. 1973b, *Ap. J.*, **186**, 69.

Talbot, R.J. & Arnett, W.D. 1975, *Ap. J.*, **197**, 551.

Tayler, R.J. 1994, *The Stars: Their Structure and Evolution*, Cambridge University Press.

Tayler, R.J. 1995, *MNRAS*, **273**, 215.

Terlevich, R.J., Davies, R.L., Faber, S.M. & Burstein, D. 1981, *MNRAS*, **196**, 381.

Thielemann, F.-K., Metzinger, J. & Klapdor, V. 1983, *A & A*, **123**, 162.

Thielemann, F.K., Nomoto, K. & Hashimoto, M. 1993, in N. Prantzos, E. Vangioni-Flam & M. Cassé (eds.), *Origin and Evolution of the Elements*, Cambridge University Press, p. 297.

Thielemann, F.-K., Nomoto, K. & Hashimoto, M. 1996, *Ap. J.*, **460**, 408.

Thielemann, F.-K., Nomoto, K. & Yokoi, K. 1986, *A & A*, **158**, 17.

Thomsen, B. & Baum, W.A. 1987, *Ap. J.*, **315**, 460.

Thorburn, J.A. 1993, *Ap. J.*, **421**, 318.

Timmes, F.X., Woosley, S.E. & Weaver, T.A. 1995, *Ap. J. Suppl.*, **98**, 617.

Tinsley, B.M. 1977, *Ap. J.*, **216**, 548.

Tinsley, B.M. 1979, *Ap. J.*, **229**, 1046.

Tinsley, B.M. 1980, *Fund. Cosmic Phys.*, **5**, 287.

Tinsley, B.M. & Larson, R.B. 1979, *MNRAS*, **186**, 503.

Tomkin, J., Lemke, M., Lambert, D.L. & Sneden, C. 1992, *Astr. J.*, **104**, 1568.

Toomre, A. 1977, in B.M. Tinsley & R.B. Larson (eds.), *The Evolution of Galaxies and Stellar Populations*, New Haven: Yale Obs., p. 401.

Trimble, V. 1987, *Ann. Rev. Astr. Ap.*, **25**, 425

Truran, J.W. 1981, *A & A*, **97**, 391.

Truran, J.W. & Cameron, A.G.W. 1971, *Ap. Sp. Sci.*, **14**, 179.

Tsujimoto, T. 1993, Thesis, Tokyo University.

Tsujimoto, T., Nomoto, K., Yoshii, Y., Hashimoto, M., Yanagida, S. & Thielemann, F.-K. 1995, *MNRAS*, **277**, 945.

Turck-Chièze, S. & Lopez, I. 1993, *Ap. J.*, **408**, 347.

Twarog, B. 1980, *Ap. J.*, **242**, 242.

Ulrich, R.K. 1973, in D.N. Schramm & W.D. Arnett (eds.), *Explosive Nucleosynthesis*, University of Texas Press, Austin, p. 139.

Uno, M., Tachibana, T. & Yamada, M. 1992, in S. Kubono & T. Kajino (eds.), *Unstable Nuclei in Astrophysics*, World Scientific, Singapore, p. 187.

Unsöld, A. 1977, *The New Cosmos*, Springer-Verlag.

Vader, J.P. 1987, *Ap. J.*, **317**, 128.

van den Bergh, S. 1962, *Astr. J.*, **67**, 486.

van den Bergh, S. 1991, *Ap. J.*, **369**, 1.

Vangioni-Flam, E. & Cassé, M. 1996, in S.S. Holt & G. Sonneborn (eds.), *Cosmic Abundances*, Astr. Soc. Pacific Conf. Series, Vol. 99, p. 366.

Vangioni-Flam, E., Cassé, M., Audouze, J. & Oberto, Y. 1990, *Ap. J.*, **364**, 568.

Vangioni-Flam, E., Cassé, M., Fields, B.D. & Olive, K.A. 1996, *Ap. J.*, **468**, 199.

Vangioni-Flam, E., Olive, K.A. & Prantzos, N. 1994, *Ap. J.*, **427**, 618.

Van Wormer, L., Görres, J., Iliadis, C., Wiescher, M. & Thielemann, F.-K. 1994, *Ap. J.*, **432**, 326.

Vidal-Madjar, A., Laurent, C., Bonnet, R.M. & York, D.G. 1977, *Ap. J.*, **211**, 91.

Vigroux, L., Chièze, J.P. & Lazareff, B. 1981, *A & A*, **98**, 119.

Vila-Costas, M.B. & Edmunds, M.G. 1993, *MNRAS*, **265**, 199.

Vilchez, J.M. & Esteban, C. 1996, *MNRAS*, **280**, 720.

Vilchez, J.M., Pagel, B.E.J., Díaz, A.I., Terlevich, E. & Edmunds, M.G. 1988, *MNRAS*, **235**, 633.

Wagoner, R.V., Fowler, W.A. & Hoyle, F. 1967, *Ap. J.*, **148**, 3.

Walker, T., Mathews, G. & Viola, V.E. 1985, *Ap. J.*, **299**, 745.

Waller, W.H., Parker, J.W. & Malamuth, E.M. 1996, in S.S. Holt & G. Sonneborn (eds.), *Cosmic Abundances*, Astr. Soc. Pacific Conf. Series, Vol. 99, p. 354.

Wallerstein, G. 1962, *Ap. J. Suppl.*, **6**, 407.

Walsh, J. & Roy, J.-R. 1989, *MNRAS*, **239**, 297.

Wasserburg, G.J. 1987, *Earth & Plan. Sci. Lett.*, **86**, 129.

Wasserburg, G.J., Busso, M., Gallino, R. & Raiteri, C.M. 1994, *Ap. J.*, **424**, 412.

Wasserburg, G.J & Papanastassiou, D.A. 1982, in C.A. Barnes, D.D. Clayton & D.N. Schramm (eds.), *Essays in Nuclear Astrophysics*, Cambridge University Press, p. 77.

Wasserburg, G.J., Papanastassiou, D.A. & Sanz, H.G. 1969, *Earth & Plan. Sci. Lett.*, **7**, 33.

Weaver, T.A. & Woosley, S.E. 1993, *Phys. Rep.*, **227**, 65.

Weinberg, S. 1972, *Gravitation & Cosmology*, John Wiley & Sons.

Weiss, A., Keady, J.J. & McGee, N.H., Jr. 1990, *Atomic Data Nuclear Data Tables*, **45**, 209.

Wheeler, J.C., Sneden, C. & Truran, J.W. 1989, *Ann. Rev. Astr. Astrophys.*, **27**, 279.

White, S.D.M. 1989, in C.S. Frenk, R.S. Ellis, T. Shanks, A. Heavens & J. Peacock (eds.), *The Epoch of Galaxy Formation*, Kluwer, Dordrecht, p. 15.

White, S.D.M. & Rees, M.J. 1978, *MNRAS*, **183**, 341.

Wielen, R., Fuchs, B. & Dettbarn, 1996, *A & A*, **314**, 438.

Wiese, W.L., Fuhr, J.R. & Dieters, T.M. 1996, *Atomic Transition Probabilities for Carbon, Nitrogen and Oxygen: A Critical Data Compilation*, *J. Phys. Chem. Ref. Data,*, Monograph 7.

Wiese, W.L. & Martin, G.A. 1980, *Wavelengths and Transition Probabilities for Atoms and Atomic Ions*, NSRDS-NBS 68, U.S. National Bureau of Standards.

Wilson, T.L. & Rood, R.T. 1994, *Ann. Rev. Astr. Astrophys.*, **32**, 191.

Witten, E. 1984, *Phys. Rev. D*, **30**, 272.

Woods, R.D. & Saxon, D.S. 1954, *Phys. Rev.*, **95**, 577.

Woosley, S.E., Fowler, W.A., Holmes, W.A. & Zimmerman, B.A. 1978, *At. Data Nucl. Data Tables*, **22**, 371.

Woosley, S.E., Hartmann, D.H., Hoffman, R.D. & Haxton, W.C. 1990, *Ap. J.*, **356**, 272.

Woosley, S.E. & Weaver, T.A. 1982, in C.A. Barnes, D.D. Clayton & D.N. Schramm (eds.), *Essays in Nuclear Astrophysics*, Cambridge University Press, p. 377.

Woosley, S.E. & Weaver, T.A. 1986, *Ann. Rev. Astr. Astrophys.*, **24**, 205.

Woosley, S.E. & Weaver, T.A. 1995, *Ap. J. Suppl.*, **101**, 181.

Woosley, S.E., Wilson, J.R., Mathews, G.J., Hoffman, R.D. & Meyer, B.S. 1994, *Ap. J.*, **433**, 229.

Worthey, G. 1994, *Ap. J. Suppl.*, **95**, 107.

Worthey, G., Faber, S.M. & Gonzales, J.J. 1992, *Ap. J.*, **398**, 69.

Wyse, R.F.G. & Gilmore, G. 1992, *Astr. J.*, **104**, 144.

Wyse, R.F.G. & Gilmore, G. 1995, *Astr. J.*, **110**, 1071.

Yang, J., Turner, M.S., Steigman, G., Schramm, D.N. & Olive, K. 1984, *Ap. J.*, **281**, 493.

Yokoi, K., Takahashi, K. & Arnould, M. 1983, *A & A*, **117**, 65.

Yoshii, Y. & Arimoto, N. 1987, *A & A*, **188**, 13.

Yoshii, Y. & Takahara, F. 1988, *Ap. J.*, **326**, 1.

Zaritsky, D., Kennicutt, R.C., Jr. & Huchra, J.P. 1994, *Ap. J.*, **420**, 87.

Zinn, R. 1985, *Ap. J.*, **293**, 424.

Index